Forest Fire Danger Prediction Using Deterministic–Probabilistic Approach

Nikolay Viktorovich Baranovskiy
National Research Tomsk Polytechnic University, Russia

IGI Global
PUBLISHER of TIMELY KNOWLEDGE

A volume in the Advances in Environmental Engineering and Green Technologies (AEEGT) Book Series

Published in the United States of America by
IGI Global
Engineering Science Reference (an imprint of IGI Global)
701 E. Chocolate Avenue
Hershey PA, USA 17033
Tel: 717-533-8845
Fax: 717-533-8661
E-mail: cust@igi-global.com
Web site: http://www.igi-global.com

Library of Congress Cataloging-in-Publication Data

Names: Baranovskiy, Nikolay Viktorovich, 1978- author.
Title: Forest fire danger prediction using deterministic-probabilistic approach / by Nikolay Viktorovich Baranovskiy.
Description: Hershey, PA : Engineering Science Reference, [2021] | Includes bibliographical references and index. | Summary: "This book is devoted to the predicting of forest fire danger using a deterministic-probabilistic approach"-- Provided by publisher.
Identifiers: LCCN 2020042221 (print) | LCCN 2020042222 (ebook) | ISBN 9781799872504 (hardcover) | ISBN 9781799872511 (paperback) | ISBN 9781799872528 (ebook)
Subjects: LCSH: Forest fire forecasting. | Forest fire forecasting--Simulation methods. | Forest fires--Prevention and control.
Classification: LCC SD421.36 .B37 2021 (print) | LCC SD421.36 (ebook) | DDC 634.9/6180112--dc23
LC record available at https://lccn.loc.gov/2020042221
LC ebook record available at https://lccn.loc.gov/2020042222

This book is published in the IGI Global book series Advances in Environmental Engineering and Green Technologies (AEEGT) (ISSN: 2326-9162; eISSN: 2326-9170)

British Cataloguing in Publication Data
A Cataloguing in Publication record for this book is available from the British Library.

Advances in Environmental Engineering and Green Technologies (AEEGT) Book Series

Sang-Bing Tsai
Zhongshan Institute, University of Electronic Science and Technology of China, China & Wuyi University, China
Ming-Lang Tseng
Lunghwa University of Science and Technology, Taiwan
Yuchi Wang
University of Electronic Science and Technology of China Zhongshan Institute, China

ISSN:2326-9162
EISSN:2326-9170

Mission

Growing awareness and an increased focus on environmental issues such as climate change, energy use, and loss of non-renewable resources have brought about a greater need for research that provides potential solutions to these problems. Research in environmental science and engineering continues to play a vital role in uncovering new opportunities for a "green" future.

The **Advances in Environmental Engineering and Green Technologies (AEEGT)** book series is a mouthpiece for research in all aspects of environmental science, earth science, and green initiatives. This series supports the ongoing research in this field through publishing books that discuss topics within environmental engineering or that deal with the interdisciplinary field of green technologies.

Coverage

- Contaminated Site Remediation
- Green Technology
- Radioactive Waste Treatment
- Alternative Power Sources
- Air Quality
- Green Transportation
- Electric Vehicles
- Renewable Energy
- Waste Management
- Pollution Management

IGI Global is currently accepting manuscripts for publication within this series. To submit a proposal for a volume in this series, please contact our Acquisition Editors at Acquisitions@igi-global.com or visit: http://www.igi-global.com/publish/.

The Advances in Environmental Engineering and Green Technologies (AEEGT) Book Series (ISSN 2326-9162) is published by IGI Global, 701 E. Chocolate Avenue, Hershey, PA 17033-1240, USA, www.igi-global.com. This series is composed of titles available for purchase individually; each title is edited to be contextually exclusive from any other title within the series. For pricing and ordering information please visit http://www.igi-global.com/book-series/advances-environmental-engineering-green-technologies/73679. Postmaster: Send all address changes to above address.

Titles in this Series

For a list of additional titles in this series, please visit:
http://www.igi-global.com/book-series/advances-environmental-engineering-green-technologies/73679

Improving Integrated Pest Management for Crop Protection and Food Security
Muhammad Haseeb (Florida A&M University, USA) Lambert H.B. Kanga (Florida A&M University, USA) and Jawwad A. Qureshi (University of Florida, USA)
Engineering Science Reference • © 2021 • 315pp • H/C (ISBN: 9781799841111) • US $195.00

Role of IoT in Green Energy Systems
Vasaki Ponnusamy (Universiti Tunku Abdul Rahman, Malaysia) Noor Zaman (Taylor's University, Malaysia) Low Tang Jung (Universiti Teknologi PETRONAS, Malaysia) and Anang Hudaya Muhamad Amin (Higher Colleges of Technology, UAE)
Engineering Science Reference • © 2021 • 405pp • H/C (ISBN: 9781799867098) • US $195.00

Combating and Controlling Nagana and Tick-Borne Diseases in Livestock
Caleb Oburu Orenge (Department of Veterinary Anatomy and Physiology, Faculty of Veterinary Medicine and Surgery, Egerton University, Kenya)
Engineering Science Reference • © 2021 • 440pp • H/C (ISBN: 9781799864332) • US $185.00

Artificial Intelligence and IoT-Based Technologies for Sustainable Farming and Smart Agriculture
Pradeep Tomar (Gautam Buddha University, India) and Gurjit Kaur (Delhi Technological University, India)
Engineering Science Reference • © 2021 • 400pp • H/C (ISBN: 9781799817222) • US $215.00

Handbook of Research on Waste Diversion and Minimization Technologies for the Industrial Sector
Ashok K. Rathoure (M/s Akone Services, Lucknow, India)
Engineering Science Reference • © 2021 • 455pp • H/C (ISBN: 9781799849216) • US $295.00

Smart Farming Techniques and Knowledge-Based Agriculture Emerging Research and Opportunities
Petraq Papajorgji (European University of Tirana, Albania) and Francois Pinet (Irstea Centre de Clermont-Ferrand, France)
Engineering Science Reference • © 2021 • 230pp • H/C (ISBN: 9781799845409) • US $195.00

Solar Concentrating Modules With Louvered Heliostats Emerging Research and Opportunities
Dmitry Strebkov (Federal State Budget Scientific Institution, Russia & Federal Scientific Agroengineering Center VIM, Russia) Natalya Filippchenkova (Federal State Budget Scientific Institution, Russia & Federal Scientific Agroengineering Center VIM, Russia) and Anatoly Irodionov (Federal State Budget Scientific Institution, Russia & Federal Scientific Agroengineering Center VIM, Russia)
Engineering Science Reference • © 2021 • 267pp • H/C (ISBN: 9781799842767) • US $190.00

IGI Global
PUBLISHER of TIMELY KNOWLEDGE

701 East Chocolate Avenue, Hershey, PA 17033, USA
Tel: 717-533-8845 x100 • Fax: 717-533-8661
E-Mail: cust@igi-global.com • www.igi-global.com

Table of Contents

Section 3
Forest Fuel Ignition Physical Modeling and Mathematical Simulation

Section 4
Deterministic-Probabilistic Criteria to Predict Forest Fire Danger

Detailed Table of Contents

the bark, current-voltage characteristics of a cloud-to-ground lightning discharge, and chemical reaction in the gas phase are taken into account. The conditions for the ignition of a coniferous tree in the typical range of the discharge effect are revealed.

Based on the one-dimensional three-layer physical and mathematical model of ignition of a coniferous tree (pine), the influence of the m-components of the cloud-to-ground lightning discharge and the appropriateness of their consideration in technical systems for monitoring the forest fire danger are evaluated. The problem is solved in a cylindrical coordinate system. Typical cloud-to-ground lightning discharges are considered. An assessment of the influence of m-components was carried out for a typical range of changes in their characteristics.

The results of numerical simulation of heating a coniferous tree (pine) by cloud-to-ground lightning discharge are presented. The problem is solved in a one-dimensional formulation in a cylindrical coordinate system. A four-layer structure of the coniferous trunk is considered taking into account the core, subcortical layer, bark and surface water layer. A parametric study of the effect of current-voltage characteristics typical of negative and positive cloud-to-ground lightning discharges on the process of heating the trunk wood has been carried out. The conditions for the ignition of a tree trunk in a typical range of variation of the parameters of the impact of the discharge are established.

A physical and mathematical formulation of the problem to simulate ignition of a deciduous tree in the approximation of large vessels is presented. The problem is considered in a flat formulation in polar coordinates. A parametric analysis of the current-voltage characteristics of cloud-to-ground lightning discharges on the process of heating the wood of a deciduous tree trunk is carried out. The conditions for the ignition of a tree trunk in the typical range of the parameters for discharge effect are established. Variants without and taking into account the surface layer of water are considered.

The results of the numerical implementation for the spatial formulation of the problem of the deciduous tree ignition by an electric current of a cloud-to-ground lightning discharge are presented. The problem was solved in a three-dimensional formulation in a cylindrical coordinate system. A symmetric formulation of the problem is considered. The localization of reactive wood, structural heterogeneity of the bark, volt-ampere characteristics of a cloud-to-ground lightning discharge, and chemical reaction in the gas phase are taken into account. The conditions of deciduous tree ignition in the typical range of the discharge

This chapter provides information on the physical and mathematical models of forest fuel layer ignition by the metal or non-metal particle heated up to high temperatures. Mathematically, the forest fuel layer ignition is described by a system of equations of heat conduction and diffusion with the corresponding initial and boundary conditions. Scenario modeling of forest fuel layer ignition was carried out in the temperature range characteristic to the occurrence of forest fires. A scenario of a catastrophic forest fire danger is considered when there is no moisture in the forest fuel layer.

This chapter is devoted to numerical simulation of forest fuel ignition by focused solar radiation. A physical model of the forest fuel ignition by focused solar radiation is presented. Three mathematical models of the studied process are presented (one-dimensional, two-dimensional, and three-dimensional). Mathematically, the ignition of the forest fuel layer by focused solar radiation is described by a system of equations of heat conduction and diffusion with the corresponding initial and boundary conditions. The results of scenario modeling are presented.

A new probabilistic criterion for forest fire danger and a new methodology for determining the probability of forest fires taking into account thunderstorm activity and the level of anthropogenic load in a controlled area are considered. The results of a parametric study of the dependence of probability on meteorological conditions, thunderstorm activity, and anthropogenic load are also presented. The fire situation corresponds to the statistics of fire accidents in the territory of the Timiryazevskiy forestry of Tomsk region.

This work is devoted to the creation of a probabilistic criterion for forest fire danger to take into account the various causes of anthropogenic load that lead to forest fires. Typical forested areas of the boreal zone are considered: Tomsk region (Russian Federation) and Vancouver Island (Canada). In addition, a description is given of a probabilistic criterion that takes into account the occurrence of a forest fire as a result of deliberate arson. The chapter presents the results of scenario modeling of forest fire danger. It is concluded that it is possible to modernize existing forest fire danger prediction systems in the USA, Canada, Southern Europe, Australia, and the Russian Federation.

Preface

Forest fires cause ecological, economic and social damage to various states of the international community (Collins et al., 2013). The causes of forest fires are rather various (Yanko, 2005), but the main factor is human activity in the neighborhood of settlements, industrial facilities, objects of transport infrastructure and in intensively developed territories (in other words, anthropogenic load) (Baranovskiy & Baranovskiy, 2020a). In turn, storm activity is basic reason in remote territories (Baranovskiy et al., 2020). Scientists of the different countries of the world develop methods, approaches and the systems to predict forest fire danger (Baranovskiy, 2020b), including, impact of the human and storm activity on forested territories.

In scientific periodicals, there are two main ways to assess the occurrence of forest fires. The first way is associated with assessing the risk of forest fires (Miller & Ager, 2013). It should be noted that many researchers make mistakes in their works devoted to the risk analysis of forest fires. The classical mathematical theory of risk analysis states that risk is the product of the probability of a catastrophic event multiplied by the amount of damage resulting from a possible catastrophic event. In our case, forest fires and their consequences are meant. All other approaches developed by various researchers do not have a rigorous mathematical basis. Such approaches can be useful in assessing the occurrence and possible consequences of forest fires, but these developments have nothing to do with risk analysis.

The second way of research is devoted to the assessment of forest fire danger (Hardy, 2005). Another term is forest fire hazard. In fact, this is the first step in assessing the risk of forest fires, but without evaluating the consequences and damage. It should be noted that it is necessary to follow the provisions of a rigorous mathematical theory of risk analysis, when a forest fire danger is expressed in the probability of a forest fire in certain conditions. In fact, all other approaches lack a rigorous mathematical basis and are poorly scientifically grounded. It should to pay tribute to the developers of various systems for predicting forest fire dangers, including those that are currently used in the practice of protecting forests from fires on the territory of various states. Some systems are designed using a large amount of factual material on forest fire incidents. However, most of the developed and developing approaches operate with abstract numerical indices corresponding to different levels of forest fire danger. Unfortunately, from the standpoint of the mathematical theory of risk analysis, these developments have nothing to do with assessing the probability of a forest fire as the only reliable, physically and mathematically justified indicator of the level of forest fire danger. Once again, it should be noted that these developments have brought practical value during the period of their use, but in fact they need to be modernized using the methods of mathematical statistics, probability theory and risk analysis.

There are various approaches to predict forest fire danger: deterministic, probabilistic, statistical, empirical, imitating (Grishin, 2002). The most perspective, according to the author, is complex deterministic-probabilistic approach which combines mathematical models of forest fuel ignition by various sources of high temperature and probabilistic criteria of forest fire occurrence (Baranovskiy, 2012a).

The present book is devoted to predicting of forest fire danger using deterministic-probabilistic approach. As practice shows, forest fires finally are result of local influence of high temperature source as the ignition spot of forest fire do not arise at a time in the territory of the large size. The main source of high temperature characteristic of anthropogenic influence, are the particles heated to high temperatures which are formed when cutting and welding metals when carrying out various technological works in the forested territory and also when cracking wood in not extinguished fires left by visitors of the forests (Baranovskiy, 2007).

As a result of lightning discharges, both trunks of trees and forest fuels at the bottom of these trees can ignite. Electric current of cloud-to-ground lightning discharge passes on the moisture carrying out ways in tree trunk and there is its heating. The following stage is growth of heat stresses in wood of tree trunk and, as a result, cracking of wood and formation of the charred or burning heated carbonaceous particles (Baranovskiy et al., 2017). Such particles drop out on layer of forest fuel under crown and lead to occurrence of surface forest fire.

Other sources of local forest fuel ignition are the concentrated fluxes of sunlight (Baranovskiy, 2015). Such conditions of sunlight influence are created when passing sunlight through glass bottles and their splinters which in large number are left by visitors of forested territories. The facts of forest fires occurrence as a result of this reason were repeatedly fixed by the staff of Emercom of Russia. The ban to scatter glass bottles and their splinters is enshrined in Rules of behavior in the forests (Russian Federation Government decree of June 30, 2007 No. 417, 2007).

Accounting of forest fuel ignition by the heated particles and the focused sunlight mathematical modeling of physical and chemical processes in layer of forest fuel at local heating is carried out. These processes are described by the systems of the nonlinear non-stationary equations of heat conductivity and diffusion with the corresponding initial and boundary conditions (Baranovskiy, 2015; Zakharevich et al., 2012).

Experiments on physical modeling of forest fuel ignition are made by heated particles and the focused sunlight to develop physical models of the considered processes and verification of the mathematical models (Baranovskiy, 2012b; Baranovskiy & Zakharevich, 2018). Special experimental installations have been developed for these purposes. Besides, experiments on identification of ignition conditions of forest fuel by group of rather small crystallizing particles which occur, for example, when welding metal works have been made.

To take into account stochastic nature of anthropogenic load and storm activity various probabilistic criteria are used. Here it is necessary to explain that various ways of definition and assessment of the members entering formulas of probabilistic criteria are possible. First, it is possible to use statistical data on forest fires retrospective for certain period of time in the controlled forested territory. Secondly, it is possible to use the generator of pseudorandom numbers and to carry out modeling of different scenarios on forest fires occurrence. And the third way is to develop the deterministic mathematical models of anthropogenic load on controlled forested territories. One of the book chapters is devoted to this new and perspective methodology.

It should be noted that simple probabilistic criterion has been used when developing software modules of forest fires predicting for the Information system of remote monitoring of forest fires ISDM-Rosleskhoz which executes within Avialesookhrana (Russian Federation) (Podolskaya et al., 2011). In 2011 the author has received the act of use of results which has been signed by Avialesookhrana's employees and the

Center for environmental problems and productivity of the forests of the Russian Academy of Sciences (the staff of the Center developed the program code on the basis of author's probabilistic criterion).

The analysis of the current state of the research area shows that various approaches are applied to the solution of problem of forest fire danger predicting now. For example, mathematical modeling (both hard and soft modeling), statistical and probabilistic estimates, empirical models, systems on the basis of knowledge and rules are used. In the author's opinion, the most perspective direction in the field of forest fire danger predicting is the methodology of deterministic-probabilistic predicting of forest fires occurrence as this methodology combines the stochastic nature of anthropogenic load and storm activity and accurate relationships of cause and effect at occurrence of forest fires as a result of influence of various high temperature sources. Mathematical models are constructed on the basis of physical mechanisms which consider the thermophysical and physical-chemical processes proceeding during forest fuel ignition.

The target audience of this book will be composed of professionals and researchers working in the field of Predicting of Forest Fire Danger disciplines, e.g. forest fire danger factors, forest fire weather, forest fuel, fire statistics, software and devices to be used in related forest fire danger researches, methods to predict forest fire danger. Moreover, book will provide insights and support executives concerned with the management of in forestry to prevent forest fires risk. Furthermore, this book can be useful for insurance companies, computer engineers who develop new system to predict forest fire danger.

Section 1 is devoted to development of mathematical models of coniferous tree ignition as a result of impact of cloud-to-ground lightning discharge on tree trunk.

Chapter 1 describes various one-dimensional mathematical models of coniferous tree ignition by the cloud-to-ground lightning discharge. Both thermophysical mathematical models and models taking into account physical and chemical processes of thermal decomposition and reaction in gas phase are considered. Estimates of ignition delays of trunk of coniferous tree at impact of cloud-to-ground lightning discharge are given. This is a deterministic part of complex deterministic-probabilistic approach. Forest fuel ignition by the cloud-to-ground discharge is the one of the reasons that leads to forest fire danger.

Chapter 2 describes various two-dimensional mathematical models of coniferous tree ignition by the cloud-to-ground lightning discharge. Both thermophysical mathematical models and models taking into account physical and chemical processes of thermal decomposition and reaction in gas phase are considered. Results on simulation of reactive wood localization are also provided. Estimates of ignition delays of trunk of coniferous tree at impact of cloud-to-ground lightning discharge are given. This is a deterministic part of complex deterministic-probabilistic approach. Forest fuel ignition by the cloud-to-ground discharge is the one of the reasons that leads to forest fire danger.

Chapter 3 describes various three-dimensional mathematical models of coniferous tree ignition by the cloud-to-ground lightning discharge. Both thermophysical mathematical models and models taking into account physical and chemical processes of thermal decomposition and reaction in gas phase are considered. Results on simulation of structural heterogeneities localization in wood are also provided. Estimates of ignition delays of trunk of coniferous tree at impact of cloud-to-ground lightning discharge are given. This is a deterministic part of complex deterministic-probabilistic approach. Forest fuel ignition by the cloud-to-ground discharge is the one of the reasons that leads to forest fire danger.

Chapter 4 is devoted to mathematical modeling of coniferous tree ignition by the cloud-to-ground lightning discharge taking into account M-components of discharge. Results of simulation, both taking into account, and without availability the M-components of cloud-to-ground lightning discharge are given. This is a deterministic part of complex deterministic-probabilistic approach. Forest fuel ignition by the cloud-to-ground discharge is the one of the reasons that leads to forest fire danger.

Section 2 is devoted to development of mathematical models of deciduous tree ignition as a result of cloud-to-ground lightning discharge on tree trunk.

Chapter 5 describes various one-dimensional mathematical models of deciduous tree ignition by the cloud-to-ground lightning discharge. Both thermophysical mathematical models and models taking into account physical and chemical processes of moisture evaporation, thermal decomposition and reaction in gas phase are considered. Estimates of ignition delays of trunk of deciduous tree at impact of cloud-to-ground lightning discharge are given. This is a deterministic part of complex deterministic-probabilistic approach. Forest fuel ignition by the cloud-to-ground discharge is the one of the reasons that leads to forest fire danger.

Chapter 6 describes various two-dimensional mathematical models of deciduous tree ignition by the cloud-to-ground lightning discharge. Both thermophysical mathematical models and models taking into account physical and chemical processes of moisture evaporation, thermal decomposition and reaction in gas phase are considered. Results on simulation of reactive wood localization and large vessels are also provided. Estimates of ignition delays of trunk of deciduous tree at impact of the cloud-to-ground lightning discharge are given. This is a deterministic part of complex deterministic-probabilistic approach. Forest fuel ignition by the cloud-to-ground discharge is the one of the reasons that leads to forest fire danger.

Chapter 7 describes various three-dimensional mathematical models of deciduous tree ignition by the cloud-to-ground lightning discharge. Both thermophysical mathematical models and models taking into account physical and chemical processes of moisture evaporation, thermal decomposition and reaction in gas phase are considered. Results on simulation of structural heterogeneities localization in wood are also provided. Estimates of ignition delays of trunk of deciduous tree at impact of the cloud-to-ground lightning discharge are given. This is a deterministic part of complex deterministic-probabilistic approach. Forest fuel ignition by the cloud-to-ground discharge is the one of the reasons that leads to forest fire danger.

Chapter 8 is devoted to mathematical modeling of deciduous tree ignition by the cloud-to-ground lightning discharge taking into account M-components of discharge. Results of modeling, both taking into account, and without availability the M-component of the cloud-to-ground lightning discharge are given. This is a deterministic part of complex deterministic-probabilistic approach. Forest fuel ignition by the cloud-to-ground discharge is the one of the reasons that leads to forest fire danger.

Section 3 is devoted to experimental study and development of mathematical models of forest fuel ignition by various sources of high temperature.

Chapter 9 describes results of physical modeling of forest fuel ignition by the particle of metal or nonmetal heated up to high temperatures. Typical forest fuels are considered: pine needles, leaves of birch and dead grass. The description of the physical mechanism of forest fuel ignition by the single heated particle is provided. Estimates of ignition delays of forest fuel layer at influence by the single particle heated to high temperatures are given. This is a basis to develop mathematical models of forest fuel ignition by the heated up to high temperatures particle. Such particles are the main sources of elevated temperature for natural and anthropogenic reasons of forest fire danger. Moreover, mathematical models should be proven by experimental results.

Chapter 10 describes results of physical modeling of forest fuel ignition by group of rather small crystallizing metal particles. Typical forest fuels are considered: pine needles, leaves of birch and dead grass. The description of the physical mechanism of forest fuel ignition by group of the crystallizing particles is provided. Estimates of ignition delays of forest fuel layer at simultaneous influence of group of the crystallizing metal particles are given. This is a basis to develop mathematical models of forest fuel ignition by the crystallizing particles. Such particles are the widely spread sources of elevated temperature for anthropogenic reasons of forest fire danger. Moreover, mathematical models should be proven by experimental results.

Chapter 11 describes results of physical modeling of forest fuel ignition by the focused sunlight. Typical forest fuels are considered: pine needles, leaves of birch and dead grass. The description of the physical mechanism of forest fuel ignition by the focused sunlight is provided. Estimates of ignition delays of forest fuel layer at impact of the focused sunlight are given. This is a basis to develop mathematical models of forest fuel ignition by the focused sunlight. Such radiation fluxes are the widely spread sources of elevated temperature for natural and anthropogenic reasons of forest fire danger. Moreover, mathematical models should be proven by experimental results.

Chapter 12 describes results of mathematical modeling of forest fuel ignition by the particle of metal or nonmetal heated up to high temperatures. The description of physical and mathematical models of forest fuel ignition by single heated particle is provided. Theoretical estimates of ignition delays of forest fuel layer at influence by the single particle heated up to high temperatures are given. This is a deterministic part of complex deterministic-probabilistic approach. Forest fuel ignition by the heated up to high temperatures particles is the one of the reasons that leads to forest fire danger.

Chapter 13 describes results of mathematical modeling of forest fuel ignition by the focused sunlight. The description of physical and mathematical models of forest fuel ignition by the focused sunlight is provided. Theoretical estimates of ignition delays of forest fuel layer at impact of the focused sunlight are given. This is a deterministic part of complex deterministic-probabilistic approach. Forest fuel ignition by the focused sunlight is the one of the reasons that leads to forest fire danger.

Section 4 is devoted to development of probabilistic criteria of forest fire danger taking into account anthropogenic load, storm activity and meteorological parameters.

Chapter 14 describes various probabilistic criteria of forest fire danger taking into account anthropogenic load, storm activity and weather conditions. The description of various scenarios of forest fire danger is provided. Scenario modeling of forest fire danger for the typical forested territory of boreal zone is carried out. This is quantitative assessment of forest fire occurrence, namely, probability of forest fire occurrence. This probability is mathematically and physically proven indicator of forest fire danger according to strong mathematical theory of risk analysis.

Chapter 15 describes probabilistic criterion of forest fire danger taking into account various reasons of anthropogenic character. The description of various scenarios of forest fire danger is provided. Scenario modeling of forest fire danger for the typical forested territory of boreal zone is carried out. This is quantitative assessment of forest fire occurrence, namely, probability of forest fire occurrence. This probability is mathematically and physically proven indicator of forest fire danger according to strong mathematical theory of risk analysis.

Chapter 16 describes results of mathematical modeling of anthropogenic load on forested territories in the context of forest fires occurrence. Point and linear sources of anthropogenic load are considered. Results of non-uniform distribution of virtual number of forest fires for point and linear sources of anthropogenic load are provided. The comparative analysis of results for various scenarios of anthropogenic load is carried out. This is a mathematical basis to compute probability of anthropogenic load over forested territory. Anthropogenic load is a widely spread reason of forest fire danger.

The relationship between the various chapters of this book should be explained in more detail. The first stage in the development of a specific forest fire danger predicting system based on the deterministic-probabilistic approach is the experimental determination of ignition conditions for specific forest fuels of the controlled area. The data obtained on the ignition delay and the physical mechanism can be further used to determine the thermokinetic constants of the pyrolysis of dry organic matter and the ignition of gaseous combustible components of the pyrolysis of forest fuels. In turn, at the next stage, it is necessary to carry out a scenario simulation of the ignition conditions for forest fuel under various conditions, taking into account a source of elevated temperature (cloud-to-ground lightning discharge, a particle heated to high temperatures, focused solar radiation). The first and second stages are a deterministic component of an integrated deterministic-probabilistic approach to predicting forest fire danger.

At the next third stage, it is necessary to assess the probability of a forest fire in various conditions, calculated using deterministic mathematical models of forest fuel ignition. It is important to note that scenario modeling should be used, since in practice some parameters and conditions of a forest fire may not be known.

The possibilities of practical application of the proposed mathematical models should be considered also. To do this, you need to follow several steps. The first step is to choose a method for solving systems of differential equations of heat conduction. In this work, for these purposes, the finite difference method and the locally one-dimensional method are used. The second step is the choice of a programming language for the software implementation of a specific mathematical model for the ignition of forest fuel. In this work, the software implementation of mathematical models is performed using the Embarcadero RAD Studio environment in the high-level programming language Delphi. Depending on the needs of the developer, console or graphical applications can be created, and multithreaded data processing tools can be used to create applications for multi-core processors. The third step is the choice of a system for visualizing the results of calculating the probability of a forest fire according to the proposed probabilistic criteria. For example, in this work, the Origin Pro software package was used to visualize the field of the virtual number of forest fires. In addition, the interaction of the developed application with host systems for the development of geographic information systems can be organized. For example, ArcGIS and QGIS systems are widely used for these purposes. In principle, step 2 and step 3 can be combined if the software implementation is executed in a high-level programming language built into the geographic information system. For example, the language constructs of the Python programming language can be used in ArcGIS and QGIS systems.

The release of the present book will cause initiation and the subsequent development of paradigm of deterministic-probabilistic to predict forest fires occurrence as a result of storm activity and anthropogenic load. The successful solution of problem of forest fires predicting for the purpose of their prevention will allow to reduce destructive consequences of forest fires on ecosystems. The present book makes contribution to development of perspective comprehensive approach of forest fire danger predicting using mathematical models of forest fuel ignition by the various sources of high temperature and probabilistic criteria which consider anthropogenic load, storm activity and meteorological parameters. In turn, expected information on the probability of forest fire occurrence can be used as input information for systems and models which assess environmental impacts and economic losses from forest fires. The present book can serve as model of carrying out complex researches of forest fire danger on the basis of probabilistic criteria and mathematical models of forest fuel ignition which are developed and verified by means of results of experiments. Perhaps, the publication of the present book will become the first step on the way of wide use of deterministic-probabilistic approach to predict forest fire danger for problem solving of forest protection from the fires.

Nikolay Viktorovich Baranovskiy
Tomsk Polytechnic University, Russia

REFERENCES

Baranovskiy, N. V. (2007). *Mathematical modeling of the most probable scenarios and conditions of forest fire occurrence* (PhD Thesis). Tomsk: Tomsk State University. (in Russian)

Baranovskiy, N. V. (2012a). *Thermophysical basis of predictive modeling of forest fire danger* (Dr.Sc. Dissertation). Tomsk: Tomsk Polytechnic University. (in Russian)

Baranovskiy, N. V. (2012b). Experimental studies of forest fuel layer ignition by the focused sunlight [in Russian]. *Fire and explosion safety.*, *21*(9), 23–27.

Baranovskiy, N. V. (2015). Mathematical modeling for the forest fuel layer ignition caused by focused solar radiation flux. In *Proceedings of SPIE*. The International Society for Optical Engineering.

Baranovskiy, N. V. (2020a). Mathematical Simulation of Anthropogenic Load on Forested Territories for Point Source. In N. Baranovskiy (Ed.), *Predicting, Monitoring, and Assessing Forest Fire Dangers and Risks* (pp. 64–88). IGI Global. doi:10.4018/978-1-7998-1867-0.ch003

Baranovskiy, N. V. (2020b). *Predicting, Monitoring, and Assessing Forest Fire Dangers and Risks*. IGI Global. doi:10.4018/978-1-7998-1867-0

Baranovskiy, N. V., Kuznetsov, G. V., & Nemova, T. N. (2017). High temperature wood particles formation caused by the cloud-to-ground lightning discharge through the coniferous tree trunk. *Far East Journal of Mathematical Sciences*, *102*(5), 1033–1044. doi:10.17654/MS102051033

Baranovskiy, N. V., & Zakharevich, A. V. (2018). Mechanical, thermophysical and aerothermochemical processes at forest fuel ignition caused by the heated to high temperatures single metal particle. *International Review of Mechanical Engineering*, *12*(6), 522–532. doi:10.15866/ireme.v12i6.15065

Collins, R. D., de Neufville, R., Claro, J., Oliveira, T., & Pacheco, A. P. (2013). Forest fire management to avoid unintended consequences: A case study of Portugal using system dynamics. *Journal of Environmental Management*, *130*, 1–9. doi:10.1016/j.jenvman.2013.08.033 PMID:24036501

Grishin, A. M. (2002). Modeling and forecast of catastrophes. Tomsk: Publishing house of Tomsk University. (in Russian)

Hardy, C. C. (2005). Wildland fire danger and risk: Problems, definitions, and context. *Forest Ecology and Management*, *211*(1-2), 73–82. doi:10.1016/j.foreco.2005.01.029

Karanina, S. Y., Kocheeva, N. A., Karanin, A. V., & Belikova, M. Y. (2020). Assessing the Territory According to the Degree of Thunderstorm Danger. In N. Baranovskiy (Ed.), *Predicting, Monitoring, and Assessing Forest Fire Dangers and Risks* (pp. 122–150). IGI Global. doi:10.4018/978-1-7998-1867-0.ch006

Miller, C., & Ager, A. A. (2013). A review of recent advances in risk analysis for wildfire management. *International Journal of Wildland Fire*, *22*(1), 1–14. doi:10.1071/WF11114

Podolskaya, A. S., Ershov, D. V., & Shulyak, P. P. (2011). Application of method to assess the probability of forest fires occurrence in ISDM-Rosleskhoz [in Russian]. *Modern problems of remote sensing of Earth from space.*, *8*(1), 118–126.

About the approval of Fire safety regulations in forests: Russian Federation Government decree of June 30, 2007 No. 417 Fire safety. 2007. N 4. (in Russian)

Yanko, I. V. (2005). *Pyrological assessment of the territory of the Tomsk Region* (PhD Thesis). Tomsk: Tomsk State Pedagogical University. (in Russian)

Zakharevich, A. V., Baranovsky, N. V., & Maksimov, V. I. (2012). Ignition of forest combustible materialby by the carbonaceous particle heated up to high temperatures [in Russian]. *Butlerov Communications.*, *29*(2), 102–108.

Acknowledgment

The author considers a duty to express gratitude to the chief researcher of the I.N. Butakov Scientific Educational Center of the School of Energy and Power Engineering of the Tomsk Polytechnic University, Dr. Sc., professor Kuznetsov Geniy Vladimirovich for long-term cooperation and the leadership in works on preparation of the thesis for PhD degree. This thesis became the first link in the subsequent intensive development of deterministic-probabilistic approach to the forecast of forest fire danger.

The author expresses gratitude to the corresponding member of the Russian Academy of Sciences, director of the Research Computing Center of the Lomonosov Moscow State University Voevodin Vladimir Valentinovich who was the training head of the book author in the period of Summer Supercomputer Academy of 2015.

The author expresses gratitude to Dr.Sc., head of the laboratory at Institute of Thermophysics of the Siberian Branch of the Russian Academy of Sciences Berdnikov Vladimir Stepanovich for the provided information on real ignitions of forest fuels under the influence of the focused sunlight. This information became the basis for cycle of researches on studying mechanisms of forest fuel ignition and methods to assess the forest fire danger caused by action of the focused sunlight.

The author expresses gratitude to the leading researcher of the Institute of Computational Mathematics of the Russian Academy of Sciences, Dr.Sc. Tolstykh Mikhail Andreevich for valuable advice which promoted generation of the new ideas when developing mathematical models of anthropogenic load on forested territories.

The author expresses gratitude to the Dr.Sc., professor of the Tomsk State University of Architecture and Building Nemova Tatyana Nikolaevna for the invaluable help at research of forest fire occurrence as a result of storm activity.

The author expresses gratitude to the staff of Timiryazevskiy forestry of the Tomsk Region for consultations and the help in the organization of field researches on quantitative assessment of probabilities of solar energy concentrator availability.

The author expresses to gratitude to drivers of cab company "Maxim" (Tomsk) for the help in the organization of transport for movement between control points on research of anthropogenic load on forested territories (especially, to the driver Lyubushin Alexander Vladimirovich).

The author expresses gratitude to the employee of the Tomsk bituminous concrete plant Akimov Vladislav Viktorovich for the help in carrying out production experiments on forest fuel ignition by group of the crystallizing metal particles which have allowed to develop technique of carrying out experiment and the experimental stand for physical modeling of the above specified processes.

Work with this book was partially supported by the Russian Foundation for Basic Researches within scientific project 17-29-05093 "Development of methods to monitor forest fire danger caused by human activity in Baikal Lake basin" (Research supervisor: Nikolay Viktorovich Baranovskiy) and within scientific project 16-41-700831 "Development of forest fire conditions monitoring and prediction technologies taking into account storm activity: Case study on Timiryazevskiy forestry of Tomsk region" (Research supervisor: Nikolay Viktorovich Baranovskiy) and also with assistance of the CE Program of the Tomsk Polytechnic University.

Section 1

Mathematical Simulation of the Coniferous Tree Ignition by the Cloud-to-Ground Lightning Discharge

Chapter 1
One-Dimensional Mathematical Models to Simulate Coniferous Tree Ignition

ABSTRACT

The results of numerical simulation of heating a coniferous tree (pine) by cloud-to-ground lightning discharge are presented. The problem is solved in a one-dimensional formulation in a cylindrical coordinate system. A four-layer structure of the coniferous trunk is considered taking into account the core, subcortical layer, bark, and surface water layer. A parametric study of the effect of current-voltage characteristics typical of negative and positive cloud-to-ground lightning discharges on the process of heating the trunk wood has been carried out. The conditions for the ignition of a tree trunk in a typical range of variation of the parameters of the impact of the discharge are established.

INTRODUCTION

Currently, the occurrence of forest fires is mainly due to anthropogenic causes (Baranovskiy, 2020a). But in sparsely populated (Ivanov, 2006) and high-mountain (Conedera et al., 2006) regions, the occurrence of fires in forests during the passage of a thunderstorm as a result of cloud-to-ground lightning discharge is of great importance (Latham and Williams, 2001). There are various variants of the determinate-probability criterion for predicting forest fire danger (Baranovskiy, 2007; 2017), which includes a subsystem for assessing the probability of forest fires from thunderstorms. But existing criteria do not take into account the physical mechanism of wood ignition as a result of cloud-to-ground lightning discharge. The main characteristics of cloud-to-ground lightning discharges are polarity, peak shock current and voltage, as well as duration of action (Burke and Jones, 1996). Therefore, it is advisable to create a methodology for predicting forest fires based on the mathematical formulation of the problem of igniting a tree with a cloud-to-ground lightning discharge. Mathematical modeling of such a complex process can avoid expensive experimental studies. Ignition of a tree by an electric lightning discharge is characterized by high energy. To date, the results of modeling the processes of ignition of trunk wood

DOI: 10.4018/978-1-7998-7250-4.ch001

due to cloud-to-ground lightning discharge have not been published. There are also no experimental data on the dependences of the parameters of this process on the exposure conditions.

The purpose of the study is a mathematical simulation of the ignition of a coniferous tree by a cloud-to-ground lightning discharge.

MATHEMATICAL STATEMENT

The processes of current flow have their own characteristics in the event of a lightning strike in the trunk of a coniferous tree (belong to gymnosperms), for example, pine. In the structure of angiosperms, a significant role is played by the vessels along which moisture moves (Esau, 1980). The inner part of the trunk, penetrated by such transport channels, is a good conductor of electric current. One of the differences in the structure of coniferous wood is the absence of vessels (Esau, 1980). Therefore, the central part of the trunk of resinous conifers has much greater resistance than the bark and subcortical layer (Fig. 1). As a result of this, in a pine, the electric lightning discharge current passes mainly through the outer layers saturated with moisture (Esau, 1980).

The following physical model is considered. A separate coniferous tree grows on the surface of the land. A terrestrial lightning discharge of a certain polarity hits a tree trunk. An electric current from a ground lightning discharge flows through the trunk. It is assumed that in different sections of the trunk the current parameters are the same and it flows in the subcortical zone of the coniferous tree. As a result, the wood is heated due to the Joule heat, and when certain heat fluxes from the subcortical zone of the trunk and the critical temperature are reached, the wood ignites. The effect of wood moisture on the ignition process is neglected. The last assumption is justified enough for short-term rains with thunderstorms, as well as for the initial period of rain, when thunderstorm activity is usually maximum. It should be noted that the scenario of current flow in the surface water layer is also considered. This is consistent with a thunderstorm scenario, accompanied by precipitation.

The problem is solved for a cylinder that simulates a tree trunk. A certain section of the trunk is considered. The scheme of the solution domain is presented in Fig. 1, where 1 is the core, 2 is the subcortical zone, 3 is the bark of the tree trunk, 4 is the surface layer of water; R_w is ste water layer, R_s is the outer radius of the trunk, R_1 is the interface between the subcortical zone and the bark, R_2 is the interface between the core and the subcortical zone.

Mathematically, the process of heating a tree with a cloud-to-ground lightning discharge before ignition is described by a system of non-stationary differential heat equations:

$$\rho_1 c_1 \frac{\partial T_1}{\partial t} = \frac{\lambda_1}{r}\frac{\partial}{\partial r}\left(r\frac{\partial T_1}{\partial r}\right), \quad (1)$$

$$\rho_2 c_2 \frac{\partial T_2}{\partial t} = \frac{\lambda_2}{r}\frac{\partial}{\partial r}\left(r\frac{\partial T_2}{\partial r}\right) + JU, \quad (2)$$

Figure 1. Area of the solution

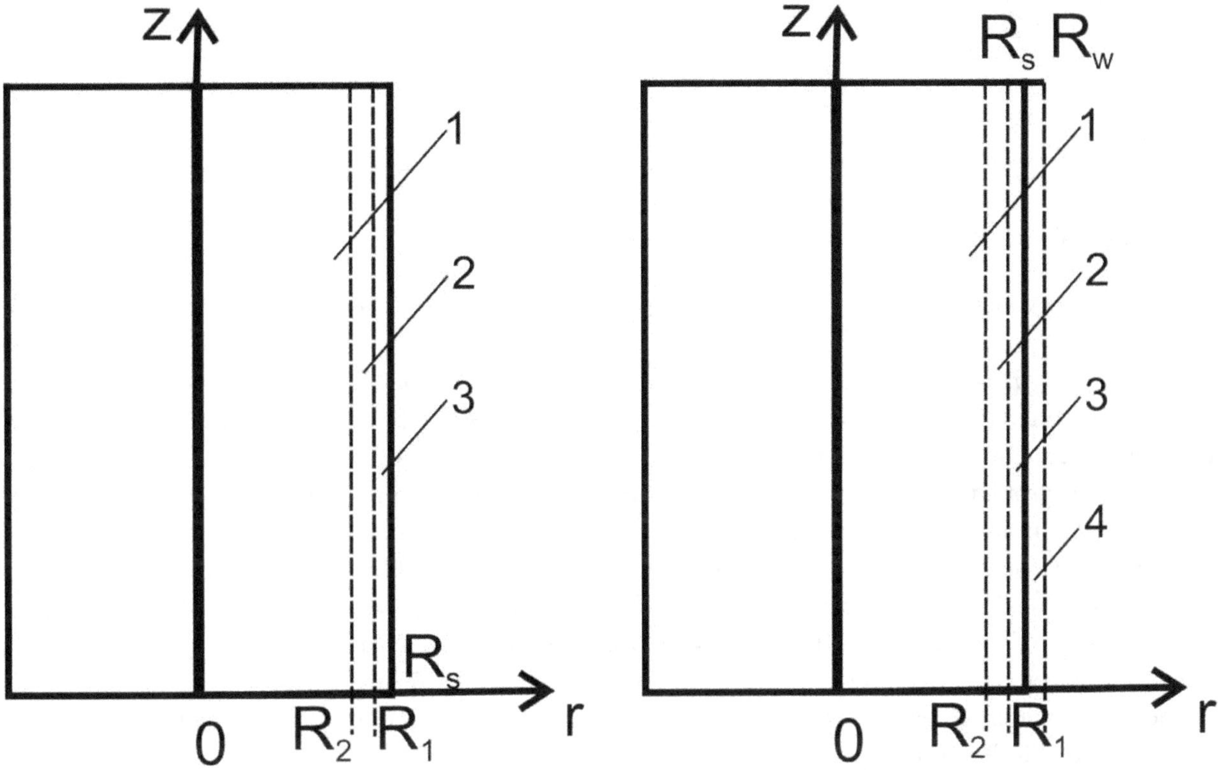

$$\rho_3 c_3 \frac{\partial T_3}{\partial t} = \frac{\lambda_3}{r}\frac{\partial}{\partial r}\left(r\frac{\partial T_3}{\partial r}\right), \tag{3}$$

$$\rho_4 c_4 \frac{\partial T_4}{\partial t} = \frac{\lambda_4}{r}\frac{\partial}{\partial r}\left(r\frac{\partial T_4}{\partial r}\right) + JU, \tag{4}$$

The boundary conditions for equations (1) - (4):

$$r=0,\ \lambda_1 \frac{\partial T_1}{\partial r} = 0, \tag{5}$$

$$r=R_3,\ \lambda_1 \frac{\partial T_1}{\partial r} = \lambda_2 \frac{\partial T_2}{\partial r},\ T_1=T_2, \tag{6}$$

$$r=R_2,\ \lambda_2 \frac{\partial T_2}{\partial r} = \lambda_3 \frac{\partial T_3}{\partial r},\ T_2=T_3, \tag{7}$$

$$r=R_l,\ \lambda_3\frac{\partial T_3}{\partial r}=\lambda_4\frac{\partial T_4}{\partial r},\ T_3=T_3, \quad (8)$$

$$r=R_s,\ \lambda_4\frac{\partial T_4}{\partial r}=\alpha(T_e-T_4), \quad (9)$$

The initial conditions for equations (1) - (4):

$$t=0, T_i(r)=T_{i0}(r),\ i=1,2,3,4 \quad (10)$$

where T_i, ρ_i, c_i, λ_i — temperature, density, heat capacity and thermal conductivity of the core (i=1), subcortical zone (i=2), bark (i=3) of the trunk and surface water layer (i=4), respectively; α - heat transfer coefficient; J is the current strength of a cloud-to-ground lightning discharge; U is the voltage of a cloud-to-ground lightning discharge; r is spatial coordinate, t is time. The indices "e" and "0" correspond to environmental parameters and wood parameters at the initial time.

Initial data (pine wood, core) (Zabolotny et al., 1995): ρ=500 kg/m^3; c=1670 J/(kg·K); λ=0.12 W/(m·K). Subcortical layer parameters: ρ=500 kg/m^3; c=2600 J/(kg·K); λ=0.35 W/(m·K). Thermophysical characteristics of the bark: ρ=500 kg/m^3; c=1670 J/(kg·K); λ=0.12 W/(m·K). Thermophysical properties of the water layer: ρ=1000 kg/m^3; c=4215 J/(kg·K); λ=0.569 W/(m·K). Geometric characteristics of the solution area: R_w=0.255 m; R_s=0.25 m; R_1=0.245 m; R_2=0.235 m. Environmental parameters: T_e=300 K, α=80 W/(m^2·K).

RESULTS AND DISCUSSION

The formulated mathematical model with the corresponding boundary and initial conditions was solved by the finite difference method (Samarskiy, 1983). To solve the difference analogues of one-dimensional differential equations, the marching method (Samarskiy, 1983) was used.

Quite extensive information is known on the parameters of cloud-to-ground-based lightning discharges. Average peak shock current (Soriano et al., 2005): J = 23.5 kA for a negative discharge and J = 35.3 kA for a positive discharge. About 16.5% of positive discharges have a current of less than 10 kA (Cummins et al., 1998).

A parametric study of the influence of the cloud-to-ground lightning discharge on the process of heating the trunk wood during a lightning pulse is carried out. It should be noted that so far no experimental data on the kinetics of the ignition process of large wood masses have been published. Obviously, this is due to the difficulties of a real experimental study of this process. But an approach is known (Zabolotny et al., 1995), in which the ignition conditions of coniferous wood (pine) are described by two parameters (heat flux and ignition surface temperature). In fact, in (Zabolotny et al., 1995) the process of gas-phase ignition of a condensed substance is simulated under conditions of high heat fluxes and relatively short times of exposure to a heating source with an excess of oxidizing agent. The experimental data (Zabolotny et al., 1995) on the critical temperature and heat flux to the crustal surface are used in the present work as ignition criteria.

When conducting numerical studies, it was assumed that a negative thunderstorm discharge lasting 500 ms with a peak current of 23.5 kA and a voltage of 100 kV acts on the pine. These are typical discharge parameters (Soriano et al., 2005; Cummins et al., 1998). Figure 2 shows the temperature distribution along the radius of a tree trunk at various points in time before and at the time of ignition of a tree trunk by electric current: a) - t = 0.01 s; b) - 0.1 s; c) 0.3 s; d) - 0.5 s.

Figure 3 shows the time dependence of the heat flux from the subcortical zone to the ignition surface of the tree trunk and the temperature of this boundary at various points in time. The ignition conditions of the trunk wood were determined by experimental data [9] (Table 1):

Table 1. Experimental data on the ignition conditions of pine wood (Zabolotny et al., 1995)

Ignition delay, s	Heat flux, kw/m²	Surface temperature, K
63.5	12.5	658
45.0	21	700
11.1	42	726
2.6	84	773
0.4	210	867

Table 2 shows the results of numerical calculations of the ignition conditions (depending on the voltage of a cloud-to-ground lightning discharge with a flash duration of 500 ms).

Table 2. The conditions of ignition of a tree depending on the discharge voltage at a current of J=23.5 kA

Voltage, U, kV	Ignition conditions (Zabolotny et al., 1995)	Surface temperature, K	Heat flux, kW/m²
1 – 85	No	<867	<210
90	Yes	867	242
95	Yes	867	246
100	Yes	867	249
105	Yes	867	252
110	Yes	867	255

Table 3 shows the ignition conditions depending on the current of a cloud-to-ground lightning discharge at a voltage of U=100 kV.

Analysis of the results presented in Fig. 2, shows that as a result of the action of the considered ground lightning discharge, the tree trunk in the subcortical zone is heated to temperatures at which natural combustible materials burn (more than 1200 K). The analyzed results allow us to conclude that the tree trunk is ignited and, at least, carbonized. In addition, at such temperatures, wood material is ablated to form furrows on the outside of the trunk. This conclusion is consistent with observations of

Table 3. The ignition conditions of the tree trunk, depending on the current at a voltage of U=100 kV

Current, J, kA	Ignition conditions (Zabolotny et al., 1995)	Surface temperature, K	Heat flux, kW/m²
1 – 20	No	<867	<210
23.5	Yes	867	249
30	Yes	867	264
35	Yes	867	274

thunderstorms (Plummer, 1912), where it is indicated that as a result of a cloud-to-ground lightning discharge, the tree was charred.

An analysis of the dependences of the heat flux and the temperature of the boundary of the subcortical zone (Fig. 3) shows that, in terms of temperature (867 K) and heat flux (249 kW / m2), the ignition conditions for the discharge under consideration are achieved for fairly typical parameters of a lightning discharge.

Figure 4 shows the temperature distribution along the radius of a tree trunk in the case of a surface water layer. In this case, two regions of elevated temperature can be observed. The first area is the subcortical zone. The second area is the surface layer of water and the surface layer of the bark. In the second region, the temperature is about two times lower, since part of the heat is removed to the environment.

An analysis of Figure 5 shows that in the second zone, over a long period of exposure, temperatures and heat fluxes are reached, which can lead to ignition of the bark layer. According to table 1, the ignition delay in this case will increase by about 10 seconds relative to the subcortical layer.

Figure 2. The temperature distribution along the radius of the tree trunk at various points in time (duration of the discharge is 500 ms): a) - t = 0.01 s; b) - 0.1 s; c) 0.3 s; d) 0.5 s

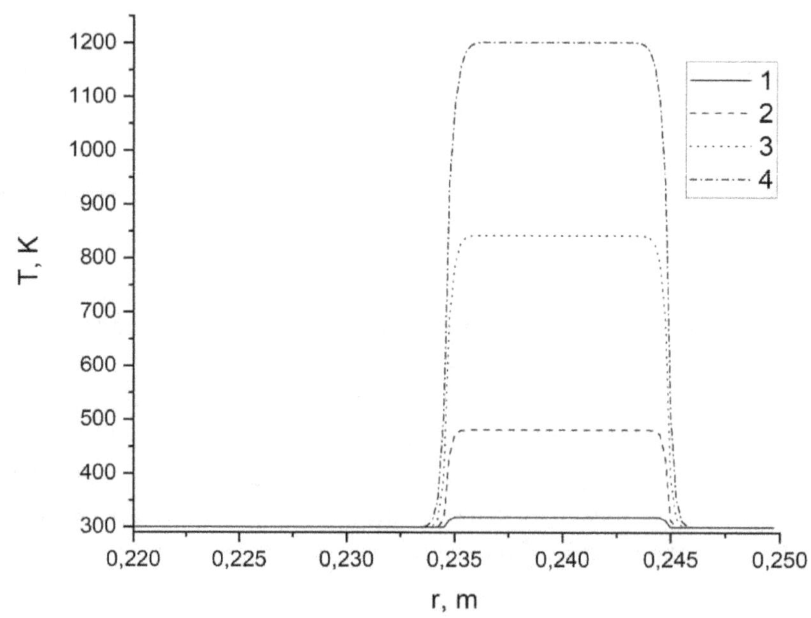

Figure 3. The heat flux from the subcortical zone (2) to the ignition surface and its temperature (1) at various points in time

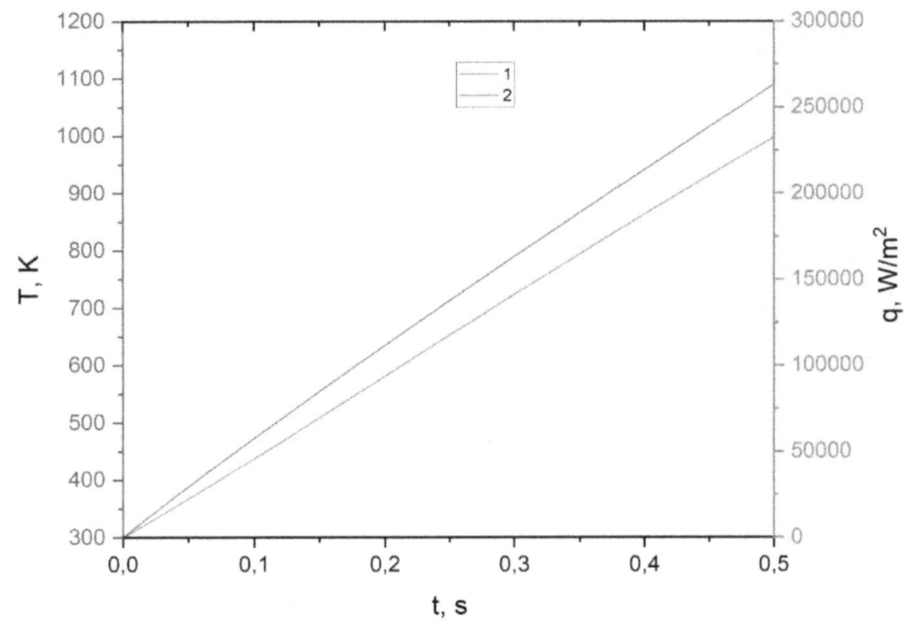

The results also show that the assessment of the conditions for the occurrence of forest fires as a

Figure 4. The temperature distribution along the radius of the tree trunk, taking into account the water layer at different points in time (duration of the discharge 500 ms): a) - t = 0.01 s; b) - 0.1 s; c) 0.3 s; d) 0.5 s

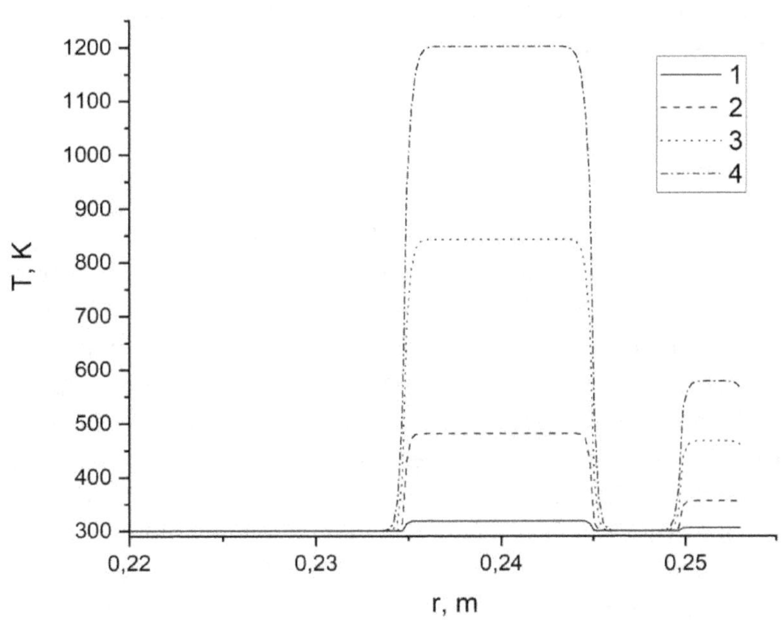

Figure 5. The heat flux to the ignition surface and its temperature (1) at various points in time from the subcortical zone (2) and the surface layer of water (3)

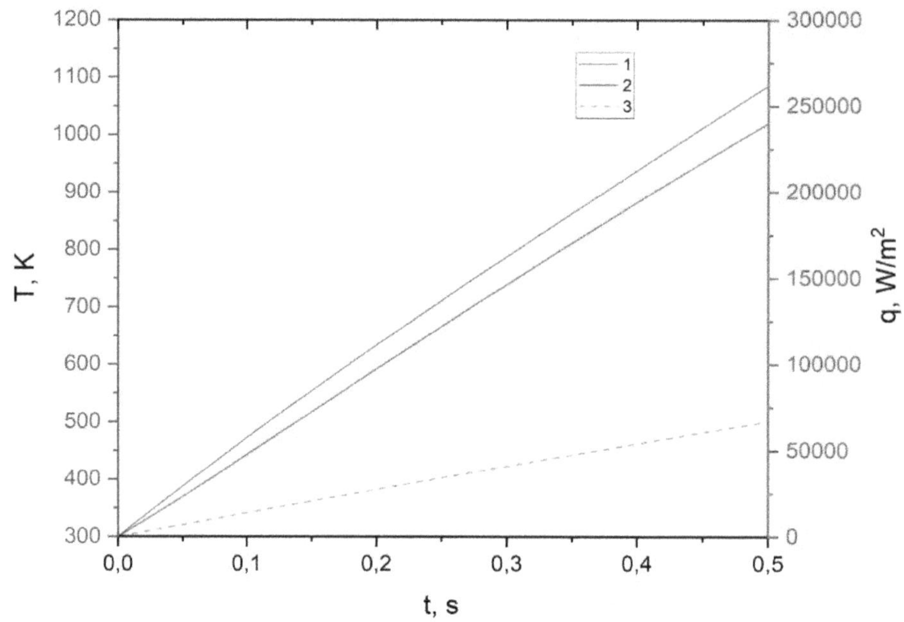

result of lightning discharges can be carried out using a fairly simple mathematical model (1) - (10). This model can be easily implemented as part of forest fire danger prediction systems (Baranovskiy, 2020b; Adab et al., 2013). The initial data for the operation of such a model are the parameters of a cloud-to-ground lightning discharge and the characteristics of wood. The voltage, current, and duration of a specific discharge can be estimated or even fixed by means of modern systems for recording thunderstorm activity (Burke and Jones, 1996; Soriano et al., 2005; Cummins et al., 1998). The thermophysical characteristics of any type of wood can be determined by fairly simple methods (Zabolotny et al., 1995) for each moisture level.

CONCLUSION

The result of this study is the numerical implementation of a mathematical model of conifer warming up as a result of the flow of electric lightning current through it. In the process of parametric research, the conditions for the implementation of the phenomenon under consideration are identified, which are characteristic of a typical range of changes in the parameters of a cloud-to-ground lightning discharge. The possibility of coniferous tree ignition under conditions of a thunderstorm discharge by a cloud-to-ground discharge is shown. The presented physical and mathematical model can be included in the subsystem for assessing the probability of occurrence of forest fire accidents. In addition, the results complement the theoretical basis for the further development of ignition models for fire danger prediction and assessment (Al Janabi et al., 2017; Bui et al., 2017; CWFIS, 2020; Cardille et al., 2001; Chen et al., 2017; Curt et al., 2016; Ganteaume and Guerra, 2018; Chen et al., 2020; Glagolev et al., 2020; Papagiannaki et al., 2020; Silva et al., 2020; Wierzchowski et al., 2002; Yankovich and Yankovich, 2020)[16-28].

ACKNOWLEDGMENT

This work was implemented with the financial support of the Russian Foundation for Basic Research. [Project N 17-29-05093].

REFERENCES

Adab, H., Kanniah, K. D., & Solaimani, K. (2013). Modeling forest fire in the northeast of Iran using remote sensing and GIS techniques. *Natural Hazards*, *65*(3), 1723–1743. doi:10.100711069-012-0450-8

Al Janabi, S., Al Shourbaji, I., & Salman, M. A. (2017). *Assessing the suitability of soft computing approaches for forest fires prediction*. Applied Computing and Informatics. doi:10.1016/j.aci.2017.09.006

Baranovskiy, N. V. (2007). *Mathematical modeling of the most likely scenarios and conditions for forest fires* (PhD Thesis). Tomsk: Tomsk State University. (In Russian)

Baranovskiy, N. V. (2017). Forest fire danger assessment using SPMD-model of computation for massive parallel system. *International Review on Modeling and Simulations*, *10*(3), 193–201. doi:10.15866/iremos.v10i3.10570

Baranovskiy, N. V. (2020a). Mathematical Simulation of Anthropogenic Load on Forested Territories for Point Source. In N. V. Baranovskiy (Ed.), *Predicting, Monitoring, and Assessing Forest Fire Dangers and Risks* (pp. 64–88). IGI Global. doi:10.4018/978-1-7998-1867-0.ch003

Baranovskiy, N. V. (2020b). Predicting Forest Fire Numbers Using Deterministic-Probabilistic Approach. In N. V. Baranovskiy (Ed.), *Predicting, Monitoring, and Assessing Forest Fire Dangers and Risks* (pp. 89–100). IGI Global. doi:10.4018/978-1-7998-1867-0.ch004

Bui, D. T., Bui, Q.-T., Nguyen, Q.-P., Pradhan, B., Nampak, H., & Trinh, P. T. (2017). A hybrid artificial intelligence using GIS-based neural-fuzzy inference system and particle swarm optimization for forest fire susceptibility modeling at a tropical area. *Agricultural and Forest Meteorology*, *233*, 32–44. doi:10.1016/j.agrformet.2016.11.002

Burke, C. P., & Jones, D. L. (1996). On the polarity and continuing current in unusually large lightning flashes deduced from ELF events. *Journal of Atmospheric and Solar-Terrestrial Physics*, *58*(5), 531–548. doi:10.1016/0021-9169(95)00054-2

Canadian Wildland Fire Information System. (n.d.). Access: http://cwfis.cfs.nrcan.gc.ca/home

Cardille, J. A., Ventura, S. J., & Turner, M. G. (2001). Environmental and Social Factors of Influencing Wildfires in the Upper Midwest, United States. *Ecological Applications*, *11*(1), 111–127. doi:10.1890/1051-0761(2001)011[0111:EASFIW]2.0.CO;2

Chen, C., Xu, T., Zhao, D., & Lei, P. (2020). Study on the Wildfires Occurring Risk Based on Fuzzy FTA. *IOP Conference Series. Earth and Environmental Science*, *440*(5), 052100. Advance online publication. doi:10.1088/1755-1315/440/5/052100

Chen, W., Xie, X., Wang, J., Pradhan, B., Hong, H., Tien Bui, D., Duan, Z., & Ma, J. (2017). A comparative study of logistic model tree, random forest and classification and regression tree models for spatial prediction of landslide susceptibility. *Catena*, *151*, 147–160. doi:10.1016/j.catena.2016.11.032

Conedera, M., Cesti, G., Pezzatti, G. B., Zumbrunnen, T., & Spinedi, F. (2006). Lightning-induced fires in the Alpine region: An increasing problem. *Forest Ecology and Management*, *234*(Supplement 1), S68. doi:10.1016/j.foreco.2006.08.096

Cummins, K. L., Murphy, M. J., Bardo, E. A., Hiscox, W. L., Pyle, R. B., & Pifer, A. E. (1998). A combined TOA/MDF technology upgrade of the U.S. national lightning detection network. *Journal of Geophysical Research*, *103*(D8), 9035–9044. doi:10.1029/98JD00153

Curt, T., Frejaville, T., & Lahaye, S. (2016). Modelling the spatial patterns of ignition causes and fire regime features in southern France: Implications for the prevention policy. *International Journal of Wildland Fire*, *25*(7), 785–796. doi:10.1071/WF15205

Esau, K. (1980). Anatomy of seed plants. Moscow: Mir. (in Russian)

Ganteaume, A., & Guerra, F. (2018). Explaining the spatio-seasonal variation of fires by their causes: The case of southeastern France. *Applied Geography (Sevenoaks, England)*, *90*, 69–81. doi:10.1016/j.apgeog.2017.11.012

Glagolev, V., Bazhenov, R., Povkh, I., Chuyko, O., & Dolmatova, O. (2020). Designing and producing a geographic information system to forecast vegetation fire danger according to weather conditions. *IOP Conference Series. Materials Science and Engineering*, *862*(5), 052019. Advance online publication. doi:10.1088/1757-899X/862/5/052019

Ivanov, V. A. (2006). *Methodological basis for the classification of forests of Central Siberia by the degree of fire hazard from thunderstorms* (PhD dissertation). Krasnoyarsk: SibGTU. (in Russian)

Latham, D., & Williams, E. (2001). *Lightning and forest fires // Forest fires: Behavior and Ecological Effects*. Elsevier.

Papagiannaki, K., Giannaros, T. M., Lykoudis, S., Kotroni, V., & Lagouvardos, K. (2020). Weather-related thresholds for wildfire danger in a Mediterranean region: The case of Greece. *Agricultural and Forest Meteorology*, *291*, 108076. Advance online publication. doi:10.1016/j.agrformet.2020.108076

Plummer, F. G. (1912). *Lightning in Relation to Forest Fires. Bulletin 111. USDA Forest Service*. Government Printing Office.

Samarsky, A. A. (1983). *Theory of difference schemes*. Nauka. (in Russian)

Silva, I. D. B., Valle, M. E., Barros, L. C., & Meyer, J. F. C. A. (2020). A wildfire warning system applied to the state of Acre in the Brazilian Amazon. *Applied Soft Computing*, *89*, 106075. Advance online publication. doi:10.1016/j.asoc.2020.106075

Soriano, L. R., De Pablo, F., & Tomas, C. (2005). Ten-year study of cloud-to-ground lightning activity in the Iberian Peninsula. *Journal of Atmospheric and Solar-Terrestrial Physics*, *67*(16), 1632–1639. doi:10.1016/j.jastp.2005.08.019

Wierzchowski, J., Heathcott, M., & Flannigan, M. D. (2002). Lightning and lightning fire, central Cordillera, Canada. *International Journal of Wildland Fire*, *11*(1), 41–51. doi:10.1071/WF01048

Yankovich, E. P., & Yankovich, K. S. (2020). Classification of Territory on Forest Fire Danger Level Using GIS and Remote Sensing. In N. V. Baranovskiy (Ed.), *Predicting, Monitoring, and Assessing Forest Fire Dangers and Risks* (pp. 258–267). IGI Global. doi:10.4018/978-1-7998-1867-0.ch011

Zabolotny, A.E., Zabolotnaya, M.M., Zabolotnaya, Y.A., & Timoshin, V.N. (1995). Determination of safe use zones of solid fuel generators of fire extinguishing aerosols. *Issues of Special Engineering*, (7-8), 15 - 21. (in Russian)

ADDITIONAL READING

Azadifar, M., Rubinstein, M., Rachidi, F., Rakov, V.A., Diendorfer, G., Schulz, W., Pavanello, D. A (2019) Study of a Large Bipolar Lightning Event Observed at the Säntis Tower, IEEE Transactions on Electromagnetic Compatibility, 61 (3), 796-806. doi:10.1109/TEMC.2019.2913220

De Paiva, A. R., Saba, M. M. F., Naccarato, K. P., Schumann, C., Jaques, R., Da Silva Ferro, M. A., & Warner, T. A. (2014) Detection of upward lightning by lightning location systems, 2014 International Conference on Lightning Protection, ICLP 2014, ArticleN° 6973425, pp. 1824-1826. DOI: 10.1109/ICLP.2014.6973425

Lu, G., Cummer, S. A., Tian, Y., Zhang, H., Lyu, F., Wang, T., Stanley, M. A., Yang, J., & Lyons, W. A. (2016). Sprite produced by consecutive impulse charge transfers following a negative stroke: Observation and simulation. *Journal of Geophysical Research*, *121*(8), 4082–4092. doi:10.1002/2015JD024644

Nag, A., Murphy, M. J., Schulz, W., & Cummins, K. L. (2014) Lightning locating systems: Characteristics and validation techniques, 2014 International Conference on Lightning Protection, ICLP 2014, Article N° 6973283, pp. 1070-1082. DOI: 10.1109/ICLP.2014.6973283

Nag, A., Rakov, V. A., & Murphy, M. J. (2014) Measurement of preliminary breakdown pulse trains in cloud-to-ground lightning using lightning locating systems, 2014 International Conference on Lightning Protection, ICLP 2014, Article N° 6973356, pp. 1437-1444. DOI: 10.1109/ICLP.2014.6973356

Saba, M. M. F., Shigihara, M., & Kodaira, C. Y. (2018) Observation of a lightning flashover with a high-speed video camera, 34th International Conference on Lightning Protection, ICLP 2018, Article N° 8503282. DOI: 10.1109/ICLP.2018.8503282

Stolzenburg, M., Marshall, T. C., Karunarathne, S., Karunarathna, N., & Orville, R. E. (2015). An M component with a concurrent dart leader traveling along different paths during a lightning flash. *Journal of Geophysical Research*, *120*(19), 10267–10284. doi:10.1002/2015JD023417

Wang, X., Yuan, P., Cen, J., & Xue, S. (2016). Correlation between the spectral features and electric field changes for natural lightning return stroke followed by continuing current with M-components. *Journal of Geophysical Research*, *121*(14), 8615–8624. doi:10.1002/2016JD025314

Xue, S., Yuan, P., Cen, J., Li, Y., & Wang, X. (2015). Spectral observations of a natural bipolar cloud-to-ground lightning. *Journal of Geophysical Research, 120*(5), 1972–1979. doi:10.1002/2014JD022598

Yuan, P., Chen, R., Su, Y., Fan, B., & Wang, X. (2019). Characteristic parameters of positive cloud-to-ground lightning channel. *Journal of Earth System Science, 128*(6), 164. Advance online publication. doi:10.100712040-019-1196-4

KEY TERMS AND DEFINITIONS

Cloud-to-Ground Lightning Discharge: An electrical discharge during a thunderstorm that occurs between a cloud and the earth's surface. It is a natural source of forest fires.

Forest Fire: Uncontrolled aerothermochemical phenomenon characterized by step-by-step mechanism which includes following stages: inert heating, moisture evaporation, high temperature terpens evaporation, dry organic matter pyrolysis, flammable combustion and smoldering.

Forest Fuel: It can be considered like dead and live forest fuel. Main types of forest fuel which can be involved in combustion during forest fire: ground forest fuel (needles, leaves and dry grass, small branches) and crown forest fuel (needles, small branches).

Ignition: Inflammation of forest fuel caused by definite source of high temperature or energy.

Ignition Delay: Time before flame flash after forest fuel heating.

Lightning Activity: An atmospheric phenomenon characterized by discharges of the cloud-to-cloud and cloud-to-ground class.

Mathematical Simulation: The production of a computer model of forest fire conditions and prerequisites, especially for the purpose of study.

Meteorological Parameters: Physical characteristics of local weather conditions in the forested area under consideration. Key parameters include ambient temperature, soil temperature, precipitation, wind speed, solar radiation, cloud cover, dew point temperature. These parameters are used for mathematical modeling of the drying of a layer of forest fuel.

Monitoring: Monitoring refers to the periodic calculation of the parameters of forest fire danger with a portion of information available in real time.

Prediction: Under the prediction of forest fires is the calculation of the parameters of forest fire danger with a certain projection in advance in order to have enough time to anticipate an emergency. The calculation in this case is carried out in a mode ahead of the real time of the development of the catastrophe - the occurrence of a forest fire.

Chapter 2
Two-Dimensional Mathematical Models to Simulate Coniferous Tree Ignition

ABSTRACT

The physical and mathematical formulation and results of the numerical solution of the problem for ignition of a coniferous tree (pine) by a cloud-to-ground lightning discharge are presented. The problem is considered in a flat formulation in a cylindrical coordinate system. The features of current flow and heat transfer are considered taking into account the localization of reactive wood and the presence of a surface layer of water. A parametric analysis was carried out, and the conditions for the ignition of a tree trunk were determined in a typical range of the exposure to negative and positive discharges.

INTRODUCTION

The natural phenomenon of thunderstorm activity (Kozlov and Mullayarov, 2004) is one of the causes of many forest fires (Flannigan and Wotton, 1991). Thunderstorms create a particularly tense situation in sparsely populated (Ivanov, 2006) and highland (Conedera et al., 2006) regions. As a rule, a cloud-to-ground lightning discharge is a source of fire (Latham and Williams, 2001). Recently, attempts have been made to assess forest fire danger (including thunderstorms) using empirical (Kurbatskiy and Kostyrina, 1977; Larjavaara et al., 2005) and deterministic-probabilistic (Baranovskiy, 2007; 2017) approaches.

In addition, to date, no mathematical models have been created that are adequate to the real physical mechanism of ignition of a tree as a result of the cloud-to-ground lightning discharge. Polarity, peak current and voltage, and duration of action are the main characteristics of cloud-to-ground lightning discharges (Burke and Jones, 1996). The average peak current can reach (Soriano et al., 2005): J = 23.5 kA for a negative discharge and J = 35.3 kA for a positive discharge. At the same time, 16.5% of positive discharges have a current of less than 10 kA (Cummins et al., 1998).

It is known (Zabolotny et al., 1995) that ignition of wood by an energy source is possible when a certain level of heat fluxes in the zone of fuel gasification, temperatures of a mixture of air with wood

DOI: 10.4018/978-1-7998-7250-4.ch002

pyrolysis products and concentrations of gaseous fuel in the mixture are reached. The combustion conditions of wood are significantly influenced by the moisture content in the porous structure of the initial material. But even for dry wood, the solution of the ignition problem is, for a number of reasons, much more complicated than, for example, for polymer materials or rocket fuels (Vilyunov, 1984). One of the important factors, which, among many, must be taken into account when analyzing the ignition conditions for real wood, is its significant structural heterogeneity. The uneven distribution of branches along the length of the tree trunk should affect the conditions for the passage of the discharge and, accordingly, the intensity of heating and the achievement of ignition conditions. For this reason, it is expedient to simulate the process of heating the wood of a coniferous tree trunk under the influence of a lightning discharge in a two-dimensional setting.

The purpose of this study is to determine the conditions for the ignition of a coniferous tree by a cloud-to-ground lightning discharge, depending on the discharge parameters, taking into account the non-uniform process of heat propagation along the trunk.

MATHEMATICAL STATEMENT

The flow of electric current has its own characteristics in the event of a lightning strike into the trunk of a coniferous tree, since the resistance of the wood of the resinous tree itself is much greater than the bark and subcrustal layer. Therefore, the electric current of a lightning discharge in the trunk of a coniferous tree passes mainly through the outer layers, without penetrating inside. Of particular interest is the study of heat transfer taking into account the localization of the so-called reactive wood. Such wood is formed in the lower part of the branches of coniferous trees and is also called compression wood. Reactive wood differs from ordinary wood by its physicochemical properties (Esau, 1980).

To describe the modeled process, the following physical model is adopted. A freestanding coniferous tree is considered. At a fixed moment in time, a lightning discharge of a certain polarity and duration strikes the tree trunk. It is believed that the current-voltage characteristics of the discharge are the same for different sections of the tree trunk. The heating of the trunk wood occurs due to the Joule heat generated in the subcrustal zone of the tree trunk and in the surface layer of water. As a result of the electric current, the wood heats up and when critical heat fluxes from the subcrustal zone of the trunk to the ignition surface and its temperature are reached, the wood is ignited. The influence of wood moisture on the ignition process is neglected.

The solution area is shown in Fig. 1.a, 1.c, where the numbers indicate the areas: 1 - the core of the tree trunk; 2 - subcortical zone; 3 - tree bark; 4 - wood from the upper part of the branches; 5 - reactive wood of the lower part of the branches; 6 - part of the subcortical zone, which has the same properties as area 4; 7 - part of the subcortical zone, which has the same properties as area 5; 8 - part of the core, which has the same properties as area 4; 9 - part of the core, which has the same properties as area 5; 10, 11 - air; 12.13 - surface layer of water. The boundaries of the regions are indicated in Fig. 1.b, 1.d.

Heat transfer in the system under consideration is described using nonstationary heat conduction equations:

Figure 1. Scheme of the solution domain (a, c) and the boundaries of subdomains (b, d)

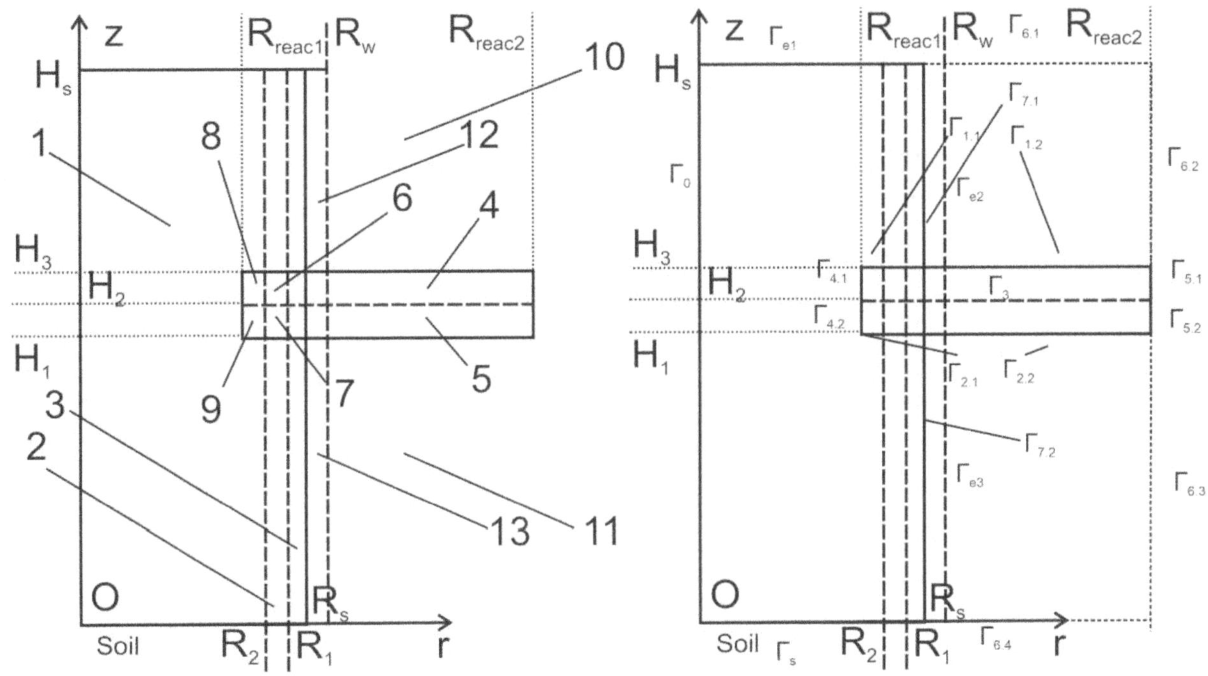

$$\rho_1 c_1 \frac{\partial T_1}{\partial t} = \frac{\lambda_1}{r}\frac{\partial}{\partial r}\left(r\frac{\partial T_1}{\partial r} \right) + \lambda_1 \frac{\partial^2 T_1}{\partial z^2}, \qquad \begin{matrix} z_0 \le z \le H_1 \\ 0 \le r \le R_2 \end{matrix}, \quad \begin{matrix} H_1 \le z \le H_3 \\ 0 \le r \le R_{reac1} \end{matrix}, \quad \begin{matrix} H_3 \le z \le H_s \\ 0 \le r \le R_2 \end{matrix}, \tag{1}$$

$$\rho_2 c_2 \frac{\partial T_2}{\partial t} = \frac{\lambda_2}{r}\frac{\partial}{\partial r}\left(r\frac{\partial T_2}{\partial r} \right) + \lambda_2 \frac{\partial^2 T_2}{\partial z^2} + JU, \qquad \begin{matrix} z_0 \le z \le H_1 \\ R_2 \le r \le R_1 \end{matrix}, \quad \begin{matrix} H_3 \le z \le H_s \\ R_2 \le r \le R_1 \end{matrix}, \tag{2}$$

$$\rho_3 c_3 \frac{\partial T_3}{\partial t} = \frac{\lambda_3}{r}\frac{\partial}{\partial r}\left(r\frac{\partial T_3}{\partial r} \right) + \lambda_3 \frac{\partial^2 T_3}{\partial z^2}, \qquad \begin{matrix} z_0 \le z \le H_1 \\ R_1 \le r \le R_s \end{matrix}, \quad \begin{matrix} H_3 \le z \le H_s \\ R_1 \le r \le R_s \end{matrix},, \tag{3}$$

$$\rho_4 c_4 \frac{\partial T_4}{\partial t} = \frac{\lambda_4}{r}\frac{\partial}{\partial r}\left(r\frac{\partial T_4}{\partial r} \right) + \lambda_4 \frac{\partial^2 T_4}{\partial z^2}, \qquad \begin{matrix} H_2 \le z \le H_3 \\ R_1 \le r \le R_{reac2} \end{matrix},, \tag{4}$$

$$\rho_4 c_4 \frac{\partial T_4}{\partial t} = \frac{\lambda_4}{r}\frac{\partial}{\partial r}\left(r\frac{\partial T_4}{\partial r}\right) + \lambda_4 \frac{\partial^2 T_4}{\partial z^2} + JU, \qquad \begin{matrix} H_2 \le z \le H_3 \\ R_2 \le r \le R_1 \end{matrix}, \tag{5}$$

$$\rho_4 c_4 \frac{\partial T_4}{\partial t} = \frac{\lambda_4}{r}\frac{\partial}{\partial r}\left(r\frac{\partial T_4}{\partial r}\right) + \lambda_4 \frac{\partial^2 T_4}{\partial z^2}, \qquad \begin{matrix} H_2 \le z \le H_3 \\ R_{reac1} \le r \le R_2 \end{matrix}, \tag{6}$$

$$\rho_5 c_5 \frac{\partial T_5}{\partial t} = \frac{\lambda_5}{r}\frac{\partial}{\partial r}\left(r\frac{\partial T_5}{\partial r}\right) + \lambda_5 \frac{\partial^2 T_5}{\partial z^2}, \qquad \begin{matrix} H_1 \le z \le H_2 \\ R_1 \le r \le R_{reac2} \end{matrix}, \tag{7}$$

$$\rho_5 c_5 \frac{\partial T_5}{\partial t} = \frac{\lambda_5}{r}\frac{\partial}{\partial r}\left(r\frac{\partial T_5}{\partial r}\right) + \lambda_5 \frac{\partial^2 T_5}{\partial z^2} + JU, \qquad \begin{matrix} H_1 \le z \le H_2 \\ R_2 \le r \le R_1 \end{matrix}, \tag{8}$$

$$\rho_5 c_5 \frac{\partial T_5}{\partial t} = \frac{\lambda_5}{r}\frac{\partial}{\partial r}\left(r\frac{\partial T_5}{\partial r}\right) + \lambda_5 \frac{\partial^2 T_5}{\partial z^2}, \qquad \begin{matrix} H_1 \le z \le H_2 \\ R_{reac1} \le r \le R_1 \end{matrix}, \tag{9}$$

$$\rho_6 c_6 \frac{\partial T_6}{\partial t} = \frac{\lambda_6}{r}\frac{\partial}{\partial r}\left(r\frac{\partial T_6}{\partial r}\right) + \lambda_6 \frac{\partial^2 T_6}{\partial z^2}, \qquad \begin{matrix} z_0 \le z \le H_1 \\ R_w \le r \le R_{reac2} \end{matrix}, \quad \begin{matrix} H_3 \le z \le H_s \\ R_w \le r \le R_{reac2} \end{matrix}, \tag{10}$$

$$\rho_7 c_7 \frac{\partial T_7}{\partial t} = \frac{\lambda_7}{r}\frac{\partial}{\partial r}\left(r\frac{\partial T_7}{\partial r}\right) + \lambda_7 \frac{\partial^2 T_7}{\partial z^2} + JU, \qquad \begin{matrix} H_3 \le z \le H_s \\ R_s \le r \le R_w \end{matrix}, \quad \begin{matrix} z_0 \le z \le H_1 \\ R_s \le r \le R_w \end{matrix}, \tag{11}$$

At the initial moment of time, the temperature field is constant:

$$T_i = T_{i\mathcal{H}}, \tag{12}$$

Boundary conditions for equations (1) - (11):

$$\Gamma_0\colon \lambda_1 \frac{\partial T_1}{\partial r} = 0, \tag{13}$$

$$\Gamma_{e1}\text{: } \lambda_i \frac{\partial T_i}{\partial z} = 0\text{, i=1,2,3,} \tag{14}$$

$$\Gamma_s\text{: } -\lambda_i \frac{\partial T_i}{\partial z} = \alpha_s (T_s - T_i)\text{, i=1,2,3,} \tag{15}$$

$$\Gamma_{e2}\text{: } \lambda_7 \frac{\partial T_7}{\partial r} = \alpha_e (T_e - T_7), \tag{16}$$

$$\Gamma_{e3}\text{: } \lambda_7 \frac{\partial T_7}{\partial r} = \alpha_e (T_e - T_7), \tag{17}$$

$$\Gamma_{1.1}\text{: } \lambda_4 \frac{\partial T_4}{\partial z} = \lambda_1 \frac{\partial T_1}{\partial z}, T_4{=}T_1, \tag{18}$$

$$\Gamma_{1.2}\text{: } \lambda_4 \frac{\partial T_4}{\partial z} = \alpha_e (T_e - T_4), \tag{19}$$

$$\Gamma_{2.1}\text{: } \lambda_5 \frac{\partial T_5}{\partial z} = \lambda_1 \frac{\partial T_1}{\partial z}, T_5{=}T_1 \tag{20}$$

$$\Gamma_{2.2}\text{: } -\lambda_5 \frac{\partial T_5}{\partial z} = \alpha_e (T_e - T_5), \tag{21}$$

$$\Gamma_3\text{: } \lambda_4 \frac{\partial T_4}{\partial z} = \lambda_5 \frac{\partial T_5}{\partial z}, T_4{=}T_5 \tag{22}$$

$$\Gamma_{4.1}\text{: } \lambda_1 \frac{\partial T_1}{\partial r} = \lambda_4 \frac{\partial T_4}{\partial r}, T_1{=}T_4 \tag{23}$$

$$\Gamma_{4.2}: \lambda_1 \frac{\partial T_1}{\partial r} = \lambda_5 \frac{\partial T_5}{\partial r}, T_1 = T_5 \quad (24)$$

$$\Gamma_{5.1}: \lambda_4 \frac{\partial T_4}{\partial r} = \alpha_e (T_e - T_4), \quad (25)$$

$$\Gamma_{5.2}: \lambda_5 \frac{\partial T_5}{\partial r} = \alpha_e (T_e - T_5), \quad (26)$$

$$\Gamma_{6.1}: T_6 = T_e, \quad (27)$$

$$\Gamma_{6.2}: T_6 = T_e, \quad (28)$$

$$\Gamma_{6.3}: T_6 = T_e, \quad (29)$$

$$\Gamma_{6.4}: -\lambda_6 \frac{\partial T_6}{\partial z} = \alpha_s (T_s - T_6), \quad (30)$$

$$\Gamma_{7.1}: \lambda_3 \frac{\partial T_3}{\partial r} = \lambda_7 \frac{\partial T_7}{\partial r}, T_3 = T_7 \quad (31)$$

$$\Gamma_{7.2}: \lambda_3 \frac{\partial T_3}{\partial r} = \lambda_7 \frac{\partial T_7}{\partial r}, T_3 = T_7 \quad (32)$$

where T_i, ρ_i, λ_i, c_i are the temperature, density, thermal conductivity and heat capacity of the corresponding subdomains (i = 1, ..., 7), J is the current, U is the voltage, α_e, α_s are the heat transfer coefficients, r, z are spatial coordinates, t is time, R_w is the surface layer of water, R_s is the outer radius of the trunk, R_1 is the interface between the bark and the subcortical zone, R_2 is the interface between the core of the trunk and the subcortical zone, H_s is the height of the tree trunk, H_1H_2 - thickness of the zone of reactive wood (lower zone of the branch), H_2H_3 - thickness of the upper zone of the branch, Γ_i, $\Gamma_{i.j}$ - designations of the boundaries of areas

Numerical studies were carried out with the following initial data (pine wood, core) (Zabolotny et al., 1995): ρ=500 kg/m^3; c=1670 J/(kg·K); λ=0.12 W/(m·K). Subcrustal layer parameters: ρ=500 kg/m^3; c=2600 J/(kg·K); λ=0.35 W/(m·K). Thermophysical characteristics of the bark: ρ=500 kg/m^3; c=1670 J/(kg·K); λ= 0.12 W/(m·K). Thermophysical characteristics of reactive wood: ρ=550 kg/m^3; c=1670 J/(kg·K); λ=0.12 W/(m·K). Thermophysical properties of the water layer: ρ=1000 kg/m^3; c=4215 J/(kg·K); λ=0.569 W/(m·K). Geometrical characteristics of the solution area: R_w=0.255 m; R_s=0.25 m; R_1=0.245 m; R_2=0.235 m; R_{reac1}=0.225 m; R_{reac2}=0.5 m; H_s=17 m; H_1H_2=0.05 m; H_2H_3=0.05 m. Environmental parameters: T_e=300 K, T_s=297 K, α_e=80 W/(m^2·K), α_s=20 W/(m^2·K).

RESULTS AND DISCUSSION

The formulated system of equations (1) - (11) with boundary and initial conditions (12) - (32) was solved locally by a one-dimensional finite-difference method [16]. To solve difference analogs of one-dimensional equations, the marching method was used (Samarskiy, 1983).

Figure 2. Temperature distribution along the radius and height of the tree trunk at different times: a) - t = 0.01 s; b) - 0.1 s; c) - 0.3 s; d) - 0.5 s

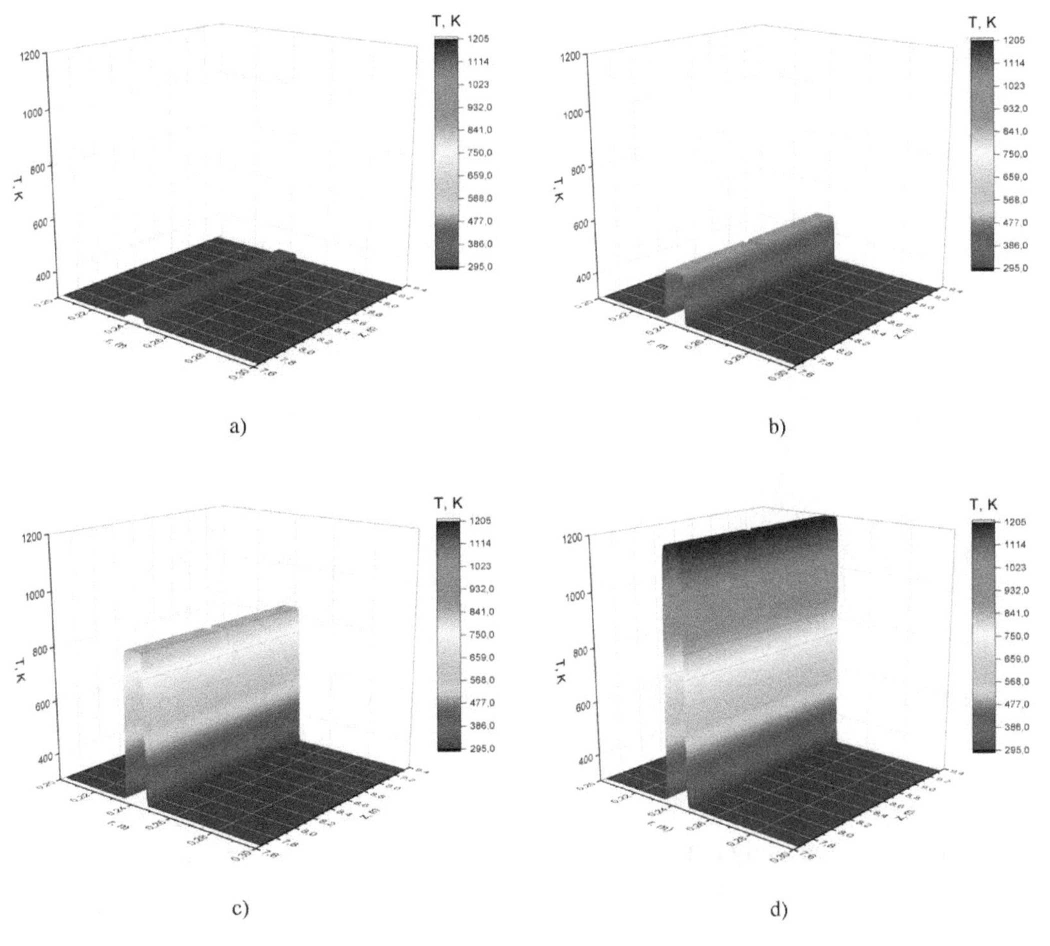

A typical ignition scenario is considered. A cloud-to-ground lightning discharge of negative polarity with a duration of 500 ms with a peak current of 23.5 kA and a voltage of 100 kV strikes a pine trunk (Popov et al., 1977). Fig. 2 shows the temperature distribution along the radius and height of the trunk of a coniferous tree at different points in time: a) - t = 0.01 s; b) - 0.1 s; c) - 0.3 s; d) - 0.5 s. Fig. 4 shows the dependences of the heat flux from the subcrustal zone of the tree trunk on time at different heights from the ground surface. Fig. 5 shows the dependence of the temperature of the border of the subcrustal zone of the tree trunk on time at certain levels above the earth's surface. In fig. 4,5, numbers 1 and 2 indicate the curves for ordinary (z = 8.2 m) and reactive (z = 8.47 m) wood, respectively.

Figure 3. Temperature distribution along the radius and height of the tree trunk at different points in time, taking into account the surface layer of water: a) - t = 0.01 s; b) - 0.1 s; c) - 0.3 s; d) - 0.5 s

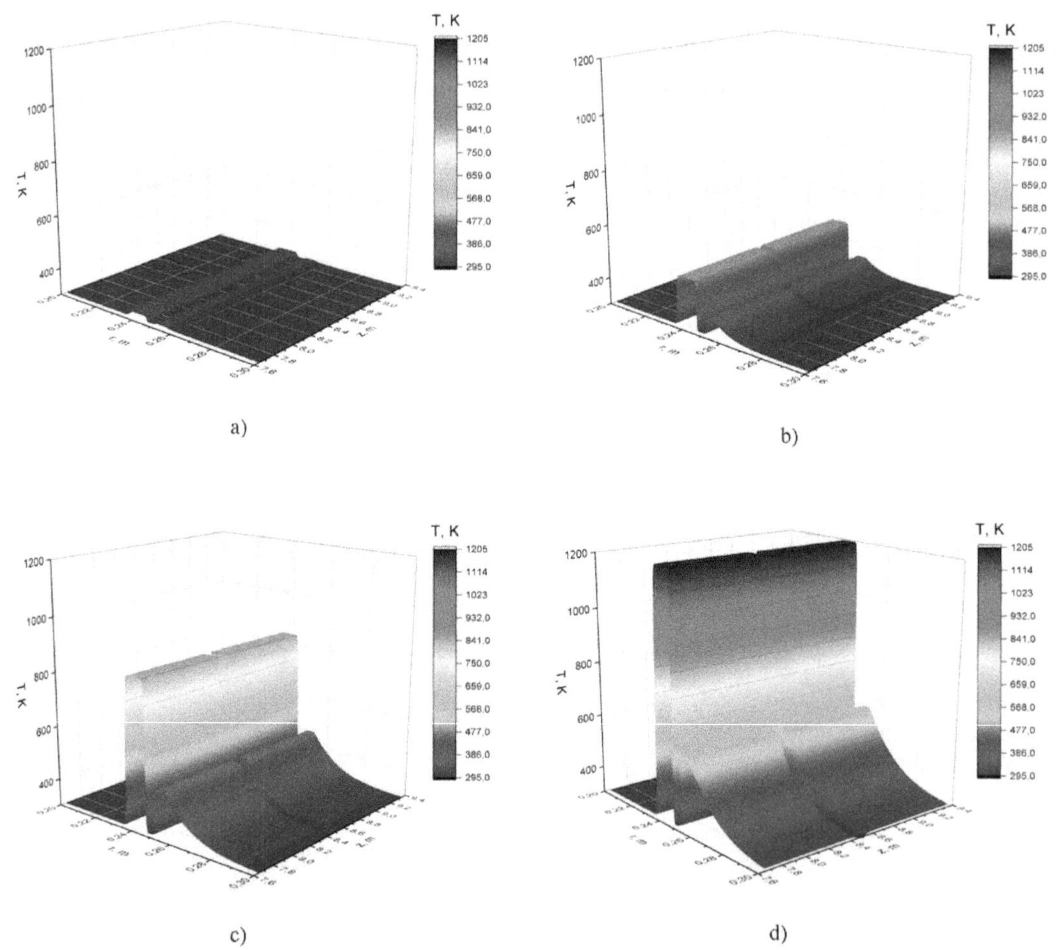

In this work, it is proposed to use experimentally determined ignition conditions as ignition criteria. Within the framework of the study under consideration, these conditions were numerically determined for ordinary and reactive wood. Conditions (Zabolotny et al., 1995) were used as ignition criteria (Table

1). A similar approach in determining the conditions of ignition was used in (Beltsova and Korolchenko, 2008).

Figure 3 shows the distribution of temperature along the radius and height of the trunk of a coniferous tree at different times, taking into account the surface layer of water: a) - t = 0.01 s; b) - 0.1 s; c) - 0.3 s; d) - 0.5 s.

Table 1. Experimentally determined pine ignition conditions (Zabolotny et al., 1995)

Ignition delay, s	Heat flux, kW/m^2	Surface temperature, K
63.5	12.5	658
45.0	21	700
11.1	42	726
2.6	84	773
0.4	210	867

Research has been carried out on the influence of the current-voltage characteristics of a cloud-to-ground lightning discharge on the ignition of a coniferous tree trunk. Table 2 shows the ignition condi-

Table 2. Numerically determined ignition conditions for a coniferous tree depending on the discharge voltage at a current J=23.5 kA (z=8.2 m)

Voltage, U, kV	Ignition conditions (Zabolotny et al., 1995)	Surface temperature, K	Heat flux, kW/m^2
1 – 85	No	<867	<210
90	Yes	867	242
95	Yes	867	246
100	Yes	867	249
105	Yes	867	252
110	Yes	867	255

Table 3. Numerically determined conditions for the ignition of a tree trunk depending on the current at a voltage of U=100 kV (z=8.2 m)

Current, J, kA	Ignition conditions (Zabolotny et al., 1995)	Surface temperature, K	Heat flux, kW/m^2
1 – 20	No	<867	<210
23.5	Yes	867	249
30	Yes	867	264
35	Yes	867	274

tions, determined numerically, depending on the voltage of the ground lightning discharge at the current J=23.5 kA for a representative section of the tree trunk (z=8.2 m).

Table 3 shows the ignition conditions, determined numerically, depending on the current of the cloud-to-ground lightning discharge at a voltage of U=100 kV for a representative section of a tree trunk.

Analysis of the results presented in Tables 2 and 3 shows that a typical cloud-to-ground lightning discharge with parameters (U=100 - 110 kV and J=23.5 - 35 kA) causes ignition of ordinary coniferous wood. These conditions were established earlier in the approximation of a one-dimensional mathematical model of ignition of a coniferous tree trunk by a cloud-to-ground lightning discharge (Kuznetsov and Baranovskiy, 2008). It can be concluded that most cloud-to-ground lightning discharges cause ignition of ordinary coniferous wood.

Table 4 shows the theoretically determined ignition conditions for reactive wood depending on the voltage of a cloud-to-ground lightning discharge at a current J=23.5 kA (z=8.47 m)

Table 4. Numerically determined conditions for the ignition of reactive wood depending on the discharge voltage at a current J=23.5 kA (z=8.47 m)

Voltage, U, kV	Ignition conditions (Zabolotny et al., 1995)	Surface temperature, K	Heat flux, kW/m^2
1 – 85	No	<867	<210
90		<867	215
95		<867	227
100		<867	238
105		<867	250
110	Yes	867	255

Table 5 shows the theoretically determined conditions for the ignition of reactive wood depending on the current of a cloud-to-ground lightning discharge at a voltage of U=100 kV (z=8.47 m).

Analysis of the results of numerical modeling allows us to conclude that the ignition conditions (Zabolotny et al., 1995) are not always satisfied for reactive wood, while they are simultaneously fulfilled for ordinary wood. In a parametric study of the influence of the lightning discharge voltage, it was found that only at a voltage of 110 kV and higher is the fulfillment of the ignition conditions observed (Zabolotny

Table 5. Numerically determined ignition conditions for reactive wood depending on the current at a voltage of U=100 kV (z=8.47 m)

Current, J, kA	Ignition conditions (Zabolotny et al., 1995)	Surface temperature, K	Heat flux, kW/m^2
1 – 20	No	<867	<210
23.5		<867	238
30	Yes	867	264
35	Yes	867	274

Figure 4. Dependence of the heat flux from the subcrustal zone to the ignition surface on time: 1 - ordinary wood (z = 8.2 m, r = 0.244 m), 2 - reactive wood (z = 8.47 m, r = 0.244 m)

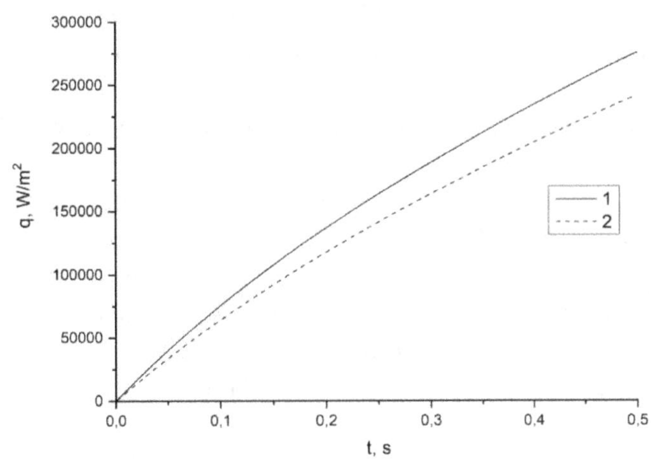

et al., 1995). A similar variation in the current of a cloud-to-ground lightning discharge showed that ignition conditions are achieved only at a current of 30 kA and higher (Zabolotny et al., 1995).

Analyzing in aggregate the results presented in Fig. 2, we can conclude that as a result of the action of the considered cloud-to-ground lightning discharge, the tree trunk in the subcrustal zone is heated to temperatures at which forest fuels burn (more than 1200 K). Therefore, it can be concluded that the tree trunk ignites under the conditions under consideration. A logical consequence of the differences in the thermophysical properties of ordinary and reactive wood is a lower temperature field in the region of reactive wood with other identical conditions. A cloud-to-ground lightning discharge with medium volt-ampere characteristics cannot ignite reactive coniferous wood.

Figure 5. Dependence of the temperature of the border of the subcrustal zone of the coniferous tree trunk on time: 1 - ordinary wood (z = 8.2 m, r = 0.244 m), 2 - reactive wood (z = 8.47 m, r = 0.244 m)

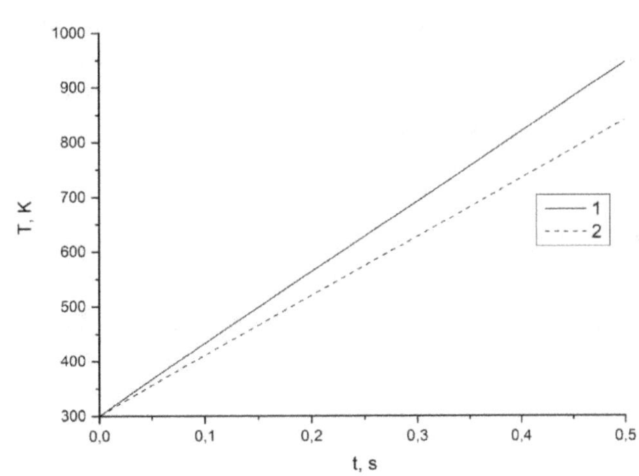

Figure 3 shows the distribution of temperature along the radius of a tree trunk in the case of a surface layer of water. In this case, two areas of increased temperature can be observed. The first area is the subcortical area. The second area is the surface layer of water and the near-surface layer of the bark. In the second area, the temperature is approximately two times lower, since part of the heat is removed to the environment.

Analysis of the results shows that in the second zone, over a long period of exposure, temperatures and heat fluxes are reached, which can lead to the ignition of the bark layer. According to the data in Table 1, the ignition delay in this case will increase by about 10 seconds relative to the subcortical zone layer. However, reactive wood cannot be ignited.

An analysis of the dependences of the heat flux and the temperature of the subcrustal zone boundary (Fig. 4 and Fig. 5) shows that in terms of temperature (867 K) and heat flux (249 kW/m^2), the ignition conditions for ordinary wood are achieved for fairly typical parameters of a lightning discharge. At the same time, there is a lag in heating the ignition surface and lower heat fluxes to it in the case of reactive wood.

The results obtained allow us to conclude that the presence of branches changes the nature of the ignition of the trunk of a coniferous tree. This effect cannot be taken into account in the one-dimensional setting. In the presence of a surface layer of water, the bark wood can also be ignited, but the branches do not ignite in the reactive wood zone.

CONCLUSION

An important scientific and practical problem was solved. A physical and mathematical model of coniferous tree ignition was developed, taking into account the localization of reactive wood. The ignition conditions are established, which are characteristic for a typical range of the lightning discharge. In the zone of reactive wood, an area of lower temperature is formed, and as a consequence, it can be expected that a lightning discharge with the same current-voltage characteristics will more likely lead to the ignition of tall trees with a small number of branches. Conversely, the ignition of trees with a developed system of branches and twigs is unlikely. In the presence of a surface layer of water, the bark wood can also be ignited, but the branches do not ignite in the reactive wood zone.

The results obtained create conditions for the further development of models for the ignition of fire hazardous materials and deterministic-probabilistic approaches to assessing the fire danger in forests (Adab et al., 2013; Al Janabi et al., 2017; Bui et al., 2017; CWFIS, 2020; Cardille et al., 2001; Chen et al., 2017; Chen et al., 2020; Curt et al., 2016; Glagolev et al., 2020; Papagiannaki et al., 2020; Wierzchowski et al., 2002; Yankovich and Yankovich, 2020).

ACKNOWLEDGMENT

This work was implemented with the financial support of the Russian Foundation for Basic Research. [Project N 17-29-05093].

REFERENCES

Adab, H., Kanniah, K. D., & Solaimani, K. (2013). Modeling forest fire in the northeast of Iran using remote sensing and GIS techniques. *Natural Hazards*, *65*(3), 1723–1743. doi:10.100711069-012-0450-8

Al Janabi, S., Al Shourbaji, I., & Salman, M. A. (2017). *Assessing the suitability of soft computing approaches for forest fires prediction.* Applied Computing and Informatics. doi:10.1016/j.aci.2017.09.006

Baranovskiy, N. V. (2007). *Mathematical modeling of the most likely scenarios and conditions for forest fires* (PhD Thesis). Tomsk: Tomsk State University. (in Russian)

Baranovskiy, N. V. (2017). Forest fire danger assessment using SPMD-model of computation for massive parallel system. *International Review on Modeling and Simulations*, *10*(3), 193–201. doi:10.15866/iremos.v10i3.10570

Beltsova, T. G., & Korolchenko, O. N. (2008). Indicators of flammability of fire retardant wood. Fire and Explosion Safety, 17(4), 31 - 33. (in Russian)

Bui, D. T., Bui, Q.-T., Nguyen, Q.-P., Pradhan, B., Nampak, H., & Trinh, P. T. (2017). A hybrid artificial intelligence using GIS-based neural-fuzzy inference system and particle swarm optimization for forest fire susceptibility modeling at a tropical area. *Agricultural and Forest Meteorology*, *233*, 32–44. doi:10.1016/j.agrformet.2016.11.002

Burke, C. P., & Jones, D. L. (1996). On the polarity and continuing current in unusually large lightning flashes deduced from ELF events. *Journal of Atmospheric and Solar-Terrestrial Physics*, *58*(5), 531–548. doi:10.1016/0021-9169(95)00054-2

Canadian Wildland Fire Information System. (n.d.). Access: http://cwfis.cfs.nrcan.gc.ca/home

Cardille, J. A., Ventura, S. J., & Turner, M. G. (2001). Environmental and Social Factors of Influencing Wildfires in the Upper Midwest, United States. *Ecological Applications*, *11*(1), 111–127. doi:10.1890/1051-0761(2001)011[0111:EASFIW]2.0.CO;2

Chen, C., Xu, T., Zhao, D., & Lei, P. (2020). Study on the Wildfires Occurring Risk Based on Fuzzy FTA. *IOP Conference Series. Earth and Environmental Science*, *440*(5), 052100. Advance online publication. doi:10.1088/1755-1315/440/5/052100

Chen, W., Xie, X., Wang, J., Pradhan, B., Hong, H., Tien Bui, D., Duan, Z., & Ma, J. (2017). A comparative study of logistic model tree, random forest and classification and regression tree models for spatial prediction of landslide susceptibility. *Catena*, *151*, 147–160. doi:10.1016/j.catena.2016.11.032

Conedera, M., Cesti, G., Pezzatti, G. B., Zumbrunnen, T., & Spinedi, F. (2006). Lightning-induced fires in the Alpine region: An increasing problem. *Forest Ecology and Management*, *234*(Supplement 1.), S68. doi:10.1016/j.foreco.2006.08.096

Cummins, K. L., Murphy, M. J., Bardo, E. A., Hiscox, W. L., Pyle, R. B., & Pifer, A. E. (1998). A combined TOA/MDF technology upgrade of the U.S. national lightning detection network. *Journal of Geophysical Research*, *103*(D8), 9035–9044. doi:10.1029/98JD00153

Curt, T., Frejaville, T., & Lahaye, S. (2016). Modelling the spatial patterns of ignition causes and fire regime features in southern France: Implications for the prevention policy. *International Journal of Wildland Fire*, *25*(7), 785–796. doi:10.1071/WF15205

Esau, K. (1980) Anatomy of seed plants. Moscow: Mir. (in Russian)

Flannigan, M. D., & Wotton, B. M. (1991). Lightning-ignited fires in northwestern Ontario. *Canadian Journal of Forest Research*, *21*(3), 277–287. doi:10.1139/x91-035

Glagolev, V., Bazhenov, R., Povkh, I., Chuyko, O., & Dolmatova, O. (2020). Designing and producing a geographic information system to forecast vegetation fire danger according to weather conditions. *IOP Conference Series. Materials Science and Engineering*, *862*(5), 052019. Advance online publication. doi:10.1088/1757-899X/862/5/052019

Ivanov, V. A. (2006). *Methodological basis for the classification of forests of Central Siberia by the degree of fire hazard from thunderstorms* (PhD dissertation). Krasnoyarsk: SibGTU. (in Russian)

Kozlov, V. I., & Mullayarov, V. A. (2004). *Thunderstorm activity in Yakutia*. YaF Publishing House SB RAS. (in Russian)

Kurbatskiy, N. P., & Kostyrina, T. V. (1977). US National Fire Hazard Calculation System. In *Detection and analysis of forest fires* (pp. 38–90). IliD SB AS USSR. (in Russian)

Kuznetsov, G. V., & Baranovskiy, N. V. (2008). Mathematical modeling of ignition of a coniferous tree by cloud-to-ground lightning discharge. Fire and Explosion Safety, 17(3), 41 - 45. (in Russian)

Larjavaara, M., Kuuluvainen, T., & Rita, H. (2005). Spatial distribution of lightning-ignited fires in Finland. *Forest Ecology and Management*, *208*(1-3), 177–188. doi:10.1016/j.foreco.2004.12.005

Latham, D., & Williams, E. (2001). *Lightning and forest fires // Forest fires: Behavior and Ecological Effects*. Elsevier.

Papagiannaki, K., Giannaros, T. M., Lykoudis, S., Kotroni, V., & Lagouvardos, K. (2020). Weather-related thresholds for wildfire danger in a Mediterranean region: The case of Greece. *Agricultural and Forest Meteorology*, *291*, 108076. Advance online publication. doi:10.1016/j.agrformet.2020.108076

Popov, B. G., Verevkin, V. N., & Bondarev, V. A. (1977). *Static electricity in the chemical industry*. Publishing Chemistry. (in Russian)

Samarskiy, A. A. (1983). *Theory of difference schemes*. Nauka. (in Russian)

Soriano, L. R., De Pablo, F., & Tomas, C. (2005). Ten-year study of cloud-to-ground lightning activity in the Iberian Peninsula. *Journal of Atmospheric and Solar-Terrestrial Physics*, *67*(16), 1632–1639. doi:10.1016/j.jastp.2005.08.019

Vilyunov, V. N. (1984). *The theory of ignition of condensed substances*. Science. (in Russian)

Wierzchowski, J., Heathcott, M., & Flannigan, M. D. (2002). Lightning and lightning fire, central Cordillera, Canada. *International Journal of Wildland Fire*, *11*(1), 41–51. doi:10.1071/WF01048

Yankovich, E. P., & Yankovich, K. S. (2020). Classification of Territory on Forest Fire Danger Level Using GIS and Remote Sensing. In N. V. Baranovskiy (Ed.), *Predicting, Monitoring, and Assessing Forest Fire Dangers and Risks* (pp. 258–267). IGI Global. doi:10.4018/978-1-7998-1867-0.ch011

Zabolotny, A.E., Zabolotnaya, M.M., Zabolotnaya, Y.A., & Timoshin, V.N. (1995). Determination of safe use zones of solid fuel generators of fire extinguishing aerosols. *Issues of Special Engineering*, (7 – 8), 15 - 21. (in Russian)

ADDITIONAL READING

Azadifar, M., Rubinstein, M., Rachidi, F., Rakov, V. A., Diendorfer, G., Schulz, W., & Pavanello, D. (2019). A Study of a Large Bipolar Lightning Event Observed at the Säntis Tower, IEEE Transactions on Electromagnetic Compatibility, 61 (3). *Article*, *61*(8715461), 796–806. doi:10.1109/TEMC.2019.2913220

De Paiva, A. R., Saba, M. M. F., Naccarato, K. P., Schumann, C., Jaques, R., Da Silva Ferro, M. A., & Warner, T. A. (2014) Detection of upward lightning by lightning location systems, 2014 International Conference on Lightning Protection, ICLP 2014, Article N° 6973425, pp. 1824-1826. DOI: 10.1109/ICLP.2014.6973425

Lu, G., Cummer, S. A., Tian, Y., Zhang, H., Lyu, F., Wang, T., Stanley, M. A., Yang, J., & Lyons, W. A. (2016). Sprite produced by consecutive impulse charge transfers following a negative stroke: Observation and simulation. *Journal of Geophysical Research*, *121*(8), 4082–4092. doi:10.1002/2015JD024644

Nag, A., Murphy, M. J., Schulz, W., & Cummins, K. L. (2014) Lightning locating systems: Characteristics and validation techniques, 2014 International Conference on Lightning Protection, ICLP 2014, Article N° 6973283, pp. 1070-1082. DOI: 10.1109/ICLP.2014.6973283

Nag, A., Rakov, V. A., & Murphy, M. J. (2014) Measurement of preliminary breakdown pulse trains in cloud-to-ground lightning using lightning locating systems, 2014 International Conference on Lightning Protection, ICLP 2014, статья N° 6973356, pp. 1437-1444. DOI: 10.1109/ICLP.2014.6973356

Saba, M. M. F., Shigihara, M., & Kodaira, C. Y. (2018) Observation of a lightning flashover with a high-speed video camera, 34th International Conference on Lightning Protection, ICLP 2018, Article N° 8503282. DOI: 10.1109/ICLP.2018.8503282

Stolzenburg, M., Marshall, T. C., Karunarathne, S., Karunarathna, N., & Orville, R. E. (2015). An M component with a concurrent dart leader traveling along different paths during a lightning flash. *Journal of Geophysical Research*, *120*(19), 10267–10284. doi:10.1002/2015JD023417

Wang, X., Yuan, P., Cen, J., & Xue, S. (2016). Correlation between the spectral features and electric field changes for natural lightning return stroke followed by continuing current with M-components. *Journal of Geophysical Research*, *121*(14), 8615–8624. doi:10.1002/2016JD025314

Xue, S., Yuan, P., Cen, J., Li, Y., & Wang, X. (2015). Spectral observations of a natural bipolar cloud-to-ground lightning. *Journal of Geophysical Research*, *120*(5), 1972–1979. doi:10.1002/2014JD022598

Yuan, P., Chen, R., Su, Y., Fan, B., Wang, X. (2019) Characteristic parameters of positive cloud-to-ground lightning channel, Journal of Earth System Science, 128 (6), статья N° 164, . Doi:10.100712040-019-1196-4

KEY TERMS AND DEFINITIONS

Cloud-to-Ground Lightning Discharge: An electrical discharge during a thunderstorm that occurs between a cloud and the earth's surface. It is a natural source of forest fires.

Forest Fire: Uncontrolled aerothermochemical phenomenon characterized by step-by-step mechanism which includes following stages: inert heating, moisture evaporation, high temperature terpens evaporation, dry organic matter pyrolysis, flammable combustion and smoldering.

Forest Fuel: It can be considered like dead and live forest fuel. Main types of forest fuel which can be involved in combustion during forest fire: ground forest fuel (needles, leaves and dry grass, small branches) and crown forest fuel (needles, small branches).

Ignition: Inflammation of forest fuel caused by definite source of high temperature or energy.

Ignition Delay: Time before flame flash after forest fuel heating.

Lightning Activity: An atmospheric phenomenon characterized by discharges of the cloud-to-cloud and cloud-to-ground class.

Mathematical Simulation: The production of a computer model of forest fire conditions and prerequisites, especially for the purpose of study.

Meteorological Parameters: Physical characteristics of local weather conditions in the forested area under consideration. Key parameters include ambient temperature, soil temperature, precipitation, wind speed, solar radiation, cloud cover, dew point temperature. These parameters are used for mathematical modeling of the drying of a layer of forest fuel.

Monitoring: Monitoring refers to the periodic calculation of the parameters of forest fire danger with a portion of information available in real time.

Prediction: Under the prediction of forest fires is the calculation of the parameters of forest fire danger with a certain projection in advance in order to have enough time to anticipate an emergency. The calculation in this case is carried out in a mode ahead of the real time of the development of the catastrophe - the occurrence of a forest fire.

Chapter 3
Three-Dimensional Mathematical Models to Simulate Coniferous Tree Ignition

ABSTRACT

The results of the numerical implementation of the spatial formulation for the problem of coniferous tree ignition by an electric current of a cloud-to-ground lightning discharge are presented. The problem is solved in a three-dimensional formulation in a cylindrical coordinate system. The axisymmetric formulation of the problem is considered. Localization of reactive wood, structural inhomogeneity of the bark, current-voltage characteristics of a cloud-to-ground lightning discharge, and chemical reaction in the gas phase are taken into account. The conditions for the ignition of a coniferous tree in the typical range of the discharge effect are revealed.

INTRODUCTION

Further development of international and domestic methods for predicting forest fire danger in the boreal zone is possible by improving the physical and mathematical models of ignition of coniferous trees and forest fuels in a thunderstorm situation (Baranovskiy and Kuznetsov, 2017; Baranovskiy et al., 2017; Latham and Williams, 2001; Zhang et al., 2020). Such models are implemented in one-dimensional and two-dimensional formulations. Ignition of the trunk wood is possible only at certain values of the parameters of a cloud-to-ground lightning discharge (polarity, peak current and voltage, as well as duration) (Soriano et al., 2005).

From experimental studies (Zabolotny et al., 1995) it is known that ignition of wood by an energy source is possible when a certain level of heat fluxes and temperature of its surface are reached. An important factor in the fire danger of trees is their complex spatial structure, in particular, the presence of branches and structural heterogeneity of the bark. In real ignitions, wood is heated and pyrolyzed with

DOI: 10.4018/978-1-7998-7250-4.ch003

the release of gaseous products, the ignition of which occurs under certain conditions. It is necessary to determine the ignition delay time of a conifer when an electric current of a cloud-to-ground lightning discharge passes through its trunk and to identify the spatial effects of the process under consideration. For this reason, it is advisable to simulate the process of ignition of a coniferous trunk under the action of a lightning discharge in a three-dimensional setting.

The aim of the study is mathematical modeling of the ignition of a coniferous tree by electric current of a cloud-to-ground lightning discharge in a spatial setting, taking into account the main factors and determining the conditions for its ignition.

MATHEMATICAL STATEMENT

According to (Esau, 1980; Yavorskiy and Seleznev, 1984), the electric current of a cloud-to-ground lightning discharge passes in the subcortical zone of the coniferous tree trunk without penetrating inside. The main assumptions and suggestions: 1) reactive wood is formed in the lower part of the branches (Esau, 1980); 2) the approximation of an "ideal" crack in the bark is used (Baranovskiy and Kuznetsov, 2017); 3) in the calculations only the part of the branch growing from the trunk is considered. The rest of the continuation is not considered, since it was previously established that during the exposure to electric current the outer part of the branch does not have time to warm up; 4) the main product of pyrolysis is carbon monoxide (Grishin, 1997); 5) the leading chemical reaction is the oxidation of carbon monoxide to carbon dioxide (Grishin and Shipulina, 2002); 6) the tree is considered as a conductor of the resistor type, for which Ohm's and Joule-Lenz's laws are valid (Yavorskiy and Seleznev, 1984).

The following physical model is adopted to describe the simulated process. A freestanding coniferous tree is considered. At a fixed moment in time, a lightning discharge of a certain polarity and duration strikes the tree trunk. It is believed that the current-voltage characteristics of the discharge are the same for different sections of the tree trunk. As a result of the flow of electric current in the subcortical zone, the wood is heated by the release of Joule heat. In the process of further heating, thermal decomposition of wood occurs with the formation of gaseous pyrolysis products. The pyrolysis products instantly enter the gas phase and mix with the oxidizing agent. At certain temperatures and concentrations of reagents, a chemical oxidation reaction of carbon monoxide occurs. It is considered that ignition occurs if the critical values of the following parameters are reached: 1) the heat flux from the chemical reaction zone exceeds the heat flux from the subcrustal zone of the tree; 2) the temperature of the gas mixture. The influence of wood moisture on the ignition process is neglected. The solution domain is shown in Fig. 1a, and the boundaries of the regions are indicated in Fig. 1.b.

The ignition of a coniferous tree by a cloud-to-ground lightning discharge is described by a system of three-dimensional nonstationary nonlinear equations of heat conduction and diffusion (1) - (26). For numerical implementation, a locally one-dimensional finite-difference method was used (Samarskiy and Vabishchevich, 2001). Difference analogs of one-dimensional heat conduction equations were solved by the marching method in combination with the simple iteration method (Samarskiy, 1983).

$$\rho_1 c_1 \frac{\partial T_1}{\partial t} = \frac{\lambda_1}{r}\frac{\partial}{\partial r}\left(r\frac{\partial T_1}{\partial r}\right) + \frac{\lambda_1}{r^2}\frac{\partial^2 T_1}{\partial \phi^2} + \lambda_1 \frac{\partial^2 T_1}{\partial z^2} - Q_p k_p \rho_1 \phi_{13} \exp\left(-\frac{E_1}{RT_1}\right), \qquad (1)$$

Figure 1. Solution domain (a) and subdomain boundaries (b)

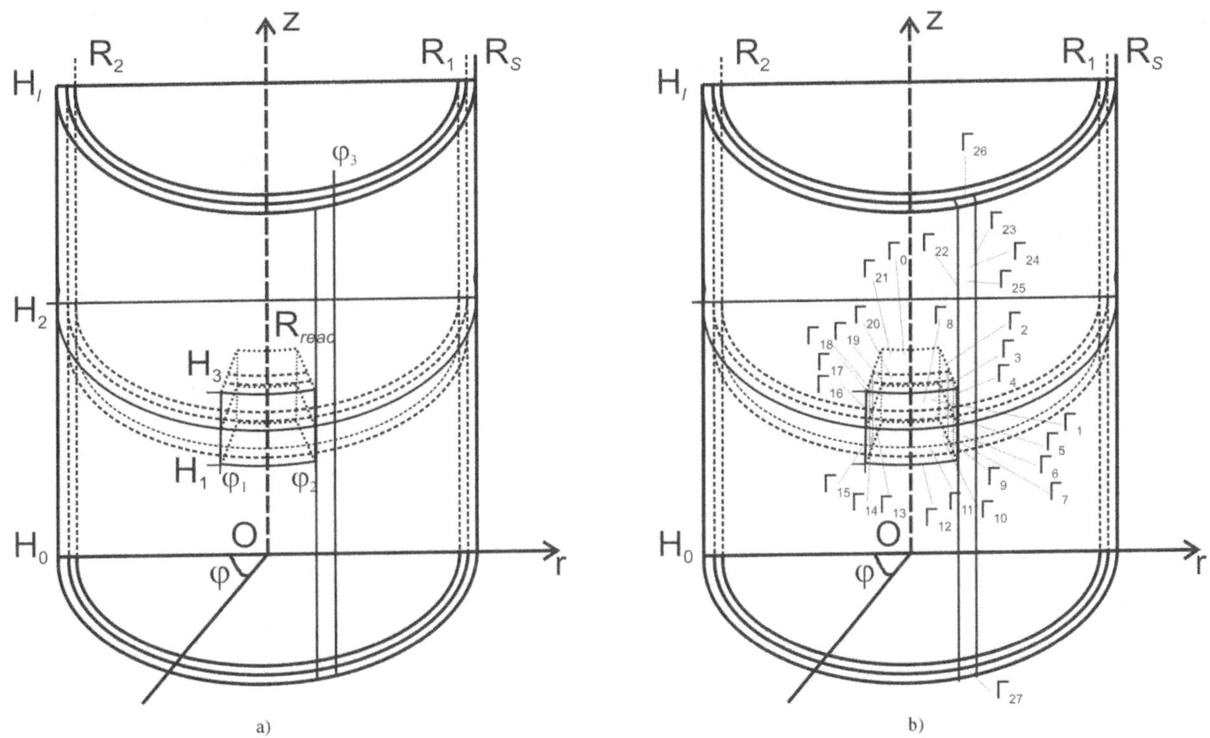

$$\begin{array}{lllll} z_0 \le z \le H_1 & H_1 < z < H_3 & H_1 < z < H_3 & H_1 < z < H_3 & H_3 \le z \le z_t \\ 0 \le r \le R_2\,, & 0 \le r \le R_2\,, & 0 \le r \le R_{reac}\,, & 0 \le r \le R_2\,, & 0 \le r \le R_2\,, \\ 0 \le \phi \le \pi & 0 \le \phi \le \phi_1 & \phi_1 \le \phi \le \phi_2 & \phi_2 \le \phi \le \pi & 0 \le \phi \le \pi \end{array}$$

$$\rho_2 c_2 \frac{\partial T_2}{\partial t} = \frac{\lambda_2}{r} \frac{\partial}{\partial r}\left(r \frac{\partial T_2}{\partial r} \right) + \frac{\lambda_2}{r^2} \frac{\partial^2 T_2}{\partial \phi^2} + \lambda_2 \frac{\partial^2 T_2}{\partial z^2} + JU - Q_p k_p \rho_2 \phi_{13} \exp\left(-\frac{E_1}{R T_2} \right), \tag{2}$$

$$\begin{array}{llll} z_0 \le z \le H_1 & H_1 < z < H_3 & H_1 < z < H_3 & H_3 \le z \le z_t \\ R_2 < r \le R_1\,, & R_2 < r \le R_1\,, & R_2 < r \le R_1\,, & R_2 < r \le R_1\,, \\ 0 \le \phi \le \pi & 0 \le \phi \le \phi_1 & \phi_2 \le \phi \le \pi & 0 \le \phi \le \pi \end{array}$$

$$\rho_3 c_3 \frac{\partial T_3}{\partial t} = \frac{\lambda_3}{r} \frac{\partial}{\partial r}\left(r \frac{\partial T_3}{\partial r} \right) + \frac{\lambda_3}{r^2} \frac{\partial^2 T_3}{\partial \phi^2} + \lambda_3 \frac{\partial^2 T_3}{\partial z^2} - Q_p k_p \rho_3 \phi_{13} \exp\left(-\frac{E_1}{R T_3} \right), \tag{3}$$

$$z_0 \le z \le H_1 \quad z_0 \le z \le H_1 \quad H_1 < z < H_3 \quad H_1 < z < H_3 \quad H_3 \le z \le z_t \quad H_3 \le z \le z_t$$
$$R_1 < r \le R_S, \quad R_1 < r \le R_S, \quad R_1 < r \le R_S, \quad R_1 < r \le R_S, \quad R_1 < r \le R_S, \quad R_1 < r \le R_S,$$
$$0 \le \phi \le \phi_2 \quad \phi_3 \le \phi \le \pi \quad 0 \le \phi \le \phi_1 \quad \phi_3 \le \phi \le \pi \quad 0 \le \phi \le \phi_2 \quad \phi_3 \le \phi \le \pi$$

$$\rho_4 c_4 \frac{\partial T_4}{\partial t} = \frac{\lambda_4}{r}\frac{\partial}{\partial r}\left(r\frac{\partial T_4}{\partial r}\right) + \frac{\lambda_4}{r^2}\frac{\partial^2 T_4}{\partial \phi^2} + \lambda_4 \frac{\partial^2 T_4}{\partial z^2} - Q_p k_p \rho_4 \phi_{13} \exp\left(-\frac{E_1}{RT_4}\right), \quad \begin{matrix} H_2 < z \le H_3 \\ R_{reac} \le r \le R_2, \\ \phi_1 \le \phi \le \phi_2 \end{matrix} \tag{4}$$

$$\rho_5 c_5 \frac{\partial T_5}{\partial t} = \frac{\lambda_5}{r}\frac{\partial}{\partial r}\left(r\frac{\partial T_5}{\partial r}\right) + \frac{\lambda_5}{r^2}\frac{\partial^2 T_5}{\partial \phi^2} + \lambda_5 \frac{\partial^2 T_5}{\partial z^2} + JU - Q_p k_p \rho_5 \phi_{13} \exp\left(-\frac{E_1}{RT_5}\right), \quad \begin{matrix} H_2 < z \le H_3 \\ R_2 < r \le R_1, \\ \phi_1 \le \phi \le \phi_2 \end{matrix} \tag{5}$$

$$\rho_6 c_6 \frac{\partial T_6}{\partial t} = \frac{\lambda_6}{r}\frac{\partial}{\partial r}\left(r\frac{\partial T_6}{\partial r}\right) + \frac{\lambda_6}{r^2}\frac{\partial^2 T_6}{\partial \phi^2} + \lambda_6 \frac{\partial^2 T_6}{\partial z^2} - Q_p k_p \rho_6 \phi_{13} \exp\left(-\frac{E_1}{RT_6}\right), \quad \begin{matrix} H_2 < z \le H_3 \\ R_1 < r \le R_S, \\ \phi_1 \le \phi \le \phi_2 \end{matrix} \tag{6}$$

$$\rho_7 c_7 \frac{\partial T_7}{\partial t} = \frac{\lambda_7}{r}\frac{\partial}{\partial r}\left(r\frac{\partial T_7}{\partial r}\right) + \frac{\lambda_7}{r^2}\frac{\partial^2 T_7}{\partial \phi^2} + \lambda_7 \frac{\partial^2 T_7}{\partial z^2} - Q_p k_p \rho_7 \phi_{13} \exp\left(-\frac{E_1}{RT_7}\right), \quad \begin{matrix} H_1 \le z \le H_2 \\ R_{reac} \le r \le R_2, \\ \phi_1 \le \phi \le \phi_2 \end{matrix} \tag{7}$$

$$\rho_8 c_8 \frac{\partial T_8}{\partial t} = \frac{\lambda_8}{r}\frac{\partial}{\partial r}\left(r\frac{\partial T_8}{\partial r}\right) + \frac{\lambda_8}{r^2}\frac{\partial^2 T_8}{\partial \phi^2} + \lambda_8 \frac{\partial^2 T_8}{\partial z^2} + JU - Q_p k_p \rho_8 \phi_{13} \exp\left(-\frac{E_1}{RT_8}\right), \quad \begin{matrix} H_1 \le z \le H_2 \\ R_2 < r \le R_1, \\ \phi_1 \le \phi \le \phi_2 \end{matrix} \tag{8}$$

$$\rho_9 c_9 \frac{\partial T_9}{\partial t} = \frac{\lambda_9}{r}\frac{\partial}{\partial r}\left(r\frac{\partial T_9}{\partial r}\right) + \frac{\lambda_9}{r^2}\frac{\partial^2 T_9}{\partial \phi^2} + \lambda_9 \frac{\partial^2 T_9}{\partial z^2} - Q_p k_p \rho_9 \phi_{13} \exp\left(-\frac{E_1}{RT_9}\right), \quad \begin{matrix} H_1 \le z \le H_2 \\ R_1 < r \le R_S, \\ \phi_1 \le \phi \le \phi_2 \end{matrix} \tag{9}$$

$$\rho_g c_g \frac{\partial T_g}{\partial t} = \frac{\lambda_g}{r}\frac{\partial}{\partial r}\left(r\frac{\partial T_g}{\partial r}\right) + \frac{\lambda_g}{r^2}\frac{\partial^2 T_g}{\partial \phi^2} + \lambda_g \frac{\partial^2 T_g}{\partial z^2} + Q_5(1-\nu_5)R_5, \quad \begin{matrix} H_0 \le z \le H_t \\ R_1 < r \le R_S, \\ \phi_2 \le \phi \le \phi_3 \end{matrix} \quad \begin{matrix} H_0 \le z \le H_t \\ R_S < r \le R_e, \\ 0 \le \phi \le \pi \end{matrix} \tag{10}$$

$$\frac{\partial C_{10}}{\partial t} = \frac{D}{r}\frac{\partial}{\partial r}\left(r\frac{\partial C_{10}}{\partial r}\right) + \frac{D}{r^2}\frac{\partial^2 C_{10}}{\partial \phi^2} + D\frac{\partial^2 C_{10}}{\partial z^2} - R_5\frac{M_4}{2M_5}, \quad \begin{matrix} H_0 \le z \le H_t \\ R_1 < r \le R_S, \\ \phi_2 \le \phi \le \phi_3 \end{matrix} \quad \begin{matrix} H_0 \le z \le H_t \\ R_S < r \le R_e, \\ 0 \le \phi \le \pi \end{matrix} \tag{11}$$

$$\frac{\partial C_{11}}{\partial t} = \frac{D}{r}\frac{\partial}{\partial r}\left(r\frac{\partial C_{11}}{\partial r}\right) + \frac{D}{r^2}\frac{\partial^2 C_{11}}{\partial \phi^2} + D\frac{\partial^2 C_{11}}{\partial z^2} - R_5, \quad \begin{matrix} H_0 \le z \le H_t \\ R_1 < r \le R_S, \\ \phi_2 \le \phi \le \phi_3 \end{matrix} \quad \begin{matrix} H_0 \le z \le H_t \\ R_S < r \le R_e, \\ 0 \le \phi \le \pi \end{matrix} \tag{12}$$

$$\sum_{i=10}^{12} C_i = 1, \tag{13}$$

$$\sum_{i=13}^{14} \phi_i = 1, \tag{14}$$

$$\rho_1 \frac{\partial \phi_{13}}{\partial t} = -k_p \rho_1 \phi_{13} \exp\left(-\frac{E_1}{RT_1}\right), \tag{15}$$

$$\rho_2 \frac{\partial \phi_{13}}{\partial t} = -k_p \rho_2 \phi_{13} \exp\left(-\frac{E_1}{RT_2}\right), \tag{16}$$

$$\rho_3 \frac{\partial \phi_{13}}{\partial t} = -k_p \rho_3 \phi_{13} \exp\left(-\frac{E_1}{RT_3}\right), \tag{17}$$

$$\rho_4 \frac{\partial \phi_{13}}{\partial t} = -k_p \rho_4 \phi_{13} \exp\left(-\frac{E_1}{RT_4}\right), \tag{18}$$

$$\rho_5 \frac{\partial \phi_{13}}{\partial t} = -k_p \rho_5 \phi_{13} \exp\left(-\frac{E_1}{RT_5}\right), \tag{19}$$

$$\rho_6 \frac{\partial \phi_{13}}{\partial t} = -k_p \rho_6 \phi_{13} \exp\left(-\frac{E_1}{RT_6}\right), \tag{20}$$

$$\rho_7 \frac{\partial \phi_{13}}{\partial t} = -k_p \rho_7 \phi_{13} \exp\left(-\frac{E_1}{RT_7}\right), \tag{21}$$

$$\rho_8 \frac{\partial \phi_{13}}{\partial t} = -k_p \rho_8 \phi_{13} \exp\left(-\frac{E_1}{RT_8}\right), \tag{22}$$

$$\rho_9 \frac{\partial \phi_{13}}{\partial t} = -k_p \rho_9 \phi_{13} \exp\left(-\frac{E_1}{RT_9}\right), \tag{23}$$

$$R_5 = k_5 M_{11} T^{-2.25} \exp\left(-\frac{E_5}{RT_g}\right) \cdot \begin{cases} x_{10}^{0.25} x_{11}, x_{10} > 0.05 \\ x_{10} x_{11}, x_{10} \leq 0.05 \end{cases}, \tag{24}$$

$$x_i = \frac{C_i}{\sum_{k=10}^{12} \frac{C_k}{M_k} M_i}, \tag{25}$$

$$P = \frac{\rho RT}{M}, \frac{1}{M} = \frac{C_{10}}{M_{10}} + \frac{C_{11}}{M_{11}} + \frac{C_{12}}{M_{12}}. \tag{26}$$

Initial and boundary conditions for the system of equations (1) - (26).

Boundary conditions outside the branch area:

$$T_i\big|_{t=0} = T_{i0}, \tag{27}$$

$$r=0,\ \lambda_i \frac{\partial T_i}{\partial r} = 0, \tag{28}$$

$$r=R_2,\ \lambda_1 \frac{\partial T_1}{\partial r} = \lambda_2 \frac{\partial T_2}{\partial r},\ T_1=T_2, \tag{29}$$

$$r=R_1,\ \lambda_2 \frac{\partial T_2}{\partial r} = \lambda_3 \frac{\partial T_3}{\partial r},\ T_2=T_3, \tag{30}$$

$$r=R_S,\ \lambda_3 \frac{\partial T_3}{\partial r} = \lambda_g \frac{\partial T_g}{\partial r},\ T_3=T_g, \tag{31}$$

$$r=R_e,\ T_g=T_e, \tag{32}$$

$$\varphi=0,\ \frac{\partial T_i}{\partial \phi} = 0, \tag{33}$$

$$\varphi=\pi,\ \frac{\partial T_i}{\partial \phi} = 0, \tag{34}$$

$$z=z_b,\ \lambda_i \frac{\partial T_i}{\partial z} = 0, \tag{35}$$

$$z=z_t,\ \lambda_i \frac{\partial T_i}{\partial z} = 0, \tag{36}$$

$$z=z_t,\ \lambda_i \frac{\partial T_i}{\partial z} = 0, \tag{37}$$

Boundary conditions on the inside of the branch:

$$\Gamma_0,\ \lambda_4 \frac{\partial T_4}{\partial r} = \lambda_1 \frac{\partial T_1}{\partial r},\ T_4=T_1, \tag{38}$$

$$\Gamma_1,\ \lambda_7 \frac{\partial T_7}{\partial r} = \lambda_1 \frac{\partial T_1}{\partial r},\ T_7=T_1, \tag{39}$$

Conditions at the boundary of the right side of a branch and a crack:

$$\Gamma_2,\ \lambda_4 \frac{\partial T_4}{\partial \phi} = \lambda_1 \frac{\partial T_1}{\partial \phi},\ T_4=T_1, \tag{40}$$

$$\Gamma_3,\ \lambda_5 \frac{\partial T_5}{\partial \phi} = \lambda_2 \frac{\partial T_2}{\partial \phi},\ T_5=T_2, \tag{41}$$

$$\Gamma_4,\ \lambda_6 \frac{\partial T_6}{\partial \phi} = \lambda_g \frac{\partial T_g}{\partial \phi},\ T_6=T_g, \tag{42}$$

$$\Gamma_5,\ \lambda_7 \frac{\partial T_7}{\partial \phi} = \lambda_1 \frac{\partial T_1}{\partial \phi},\ T_7=T_1, \tag{43}$$

$$\Gamma_6,\ \lambda_8 \frac{\partial T_8}{\partial \phi} = \lambda_2 \frac{\partial T_2}{\partial \phi},\ T_8=T_2, \tag{44}$$

$$\Gamma_7,\ \lambda_9 \frac{\partial T_9}{\partial \phi} = \lambda_g \frac{\partial T_g}{\partial \phi},\ T_9=T_g, \tag{45}$$

Boundary conditions at the outer cut of the branch:

$$\Gamma_8,\ \lambda_6 \frac{\partial T_6}{\partial r} = \lambda_g \frac{\partial T_g}{\partial r},\ T_6 = T_g, \tag{46}$$

$$\Gamma_9,\ \lambda_9 \frac{\partial T_9}{\partial r} = \lambda_g \frac{\partial T_g}{\partial r},\ T_9 = T_g, \tag{47}$$

Boundary conditions on the lower edge of the branch:

$$\Gamma_{10},\ \lambda_1 \frac{\partial T_1}{\partial z} = \lambda_7 \frac{\partial T_7}{\partial z},\ T_1 = T_7, \tag{48}$$

$$\Gamma_{11},\ \lambda_2 \frac{\partial T_2}{\partial z} = \lambda_8 \frac{\partial T_8}{\partial z},\ T_2 = T_8, \tag{49}$$

$$\Gamma_{12},\ \lambda_3 \frac{\partial T_3}{\partial z} = \lambda_9 \frac{\partial T_9}{\partial z},\ T_3 = T_9, \tag{50}$$

Boundary conditions on the left side of the branch:

$$\Gamma_{13},\ \lambda_1 \frac{\partial T_1}{\partial \phi} = \lambda_7 \frac{\partial T_7}{\partial \phi},\ T_1 = T_7, \tag{51}$$

$$\Gamma_{14},\ \lambda_2 \frac{\partial T_2}{\partial \phi} = \lambda_8 \frac{\partial T_8}{\partial \phi},\ T_2 = T_g, \tag{52}$$

$$\Gamma_{15},\ \lambda_3 \frac{\partial T_3}{\partial \phi} = \lambda_9 \frac{\partial T_9}{\partial \phi},\ T_3 = T_9, \tag{53}$$

$$\Gamma_{16},\ \lambda_1 \frac{\partial T_1}{\partial \phi} = \lambda_4 \frac{\partial T_4}{\partial \phi},\ T_1 = T_4, \tag{54}$$

$$\Gamma_{17}, \lambda_2 \frac{\partial T_2}{\partial \phi} = \lambda_5 \frac{\partial T_4}{\partial \phi}, T_2 = T_4, \tag{55}$$

$$\Gamma_{18}, \lambda_3 \frac{\partial T_3}{\partial \phi} = \lambda_6 \frac{\partial T_6}{\partial \phi}, T_3 = T_6, \tag{56}$$

Boundary conditions on the upper edge of the branch:

$$\Gamma_{19}, \lambda_3 \frac{\partial T_3}{\partial z} = \lambda_6 \frac{\partial T_6}{\partial z}, T_3 = T_6, \tag{57}$$

$$\Gamma_{20}, \lambda_2 \frac{\partial T_2}{\partial z} = \lambda_5 \frac{\partial T_5}{\partial z}, T_2 = T_5, \tag{58}$$

$$\Gamma_{21}, \lambda_1 \frac{\partial T_1}{\partial z} = \lambda_4 \frac{\partial T_4}{\partial z}, T_1 = T_4, \tag{59}$$

Conditions at the crack boundary, excluding the right edge of the branch:

$$\Gamma_{22}, \lambda_3 \frac{\partial T_3}{\partial \phi} = \lambda_g \frac{\partial T_g}{\partial \phi}, T_3 = T_g, \tag{60}$$

$$\rho D \frac{\partial C_{10}}{\partial \phi} = 0, \tag{61}$$

$$\rho D \frac{\partial C_{11}}{\partial \phi} = 0, \tag{62}$$

$$\Gamma_{23}, \lambda_g \frac{\partial T_g}{\partial \phi} = \lambda_3 \frac{\partial T_3}{\partial \phi}, T_g = T_3, \tag{63}$$

$$\rho D \frac{\partial C_{10}}{\partial \phi} = 0\,, \tag{64}$$

$$\rho D \frac{\partial C_{11}}{\partial \phi} = 0\,, \tag{65}$$

$$\Gamma_{24},\ \lambda_2 \frac{\partial T_2}{\partial r} = \lambda_g \frac{\partial T_g}{\partial r},\ T_2 = T_g, \tag{66}$$

$$\rho D \frac{\partial C_{10}}{\partial r} = 0\,, \tag{67}$$

$$\rho D \frac{\partial C_{11}}{\partial r} = Y_5\,, \tag{68}$$

$$\Gamma_{25},\ T_g = T_e, \tag{69}$$

$$\rho D \frac{\partial C_{10}}{\partial r} = 0\,, \tag{70}$$

$$\rho D \frac{\partial C_{11}}{\partial r} = 0\,, \tag{71}$$

$$\Gamma_{26},\ \lambda_g \frac{\partial T_g}{\partial z} = 0\,, \tag{72}$$

$$\rho D \frac{\partial C_{10}}{\partial z} = 0\,, \tag{73}$$

$$\rho D \frac{\partial C_{11}}{\partial z} = 0, \tag{74}$$

$$\Gamma_{27}, \ \lambda_g \frac{\partial T_g}{\partial z} = 0, \tag{75}$$

$$\rho D \frac{\partial C_{10}}{\partial z} = 0, \tag{76}$$

$$\rho D \frac{\partial C_{11}}{\partial z} = 0, \tag{77}$$

Initial conditions:

$$T_i\Big|_{t=0} = T_{i0}, \quad C_i\Big|_{t=0} = C_{i0}, \quad \phi_i\Big|_{t=0} = \phi_{i0}, \tag{66}$$

where T_i, ρ_i, c_i, λ_i are the temperature, density, heat capacity and thermal conductivity, respectively, of the inner part of the trunk (i=1), subcortical zone (i=2), bark (i=3), upper part of the branch (i=4,5,6), the lower part of the branch (i=7,8,9); C_i, M_i — concentration and molar mass of oxygen (i=10), carbon monoxide (i=11) and inert components (i=12); α - heat transfer coefficient; J is the current; U is the voltage; φ_i — volume fractions of organic matter (i=13) and gas phase (i=14); P is the pressure in the gas phase; x_i are auxiliary factors; M is molar mass; Q_p is the thermal effect of pyrolysis; k_p is the preexponential factor of the pyrolysis reaction; E_1 is the activation energy of the pyrolysis reaction; Q_5 is the heat effect of the oxidation reaction of carbon monoxide; k_5 is the preexponential factor of the oxidation reaction of carbon monoxide; E_5 is the activation energy of the oxidation reaction of carbon monoxide; ν_5 - fraction of heat absorbed by the wood layer; Y_5 - mass flow; D is the diffusion coefficient; R is the universal gas constant; r,φ,z - cylindrical coordinates; t is time. R_e is the boundary of the computational domain, R_s is the outer radius of the trunk, R_1 is the interface between the bark and the subcortical zone, R_2 is the interface between the core of the trunk and the subcortical zone, R_{reac} is the left boundary of the branch originating from the tree trunk, H_1H_2 is the thickness of the reactive wood zone (lower parts of the branch), H_2H_3 - thickness of the upper part of the branch, Γ_j - designations of the boundaries of the regions. Indices "e" and "0" correspond to the parameters of the external environment and parameters at the initial moment of time. Indexes "b", "t" correspond to the parameters at the lower and upper boundaries of the computational domain along the vertical of the wellbore. The indices in the designation of the boundaries of the structural heterogeneity (branches and cracks) are intended for their numbering.

Numerical modeling was carried out using the following initial data (pine wood, inner part): ρ=500 kg/m3; c=1670 J/(kg·K); λ=0.12 W/(m·K). Subcrustal layer parameters: ρ=500 kg/m^3; c=2600 J/(kg·K); λ=0.35 W/(m·K). Thermophysical characteristics of the bark: ρ=500 kg/m^3; c=1670 J/(kg·K); λ=0.12

W/(m·K). Thermophysical characteristics of reactive wood: ρ=650 kg/m^3; c=1670 J/(kg·K); λ=0.12 W/(m·K). Geometric characteristics of the solution area: R_e=0.3 m; R_s=0.25 m; R_1=0.245 m; R_2=0.235 m; R_{reac}=0.225 m. H_1H_2=0.05 m, H_2H_3=0.05 m. Environmental parameters: T_e=300 K.

RESULTS AND DISCUSSION

Numerical modeling in a three-dimensional setting shows that an increase in the dimension of the problem does not allow revealing new patterns. The main results coincide with the calculations obtained for a set of two-dimensional problems: (a) taking into account the localization of reactive wood; b) taking into account the chemical reactions in the gas phase and c) a one-dimensional formulation that takes into account the influence of the M-components of the cloud-to-ground lightning discharge. Thus, the three-dimensional formulation is a generalization of the coniferous tree ignition problem. Fig. 2 shows the temperature distribution in a horizontal section. Fig. 3 shows the distribution of the gas phase components at the moment of ignition in different sections. Fig. 4 illustrates the dependence of volume fractions of dry organic matter and gas in the region of the subcortical zone.

The influence of the current-voltage characteristics of a cloud-to-ground lightning discharge on the ignition of a tree trunk as a result of the passage of an electric current of a cloud-to-ground lightning discharge has been investigated. Taking into account the approximation of an "ideal" crack, the same ignition delays were obtained as for simpler one-dimensional and two-dimensional models. The use of this model made it possible to establish that it is in the places of localization of cracks that coniferous wood ignites. Moreover, the crack in the bark should deepen almost to the subcrustal layer. Since the presence of a crustal barrier even 1.5 mm thick does not allow the gas mixture in the crack to warm up to temperatures at which the oxidation reaction of carbon monoxide to carbon dioxide occurs during the period of the cloud-to-ground lightning discharge current. At the same time, the concentration of reacting gases in such a crack is sufficient (Fig. 3), so that in the case of an increase in the temperature of the gas phase to a certain value, such a reaction begins.

As a result of the action of the considered cloud-to-ground lightning discharge, the tree trunk in the subcrustal zone is heated to high temperatures and wood pyrolysis occurs. In the area of the crack, upon reaching a certain temperature of the gas mixture and the concentrations of the reacting components, they interact with the release of heat. The response area is located at a small distance from the bottom of the fracture and is characterized by a zone of elevated temperature in the trunk fracture.

It was found that the volume fraction of the gas phase in the decomposing subcrustal layer has a slight increase near the bottom of the crack. To this zone there is an influx of heat from a chemical reaction in a gas mixture and wood pyrolysis occurs more intensively. Numerical calculations have shown that as a result of the inflow of heat from the gas mixture, thermal decomposition of the side walls of the crack occurs, but the value of the mass flux of the gaseous combustible components of pyrolysis is less than 5% (the exact value depends on the radial coordinate) of the value of the mass flux of these components at the boundary passing through bottom of the crack. Therefore, the influx of carbon monoxide from these boundaries can be neglected. The boundary conditions for the diffusion equation of the combustible component presented in this article should be used. The bulk of combustible gaseous components is formed during the pyrolysis of the subcrustal zone of the tree trunk. This is evidenced by the results presented in Fig. 4. Moreover, any noticeable decomposition of wood and the transition of combustible components to the gas phase occurs at exposure times of the considered cloud-to-ground lightning discharge more

Figure 2. Temperature distributions in the horizontal section of the tree trunk: a) 0.01 s; b) 0.1 s; c) 0.3 s; 0.463 s (ignition moment)

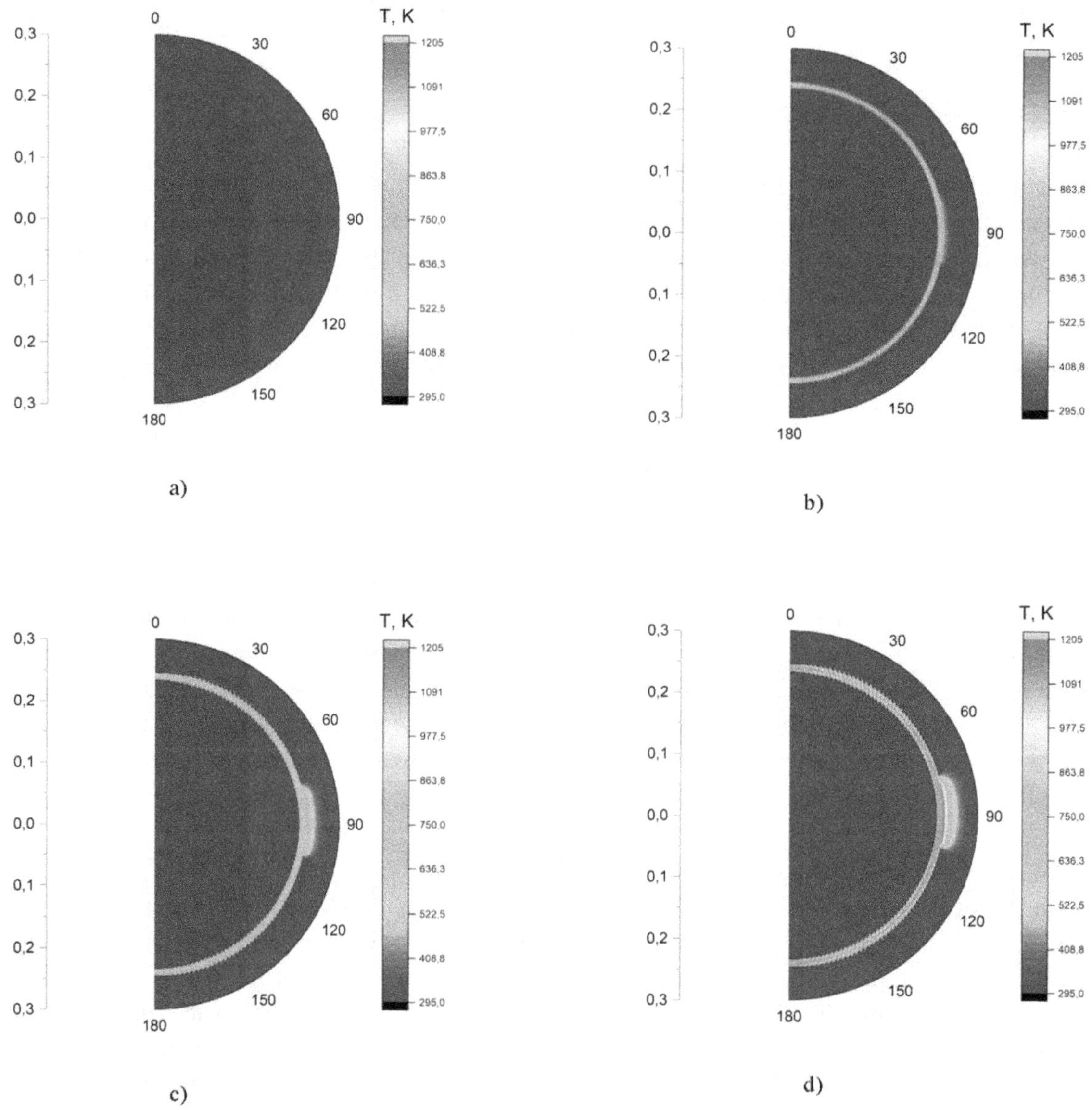

than 0.3 s. That is, a short-term discharge with the indicated volt-ampere characteristics will not lead to ignition of the wood of the tree trunk, for at least two reasons. First, the insufficient concentration of combustible components in the gas mixture. Secondly, insufficient heating of the gas mixture itself.

The development of an improved three-dimensional mathematical model allows us to evaluate the ignition conditions in coniferous forests. This model can be applied in practice and implemented as part of forest fire danger predicting systems (Adab et al., 2013; Al Janabi et al., 2017; Belikova and Glebova, 2020; Bui et al., 2017; CWFIS, 2020; Cardille et al., 2001; Chen et al., 2017; Chen et al., 2020; Curt et al., 2016; Ganteaume and Guerra, 2018; Glagolev et al., 2020; Glagolev and Zubareva, 2020; Papagiannaki

Figure 3. The dependences of the concentrations of the components in gas phase at the time of ignition: 1 - section in the crack; 2 - section outside the crack

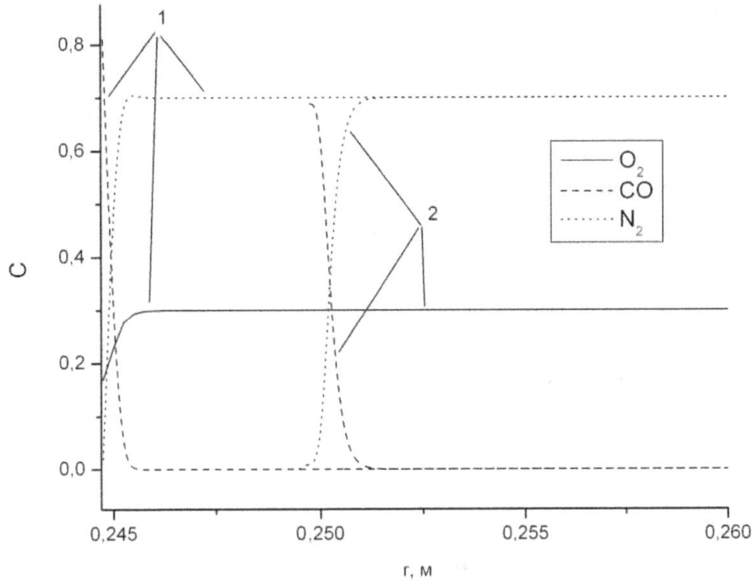

et al., 2020; Silva net al., 2020; Yankovich and Yankovich, 2020; Zharikova, 2020). The source data for such a system should be cloud-to-cloud-to-ground lightning networks (Cummins et al., 1998). They allow you to record the necessary discharge parameters. Another possible solution is to use a software pseudo-

Figure 4. The time dependence of the volume fractions of phases in the subcortical zone of the trunk: 1 - organic matter, 2 - gas phase

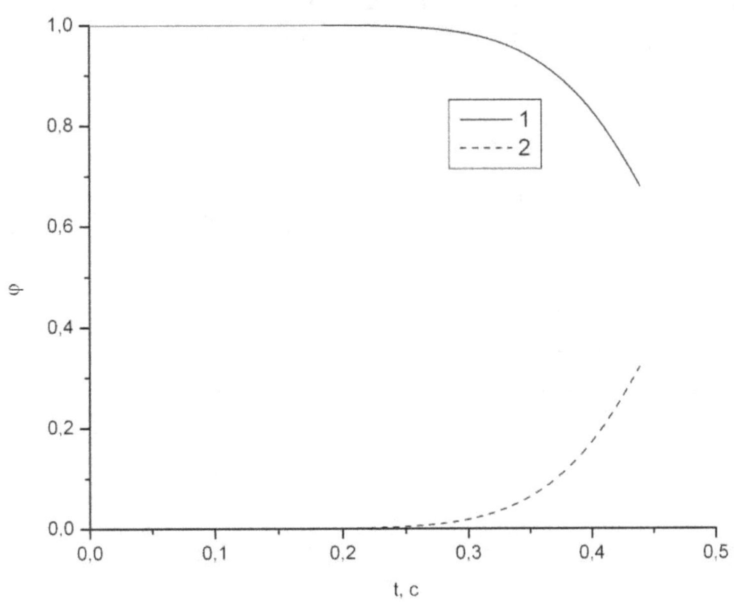

random number generator in information-computing systems for monitoring and predicting forest fires to set the discharge parameters in a certain range of values. Thermophysical characteristics of any type of wood can be determined by fairly simple methods (Zabolotny et al., 1995) for each moisture level.

The main results correspond to those obtained by simplified statements:

1) In the zone of reactive wood, a low temperature field is formed. In addition, the supply of pyrolysis products from this zone is carried out in smaller quantities. Thus, the presence of reactive wood should reduce the probability of a forest fire.
2) Wood heating occurs in a narrow subcrustal zone. This explains less damage to pines compared to deciduous trees, which sometimes tears from the inside.
3) Ignition of coniferous wood occurs in the gas phase in the area of the crack. It is here that the conditions for ignition are created. The temperature reaches a certain value at which, at certain concentrations, the components begin to react.
4) The presence of a crustal barrier in a crack even 1.5 mm thick reduces the temperature in the gas phase and ignition may not occur.
5) Any noticeable decomposition of wood and the transition of combustible components to the gas phase occurs at exposure times of the considered cloud-to-ground lightning discharge more than 0.3 s. That is, a short-term discharge with the indicated volt-ampere characteristics will not lead to ignition of the tree trunk, for at least two reasons. First, the insufficient concentration of combustible components in the gas mixture. Secondly, insufficient heating of the gas mixture itself.
6) Differences in crack thickness in real conditions do not significantly affect the ignition delay of coniferous trees.
7) The values of the ignition delay are slightly longer than as a result of the implementation of a two-dimensional formulation in the approximation of an "ideal" crack. This is explained by the closeness of the crack to the branch (this is the option considered in the calculations). The values of the ignition delay time depending on the current-voltage characteristics are presented in table. 1 and 2.
8) The presence of M-components of a cloud-to-ground lightning discharge practically does not affect the process of heating wood and ignition of coniferous trees.

Table 1. The tree ignition delay depending on the discharge voltage at the current J=23.5 kA

Voltage, U, kV	Ignition delay, s
1 – 85	No
90	0.516
95	0.486
100	0.463
105	0.441
110	0.423

Table 2. The tree trunk ignition delay depending on the current at a voltage of U=100 kV

Current, J, kA	Ignition delay, s
1 – 20	No
23.5	0.463
30	0.366
35	0.317

CONCLUSION

A generalized formulation of the problem for the ignition of a coniferous tree by a cloud-to-ground lightning discharge is presented. As a result of computational experiments, it was found that an increase in the dimension of the problem does not allow revealing new physical effects in comparison with a series of two-dimensional and one-dimensional formulations. However, this statement summarizes the statements developed earlier and allows us to consider the whole range of fire danger factors in the aggregate. The results obtained allow us to talk about the prospects of developing a module for predicting fires from thunderstorms for a system for monitoring forest fire situations. The development of such systems will reduce the negative consequences of forest fires (Baranovskiy and Kuznetsov, 2017). In practice, one should focus on the use of multiprocessor computing systems and the parallelization of computing operations. It is known that the time to obtain a forecast should be less than the catastrophe induction period (Grishin, 2002). Therefore, the use of parallel computing systems will make it possible to obtain predictive information in a mode that is ahead of the real time of the development of the disaster. Personal computers of this generation are not suitable for fire monitoring using three-dimensional setting in large forested areas.

ACKNOWLEDGMENT

This work was implemented with the financial support of the Russian Foundation for Basic Research. [Project N 17-29-05093].

REFERENCES

Adab, H., Kanniah, K. D., & Solaimani, K. (2013). Modeling forest fire in the northeast of Iran using remote sensing and GIS techniques. *Natural Hazards*, *65*(3), 1723–1743. doi:10.100711069-012-0450-8

Al Janabi, S., Al Shourbaji, I., & Salman, M. A. (2017). *Assessing the suitability of soft computing approaches for forest fires prediction*. Applied Computing and Informatics. doi:10.1016/j.aci.2017.09.006

Baranovskiy, N. V., & Kuznetsov, G. V. (2017). Coniferous tree ignition by cloud-to-cloud-to-ground lightning discharge using approximation of "ideal" crack in bark. *JP Journal of Heat and Mass Transfer*, *14*(1), 173–186. doi:10.17654/HM014010173

Baranovskiy, N. V., & Kuznetsov, G. V. (2017). *Forest fire occurrences and ecological impact prediction*. Publishing House of the Siberian Branch of the Russian Academy of Science. doi:10.15372/FOREST2017BNV

Baranovskiy, N. V., Kuznetsov, G. V., & Nemova, T. N. (2017). Mathematical computing of coniferous tree ignition by the cloud-to-cloud-to-ground lightning discharge using Joule-Lenz's law. *Iranian Journal of Electrical and Computer Engineering*, *7*(3), 1337–1346. doi:10.11591/ijece.v7i3.pp1337-1346

Belikova, M. Y., & Glebova, A. V. (2020). Cluster Analysis Algorithms for RS and WWLLN Data Processing. In N. V. Baranovskiy (Ed.), *Predicting, Monitoring, and Assessing Forest Fire Dangers and Risks* (pp. 151–181). IGI Global. doi:10.4018/978-1-7998-1867-0.ch007

Bui, D. T., Bui, Q.-T., Nguyen, Q.-P., Pradhan, B., Nampak, H., & Trinh, P. T. (2017). A hybrid artificial intelligence using GIS-based neural-fuzzy inference system and particle swarm optimization for forest fire susceptibility modeling at a tropical area. *Agricultural and Forest Meteorology*, *233*, 32–44. doi:10.1016/j.agrformet.2016.11.002

Canadian Wildland Fire Information System. (n.d.). Access: http://cwfis.cfs.nrcan.gc.ca/home

Cardille, J. A., Ventura, S. J., & Turner, M. G. (2001). Environmental and Social Factors of Influencing Wildfires in the Upper Midwest, United States. *Ecological Applications*, *11*(1), 111–127. doi:10.1890/1051-0761(2001)011[0111:EASFIW]2.0.CO;2

Chen, C., Xu, T., Zhao, D., & Lei, P. (2020). Study on the Wildfires Occurring Risk Based on Fuzzy FTA. *IOP Conference Series. Earth and Environmental Science*, *440*(5), 052100. Advance online publication. doi:10.1088/1755-1315/440/5/052100

Chen, W., Xie, X., Wang, J., Pradhan, B., Hong, H., Tien Bui, D., Duan, Z., & Ma, J. (2017). A comparative study of logistic model tree, random forest and classification and regression tree models for spatial prediction of landslide susceptibility. *Catena*, *151*, 147–160. doi:10.1016/j.catena.2016.11.032

Cummins, K. L., Murphy, M. J., Bardo, E. A., Hiscox, W. L., Pyle, R. B., & Pifer, A. E. (1998). A combined TOA/MDF technology upgrade of the U.S. national lightning detection network. *Journal of Geophysical Research*, *103*(D8), 9035–9044. doi:10.1029/98JD00153

Curt, T., Frejaville, T., & Lahaye, S. (2016). Modelling the spatial patterns of ignition causes and fire regime features in southern France: Implications for the prevention policy. *International Journal of Wildland Fire*, *25*(7), 785–796. doi:10.1071/WF15205

Esau, K. (1980). Anatomy of seed plants. Moscow: Mir. (in Russian)

Ganteaume, A., & Guerra, F. (2018). Explaining the spatio-seasonal variation of fires by their causes: The case of southeastern France. *Applied Geography (Sevenoaks, England)*, *90*, 69–81. doi:10.1016/j.apgeog.2017.11.012

Glagolev, V., Bazhenov, R., Povkh, I., Chuyko, O., & Dolmatova, O. (2020). Designing and producing a geographic information system to forecast vegetation fire danger according to weather conditions. *IOP Conference Series. Materials Science and Engineering*, *862*(5), 052019. Advance online publication. doi:10.1088/1757-899X/862/5/052019

Glagolev, V. A., & Zubareva, A. M. (2020). Using Geographic Information Systems for Study of the Natural Fire Risk on the Territory of the Jewish Autonomous Region. In N. V. Baranovskiy (Ed.), *Predicting, Monitoring, and Assessing Forest Fire Dangers and Risks* (pp. 209–231). IGI Global. doi:10.4018/978-1-7998-1867-0.ch009

Grishin, A. M. (1997). *Mathematical modeling of forest fire and new methods of fighting them. Russia.* Publishing House of the Tomsk State University.

Grishin, A. M. (2002). Modeling and forecasting disasters. Tomsk: Publishing house of Tomsk University. (in Russian)

Grishin, A. M., & Shipulina, O. V. (2002). Mathematical model for spread of crown fires in homogeneous forests and along openings. *Combustion, Explosion, and Shock Waves*, *38*(6), 622–632. doi:10.1023/A:1021127924913

Latham, D., & Williams, E. (2001). *Lightning and forest fires // Forest Fires: Behavior and Ecological Effects*. Elsevier.

Papagiannaki, K., Giannaros, T. M., Lykoudis, S., Kotroni, V., & Lagouvardos, K. (2020). Weather-related thresholds for wildfire danger in a Mediterranean region: The case of Greece. *Agricultural and Forest Meteorology*, *291*, 108076. Advance online publication. doi:10.1016/j.agrformet.2020.108076

Samarskiy, A. A. (1983). *Theory of difference schemes*. Nauka. (in Russian)

Samarskiy, A. A., & Vabishchevich, P. N. (2001). *Additive schemes for problems of mathematical physics*. Science. (in Russian)

Silva, I. D. B., Valle, M. E., Barros, L. C., & Meyer, J. F. C. A. (2020). A wildfire warning system applied to the state of Acre in the Brazilian Amazon. *Applied Soft Computing*, *89*, 106075. Advance online publication. doi:10.1016/j.asoc.2020.106075

Soriano, L. R., De Pablo, F., & Tomas, C. (2005). Ten-year study of cloud-to-cloud-to-ground lightning activity in the Iberian Peninsula. *Journal of Atmospheric and Solar-Terrestrial Physics*, *67*(16), 1632–1639. doi:10.1016/j.jastp.2005.08.019

Yankovich, E. P., & Yankovich, K. S. (2020). Classification of Territory on Forest Fire Danger Level Using GIS and Remote Sensing. In N. V. Baranovskiy (Ed.), *Predicting, Monitoring, and Assessing Forest Fire Dangers and Risks* (pp. 258–267). IGI Global. doi:10.4018/978-1-7998-1867-0.ch011

Yavorskiy, B. M., & Seleznev, Yu. A. (1984). *A reference guide to physics for university entrants and self-education*. Nauka. (in Russian)

Zabolotny, A.E., Zabolotnaya, M.M., Zabolotnaya, Y.A., & Timoshin, V.N. (1995) Determination of safe use zones of solid fuel generators of fire extinguishing aerosols. *Issues of Special Engineering*, (7-8), 15 - 21. (in Russian)

Zhang, H., Qiao, Y., Chen, H., Liu, N., Zhang, L., & Xie, X. (2020). Experimental study on flaming ignition of pine needles by simulated lightning discharge. *Fire Safety Journal*, 103029. doi:10.1016/j.firesaf.2020.103029

Zharikova, M. V. (2020). Dynamic Spatial-Distributed Fire Risk Analysis. In N. V. Baranovskiy (Ed.), *Predicting, Monitoring, and Assessing Forest Fire Dangers and Risks* (pp. 101–120). IGI Global. doi:10.4018/978-1-7998-1867-0.ch005

ADDITIONAL READING

Azadifar, M., Rubinstein, M., Rachidi, F., Rakov, V. A., Diendorfer, G., Schulz, W., & Pavanello, D. (2019). A Study of a Large Bipolar Lightning Event Observed at the Säntis Tower, IEEE Transactions on Electromagnetic Compatibility, 61 (3). *Article*, *61*(8715461), 796–806. doi:10.1109/TEMC.2019.2913220

De Paiva, A. R., Saba, M. M. F., Naccarato, K. P., Schumann, C., Jaques, R., Da Silva Ferro, M. A., & Warner, T. A. (2014) Detection of upward lightning by lightning location systems, 2014 International Conference on Lightning Protection, ICLP 2014, Article N° 6973425, pp. 1824-1826. DOI: 10.1109/ICLP.2014.6973425

Lu, G., Cummer, S. A., Tian, Y., Zhang, H., Lyu, F., Wang, T., Stanley, M. A., Yang, J., & Lyons, W. A. (2016). Sprite produced by consecutive impulse charge transfers following a negative stroke: Observation and simulation. *Journal of Geophysical Research*, *121*(8), 4082–4092. doi:10.1002/2015JD024644

Nag, A., Murphy, M. J., Schulz, W., & Cummins, K. L. (2014) Lightning locating systems: Characteristics and validation techniques, 2014 International Conference on Lightning Protection, ICLP 2014, Article N° 6973283, pp. 1070-1082. DOI: 10.1109/ICLP.2014.6973283

Nag, A., Rakov, V. A., & Murphy, M. J. (2014) Measurement of preliminary breakdown pulse trains in cloud-to-ground lightning using lightning locating systems, 2014 International Conference on Lightning Protection, ICLP 2014, статья N° 6973356, pp. 1437-1444. DOI: 10.1109/ICLP.2014.6973356

Saba, M. M. F., Shigihara, M., & Kodaira, C. Y. (2018) Observation of a lightning flashover with a high-speed video camera, 34th International Conference on Lightning Protection, ICLP 2018, Article N° 8503282. DOI: 10.1109/ICLP.2018.8503282

Stolzenburg, M., Marshall, T. C., Karunarathne, S., Karunarathna, N., & Orville, R. E. (2015). An M component with a concurrent dart leader traveling along different paths during a lightning flash. *Journal of Geophysical Research*, *120*(19), 10267–10284. doi:10.1002/2015JD023417

Wang, X., Yuan, P., Cen, J., & Xue, S. (2016). Correlation between the spectral features and electric field changes for natural lightning return stroke followed by continuing current with M-components. *Journal of Geophysical Research*, *121*(14), 8615–8624. doi:10.1002/2016JD025314

Xue, S., Yuan, P., Cen, J., Li, Y., & Wang, X. (2015). Spectral observations of a natural bipolar cloud-to-ground lightning. *Journal of Geophysical Research*, *120*(5), 1972–1979. doi:10.1002/2014JD022598

Yuan, P., Chen, R., Su, Y., Fan, B., Wang, X. (2019) Characteristic parameters of positive cloud-to-ground lightning channel, Journal of Earth System Science, 128 (6), статья N° 164, . Doi:10.100712040-019-1196-4

KEY TERMS AND DEFINITIONS

Cloud-to-Ground Lightning Discharge: An electrical discharge during a thunderstorm that occurs between a cloud and the earth's surface. It is a natural source of forest fires.

Forest Fire: Uncontrolled aerothermochemical phenomenon characterized by step-by-step mechanism which includes following stages: inert heating, moisture evaporation, high temperature terpens evaporation, dry organic matter pyrolysis, flammable combustion and smoldering.

Forest Fuel: It can be considered like dead and live forest fuel. Main types of forest fuel which can be involved in combustion during forest fire: cloud-to-ground forest fuel (needles, leaves and dry grass, small branches) and crown forest fuel (needles, small branches).

Ignition: Inflammation of forest fuel caused by definite source of high temperature or energy.

Ignition Delay: Time before flame flash after forest fuel heating.

Lightning Activity: An atmospheric phenomenon characterized by discharges of the cloud-to-cloud and cloud-to-cloud-to-ground class.

Mathematical Simulation: The production of a computer model of forest fire conditions and prerequisites, especially for the purpose of study.

Meteorological Parameters: Physical characteristics of local weather conditions in the forested area under consideration. Key parameters include ambient temperature, soil temperature, precipitation, wind speed, solar radiation, cloud cover, dew point temperature. These parameters are used for mathematical modeling of the drying of a layer of forest fuel.

Monitoring: Monitoring refers to the periodic calculation of the parameters of forest fire danger with a portion of information available in real time.

Prediction: Under the prediction of forest fires is the calculation of the parameters of forest fire danger with a certain projection in advance in order to have enough time to anticipate an emergency. The calculation in this case is carried out in a mode ahead of the real time of the development of the catastrophe - the occurrence of a forest fire.

Chapter 4
M-Components Mathematical Modeling for Coniferous Tree Ignition

ABSTRACT

Based on the one-dimensional three-layer physical and mathematical model of ignition of a coniferous tree (pine), the influence of the m-components of the cloud-to-ground lightning discharge and the appropriateness of their consideration in technical systems for monitoring the forest fire danger are evaluated. The problem is solved in a cylindrical coordinate system. Typical cloud-to-ground lightning discharges are considered. An assessment of the influence of m-components was carried out for a typical range of changes in their characteristics.

INTRODUCTION

Existing direction-finding systems for lightning discharges covering North America (Orville et al., 2010) and Europe (Diendorfer, 2007) are capable to detect cloud-to-ground lightning discharge with an efficiency of about 90% and an accuracy of 1 kilometer (Pineda et al., 2014). That is, these systems provide forest fire managers with important information about potential fires in the forest. It should be noted that only a small number of lightning discharges (one in 1000 or one in 10000) will ultimately lead to an active forest fire (Nash and Johnson, 1996; Wierzchowski et al., 2002; Larjavaara et al., 2005a). Accordingly, it is important to understand which components of a cloud-to-ground lightning discharge play a role in the ignition of forest fuel. Of great importance for the protection of forests from fires will be the ability to compare the probability of a forest fire to each cloud-to-ground lightning discharge (Pineda et al., 2014). Various assumptions have been put forward about the characteristics of a cloud-to-ground lightning discharge that lead to a forest fire. There are mainly three factors, namely polarity, multi-impact, and long continuing current (LCC) (Hall and Brown, 2006).

M-components are observed as an increase in the brightness of the channel during continuing current. M-components may have a peak in the range of kiloamperes (Campos et al., 2007). It is necessary

DOI: 10.4018/978-1-7998-7250-4.ch004

to study the influence of M-components on the ignition process of a coniferous tree trunk in order to assess the feasibility of supplementing existing technical systems and empirical (Kurbatskiy and Kostyrina, 1977; Larjavaara et al., 2005b) and deterministic-probabilistic (Baranovskiy, 2004; 2007) methods of forest fire danger prediction with subsystems for accounting for the presence of M-components in a lightning discharge. As a physical and mathematical model of coniferous tree ignition by cloud-to-ground lightning discharge, a one-dimensional three-layer model was used (Kuznetsov and Baranovskiy, 2008).

The purpose of the study is to assess the influence of M-components on the process of coniferous tree ignition by cloud-to-ground lightning discharge, as well as to determine the ignition conditions depending on the discharge parameters.

PHYSICAL AND MATHEMATICAL STATEMENT

Due to the subcortical layer of coniferous tree is more saturated with moisture (Esau, 1980) than the resinous core, the electric current of lightning discharge in the trunk of such a tree passes mainly through the outer layers, without penetrating inside. The following physical model is used. At a certain point in time, a lightning discharge of a given polarity and duration strikes the coniferous tree trunk. It were assumed that current-voltage characteristics of the discharge are the same for different sections of the tree trunk. The discharge current has M components.

Fig. 1 shows a typical discharge with M-components. The time is plotted on the abscissa, and the channel brightness in arbitrary units on the ordinate. The trunk wood is heated due to the Joule heat generated in the subcortical zone of the tree trunk. As a result of the flow of electric current, the wood is heated and when critical heat fluxes from the subcortical zone of the trunk to the ignition surface and its temperature, the wood ignites. The solution domain is shown in Fig. 2, where the numbers indicate the zones: 1 - the core of the tree trunk; 2 – the subcortical zone; 3 – the tree bark.

The mathematical model (Kuznetsov and Baranovskiy, 2008) describes heat transfer using non-stationary heat conduction equations:

$$\rho_1 c_1 \frac{\partial T_1}{\partial t} = \frac{\lambda_1}{r}\left(r\frac{\partial T_1}{\partial r}\right), \tag{1}$$

$$\rho_2 c_2 \frac{\partial T_2}{\partial t} = \frac{\lambda_2}{r}\frac{\partial}{\partial r}\left(r\frac{\partial T_2}{\partial r}\right) + JU, \tag{2}$$

$$\rho_3 c_3 \frac{\partial T_3}{\partial t} = \frac{\lambda_3}{r}\frac{\partial}{\partial r}\left(r\frac{\partial T_3}{\partial r}\right). \tag{3}$$

At the initial moment of time, the temperature field is determined by the following conditions:

Figure 1. Discharge with M-components (Campos et al., 2007)

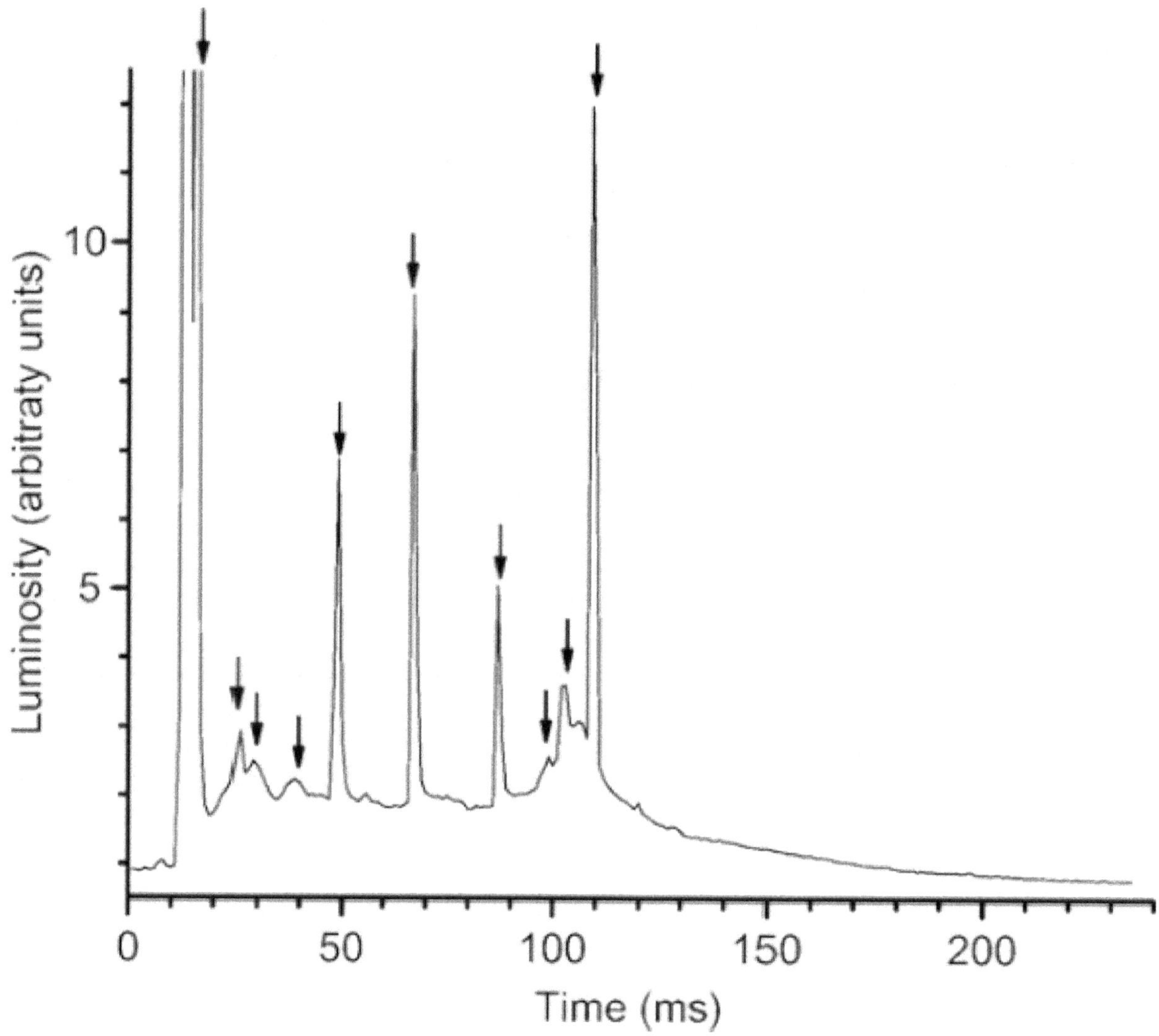

$$T_i = T_{i\mathcal{H}},\ i=1,2,3 \tag{4}$$

The boundary conditions for equations (1) - (3):

$$r=0\ \lambda_1 \frac{\partial T_1}{\partial r} = 0, \tag{5}$$

$$r=R_2\ \lambda_1 \frac{\partial T_1}{\partial r} = \lambda_2 \frac{\partial T_2}{\partial r},\ T_1=T_2 \tag{6}$$

Figure 2. Scheme of the solution area

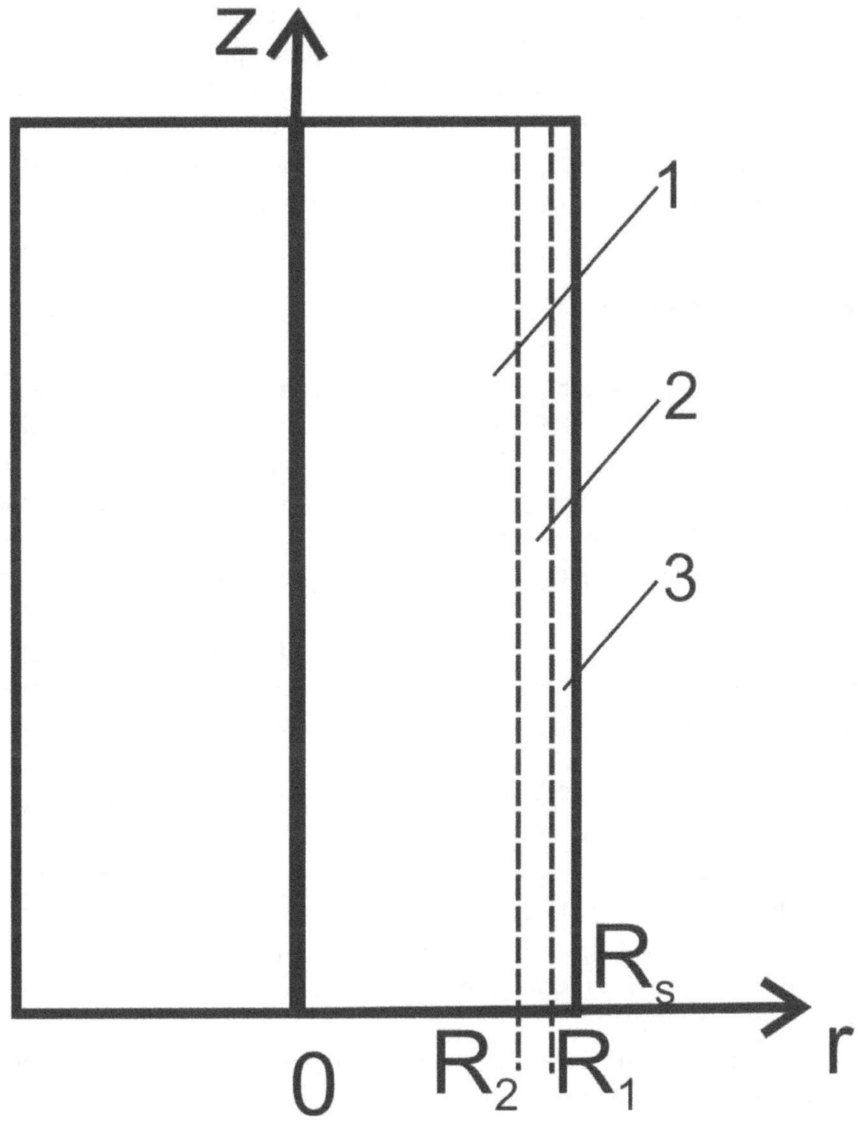

$$r{=}R_1\ \lambda_2 \frac{\partial T_2}{\partial r} = \lambda_3 \frac{\partial T_3}{\partial r},\ T_2{=}T_3 \tag{7}$$

$$r{=}R_s\ \lambda_3 \frac{\partial T_3}{\partial r} = \alpha_e (T_e - T_3), \tag{8}$$

where T_i, ρ_i, λ_i, c_i are the temperature, density, thermal conductivity and heat capacity of the corresponding subareas (i=1,2,3), J is the current, U is the voltage, α_e is the heat transfer coefficient, r is the spatial coordinate, t is time, R_s is the external radius of the trunk, R_1 is the interface between the cortex and the subcortical zone, R_2 is the interface between the core of the trunk and the subcortical zone.

In the computational experiments, the initial data were used (pine wood, core) (Zabolotny et al., 1995): $\rho=500$ kg/m^3; c=1670 J/(kg·K); $\lambda=0.12$ W/(m·K). Subcortical layer parameters: $\rho=500$ kg/m^3; c=2600 J/(kg·K); $\lambda=0.35$ W/(m·K). Thermophysical characteristics of the bark: $\rho=500$ kg/m^3; c=1670 J/(kg·K); $\lambda=0.12$ W/(m·K). The geometric characteristics of the solution area: $R_s=0.25$ m; $R_1=0.245$ m; $R_2=0.235$ m; Environmental parameters: $T_e=300$ K, $\alpha_e=80$ W/(m^2·K).

RESULTS AND DISCUSSION

The mathematical model (1) - (3) with boundary and initial conditions (4) - (8) is implemented by the finite-difference method (Samarskiy, 1983). To solve the difference analogues of one-dimensional equations, the marching method was used (Samarskiy, 1983). As a result of an experimental study (Zabolotny et al., 1995), the criteria for ignition are determined by temperature and heat flux to the surface of the ignition. A scenario is considered when a negative polarity discharge strikes a pine trunk for a duration of 500 ms with an average peak current of 23.5 kA and a voltage of 100 kV (Cummins et al., 1998). The current has 5 M components.

Figure 3 shows the time dependence of the current of a cloud-to-ground lightning discharge and M-components. In Fig. 3.a, curves 1 and 2 correspond to discharges with a duration of 500 and 300 ms, respectively. Fig. 3b shows the dependence of the current and the M component for a discharge with a large value of the average current (J = 35 kA).

Figures 4.a and 4.c show the dependences of the temperature of the ignition surface at different instants of time for discharges of 500 and 300 ms duration, respectively. Figures 4.b, 4.d show the time dependence of the heat flux to the ignition surface for discharges of 500 and 300 ms duration, respectively. Curves 1 and 2 correspond to discharges with and without M-components. Fig. 5.a illustrates the dependence of the temperature of the ignition surface, and Fig. 5.b, the dependence of the heat flux to this surface at various times for a discharge with a large average current. Ignition conditions were evaluated according to the following criteria (Table 1):

An analysis of the results (Fig. 4) shows that the ignition conditions are satisfied for a typical cloud-to-ground discharge (with a duration of about 500 ms). A short-term discharge (with a duration of less than 500 ms) with average volt-ampere characteristics does not ignite the coniferous trunk. That is, a clear dependence is established on the time of exposure of the tree trunk by the current of a cloud-to-ground lightning discharge and the process of its ignition as a result of the Joule heat release.

An analysis of the influence of the M-components of a cloud-to-ground lightning discharge revealed a weak dependence of the ignition process on their presence. The M-components of typical cloud-to-ground lightning discharges show only a slight effect on the ignition process of conifers (Fig. 4). The influence of the M-components of a cloud-to-ground lightning discharge with a high average current strength is even less noticeable (Fig. 5). Curves 1 and 2 in Fig. 5 lie almost one in one.

Figure 3. Current and M-components adopted in computational experiments

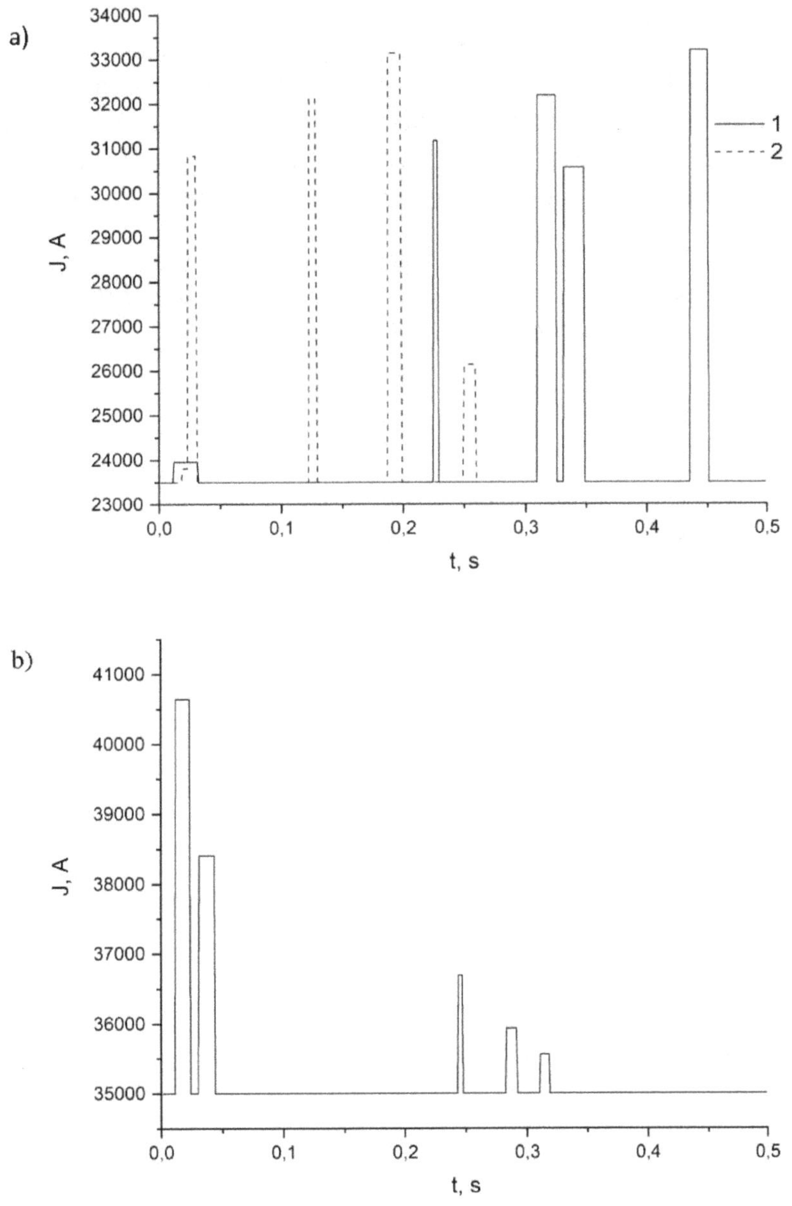

Table 1. Experimental data for ignition of pine wood (Zabolotny et al., 1995)

Ignition delay, s	Heat flux, kW/m^2	Surface temperature, K
63.5	12.5	658
45.0	21	700
11.1	42	726
2.6	84	773
0.4	210	867

CONCLUSION

The effect of the M-components was assessed on the ignition process of coniferous trees by cloud-to-ground lightning discharge. The ignition conditions that are characteristic of a typical range of lightning discharge parameters are revealed. Lightning discharge about 500 ms leads to ignition of the coniferous trunk, regardless of the presence or absence of M-components. Dependence on the presence of M-components is weak and decreases with an increase in the average current of a cloud-to-ground lightning

Figure 4. The temperature of the ignition surface and the heat flux to it at different points in time (long-term and short-term cloud-to-ground lightning discharges)

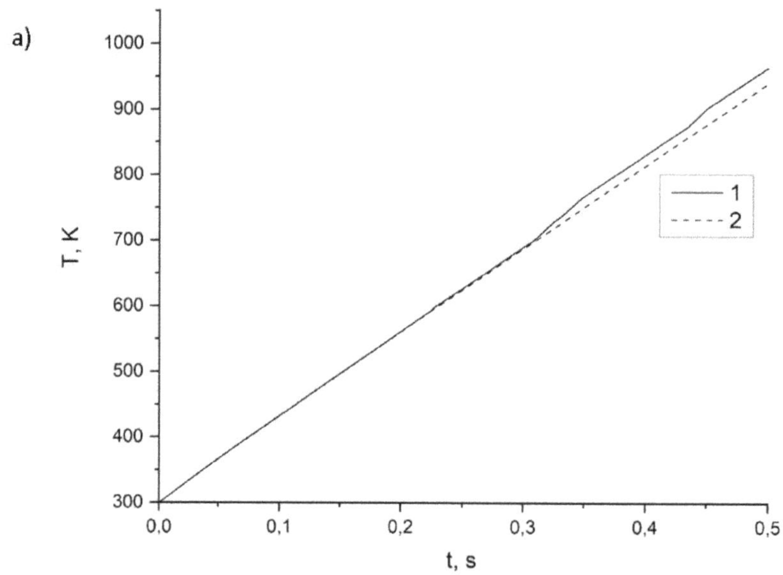

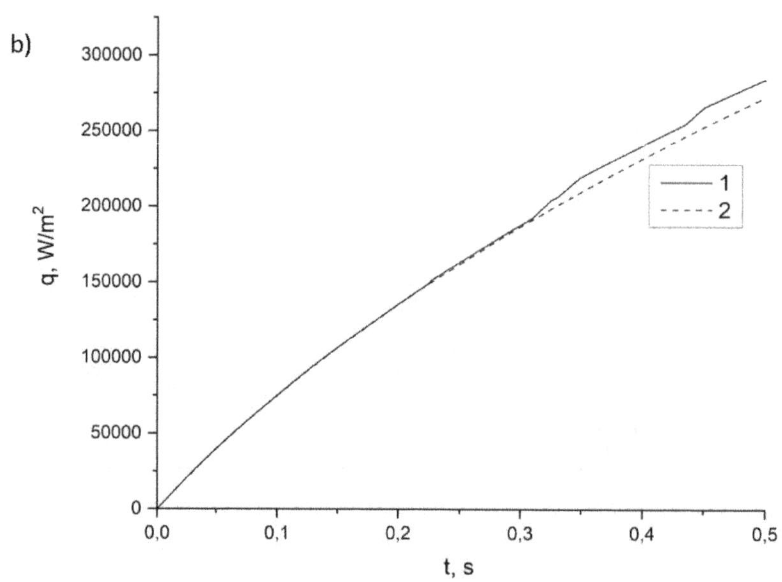

Figure 5. The temperature of the ignition surface and the heat flux to it at different points in time (high-current cloud-to-ground lightning discharge)

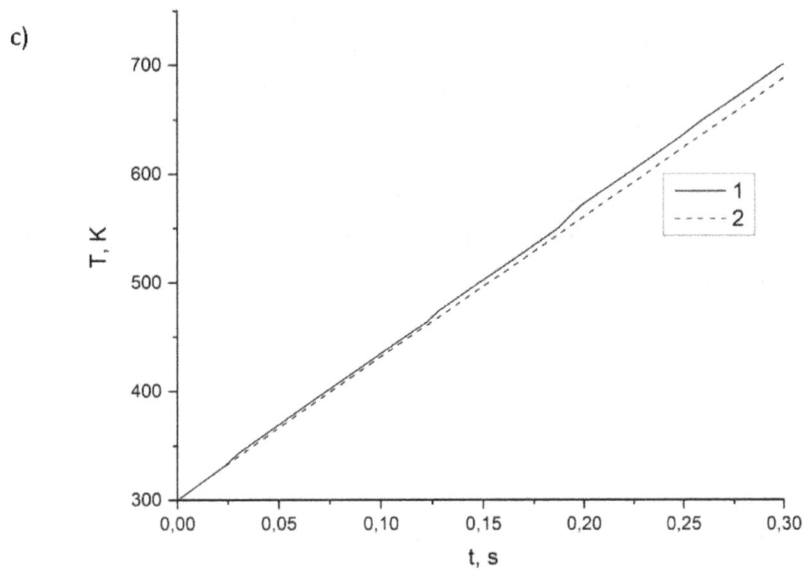

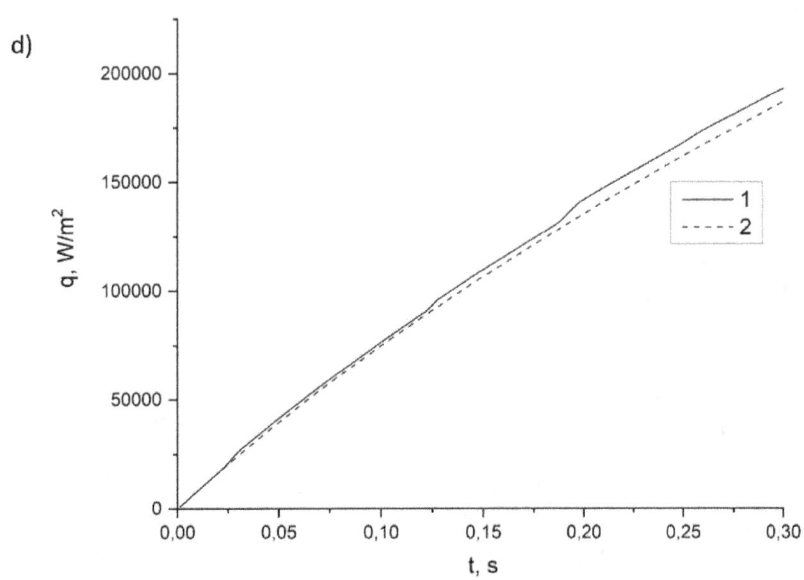

discharge. An explicit dependence is also established on the time of exposure of the tree trunk by the cloud-to-ground lightning discharge. A short-term (less than 300 ms) discharge with typical current-voltage characteristics cannot lead to ignition of conifer. Developing software for the prediction of forest fires, it is not necessary to take into account the M-components. This fact allows us to operate with simpler physical and mathematical models. In addition, the database of such a system may not contain information on the M-components of a cloud-to-ground lightning discharge. On the other hand, in the present work, coniferous wood ignition by a cloud-to-ground lightning discharge was simulated in a more complete physical formulation than in (Kuznetsov and Baranovskiy, 2008). The results obtained are of independent importance for the development of the theory of forest fires prediction (Baranovskiy, 2020; Karanina et al., 2020; Belikova and Glebova, 2020; Yankovich and Yankovich, 2020; Eskandari et al., 2020; Papagiannaki et al., 2020; Dupuy et al., 2020; Glagolev et al., 2020; Silva et al., 2020; Chen et al., 2020). In the present work, the location and duration of the M-components of the discharge were determined randomly from the range of their variation determined statistically (Campos et al., 2007). This is programmatically implemented using a random number generator. It is known that continuing current has various waveforms (Campos et al., 2007) depending on the location and duration of the M-components. In subsequent works, while conducting basic research, this issue can be worked out in more detail.

ACKNOWLEDGMENT

This work was implemented with the financial support of the Russian Foundation for Basic Research. [Project N 17-29-05093 and Project N 16-41-700831].

REFERENCES

Baranovskiy, N. V. (2004). The influence of anthropogenic load and thunderstorm activity on the probability of forest fires. [in Russian]. *Siberian Journal of Ecology*, (6), 835–842.

Baranovskiy, N. V. (2007). *Mathematical modeling of the most probable scenarios and conditions for forest fires* (PhD Thesis). Tomsk: Tomsk State University. (in Russian)

Baranovskiy, N. V. (2020). *Predicting, Monitoring, and Assessing Forest Fire Dangers and Risks*. IGI Global. doi:10.4018/978-1-7998-1867-0

Belikova, M. Y., & Glebova, A. V. (2020). Cluster Analysis Algorithms for RS and WWLLN Data Processing. In N. V. Baranovskiy (Ed.), *Predicting, Monitoring, and Assessing Forest Fire Dangers and Risks* (pp. 151–181). IGI Global. doi:10.4018/978-1-7998-1867-0.ch007

Campos, L. Z. S., Saba, M. M. F., Pinto, O. Jr, & Ballarotti, M. G. (2007). Waveshapes of continuing currents and properties of M-components in natural negative cloud-to-ground lightning from high-speed video observations. *Atmospheric Research*, *84*(9), 302–310. doi:10.1016/j.atmosres.2006.09.002

Chen, C., Xu, T., Zhao, D., & Lei, P. (2020). Study on the Wildfires Occurring Risk Based on Fuzzy FTA. *IOP Conference Series. Earth and Environmental Science*, *440*(5), 052100. Advance online publication. doi:10.1088/1755-1315/440/5/052100

Cummins, K. L., Murphy, M. J., Bardo, E. A., Hiscox, W. L., Pyle, R. B., & Pifer, A. E. (1998). A combined TOA/MDF technology upgrade of the U.S. national lightning detection network. *Journal of Geophysical Research*, *103*(D8), 9035–9044. doi:10.1029/98JD00153

Diendorfer, G. (2007) *Lightning location systems IX International Symposium on Lightning Protection*. Foz do Iguacu.

Dupuy, J.-L., Fargeon, H., Martin-StPaul, N., Pimont, F., Ruffault, J., Guijarro, M., Hernando, C., Madrigal, J., & Fernandes, P. (2020). Climate change impact on future wildfire danger and activity in southern Europe: A review. *Annals of Forest Science*, *77*(2), 35. Advance online publication. doi:10.100713595-020-00933-5

Esau, K. (1980) Anatomy of seed plants. Moscow: Mir. (in Russian)

Eskandari, S., Pourghasemi, H. R., & Tiefenbacher, J. P. (2020). Relations of land cover, topography, and climate to fire occurrence in natural regions of Iran: Applying new data mining techniques for modeling and mapping fire danger. *Forest Ecology and Management*, *473*, 118338. Advance online publication. doi:10.1016/j.foreco.2020.118338

Glagolev, V., Bazhenov, R., Povkh, I., Chuyko, O., & Dolmatova, O. (2020). Designing and producing a geographic information system to forecast vegetation fire danger according to weather conditions. *IOP Conference Series. Materials Science and Engineering*, *862*(5), 052019. Advance online publication. doi:10.1088/1757-899X/862/5/052019

Hall, B. L., & Brown, T. J. (2006). Climatology of positive polarity flashes and multiplicity and their relation to natural wildfire ignitions. *1st International Lightning Meteorology Conference*, Tucson, AZ.

Karanina, S. Y., Kocheeva, N. A., Karanin, A. V., & Belikova, M. Y. (2020). Assessing the Territory According to the Degree of Thunderstorm Danger. In N. V. Baranovskiy (Ed.), *Predicting, Monitoring, and Assessing Forest Fire Dangers and Risks* (pp. 122–150). IGI Global. doi:10.4018/978-1-7998-1867-0.ch006

Kurbatskiy, N. P., & Kostyrina, T. V. (1977). US National Fire Hazard Calculation System. In *Detection and analysis of forest fires* (pp. 38–90). IliD SB AS USSR. (in Russian)

Kuznetsov, G. V., & Baranovskiy, N. V. (2008). Mathematical modeling of ignition of coniferous trees by cloud-to-ground lightning discharge. Fire and Explosion Safety, 17(3), 41-45. (in Russian)

Larjavaara, M., Kuuluvainen, T., & Rita, H. (2005b). Spatial distribution of lightning-ignited fires in Finland. *Forest Ecology and Management*, *208*(1-3), 177–188. doi:10.1016/j.foreco.2004.12.005

Larjavaara, M., Pennanen, J., & Tuomi, T. J. (2005a). Lightning that ignites forest fires in Finland. *Agricultural and Forest Meteorology*, *123*(3-4), 171–180. doi:10.1016/j.agrformet.2005.07.005

Nash, C. H., & Johnson, E. A. (1996). Synoptic climatology of lightning caused forest fires in subalpine and boreal forests. *Canadian Journal of Forest Research*, *26*(10), 1859–1874. doi:10.1139/x26-211

Orville, R. E., Huffines, G. R., Burrows, W. R., & Cummins, K. L. (2010). The North American Lightning detection network (NALDN) – analysis of flash data: 2001-09. *Monthly Weather Review*, *139*(5), 1305–1322. doi:10.1175/2010MWR3452.1

Papagiannaki, K., Giannaros, T. M., Lykoudis, S., Kotroni, V., & Lagouvardos, K. (2020). Weather-related thresholds for wildfire danger in a Mediterranean region: The case of Greece. *Agricultural and Forest Meteorology*, *291*, 108076. Advance online publication. doi:10.1016/j.agrformet.2020.108076

Pineda, N., Montanya, J., & van der Velde, O. A. (2014). Characteristics of lightning related to wildfire ignitions. *Atmospheric Research, 135-136*, 380–387. doi:10.1016/j.atmosres.2012.07.011

Samarskiy, A. A. (1983). *Theory of difference schemes*. Nauka. (in Russian)

Silva, I. D. B., Valle, M. E., Barros, L. C., & Meyer, J. F. C. A. (2020). A wildfire warning system applied to the state of Acre in the Brazilian Amazon. *Applied Soft Computing, 89*, 106075. Advance online publication. doi:10.1016/j.asoc.2020.106075

Wierzchowski, J., Heathcott, M., & Flannigan, M. D. (2002). Lightning and lightning fire, central Cordillera, Canada. *International Journal of Wildland Fire, 11*(1), 41–51. doi:10.1071/WF01048

Yankovich, E. P., & Yankovich, K. S. (2020). Classification of Territory on Forest Fire Danger Level Using GIS and Remote Sensing. In N. V. Baranovskiy (Ed.), *Predicting, Monitoring, and Assessing Forest Fire Dangers and Risks* (pp. 258–267). IGI Global. doi:10.4018/978-1-7998-1867-0.ch011

Zabolotny, A.E., Zabolotnaya, M.M., Zabolotnaya, Y.A., & Timoshin, V.N. (1995). Determination of safe use zones of solid fuel generators of fire extinguishing aerosols. *Issues of Special Engineering*, (7-8), 15-21. (in Russian)

ADDITIONAL READING

Azadifar, M., Rubinstein, M., Rachidi, F., Rakov, V. A., Diendorfer, G., Schulz, W., & Pavanello, D. (2019). A Study of a Large Bipolar Lightning Event Observed at the Säntis Tower, IEEE Transactions on Electromagnetic Compatibility, 61 (3). *Article, 61*(8715461), 796–806. doi:10.1109/TEMC.2019.2913220

De Paiva, A. R., Saba, M. M. F., Naccarato, K. P., Schumann, C., Jaques, R., Da Silva Ferro, M. A., & Warner, T. A. (2014) Detection of upward lightning by lightning location systems, 2014 International Conference on Lightning Protection, ICLP 2014, Article N° 6973425, pp. 1824-1826. DOI: 10.1109/ICLP.2014.6973425

Lu, G., Cummer, S. A., Tian, Y., Zhang, H., Lyu, F., Wang, T., Stanley, M. A., Yang, J., & Lyons, W. A. (2016). Sprite produced by consecutive impulse charge transfers following a negative stroke: Observation and simulation. *Journal of Geophysical Research, 121*(8), 4082–4092. doi:10.1002/2015JD024644

Nag, A., Murphy, M. J., Schulz, W., & Cummins, K. L. (2014) Lightning locating systems: Characteristics and validation techniques, 2014 International Conference on Lightning Protection, ICLP 2014, Article N° 6973283, pp. 1070-1082. DOI: 10.1109/ICLP.2014.6973283

Nag, A., Rakov, V. A., & Murphy, M. J. (2014) Measurement of preliminary breakdown pulse trains in cloud-to-ground lightning using lightning locating systems, 2014 International Conference on Lightning Protection, ICLP 2014, статья N° 6973356, pp. 1437-1444. DOI: 10.1109/ICLP.2014.6973356

Saba, M. M. F., Shigihara, M., & Kodaira, C. Y. (2018) Observation of a lightning flashover with a high-speed video camera, 34th International Conference on Lightning Protection, ICLP 2018, Article N° 8503282. DOI: 10.1109/ICLP.2018.8503282

Stolzenburg, M., Marshall, T. C., Karunarathne, S., Karunarathna, N., & Orville, R. E. (2015). An M component with a concurrent dart leader traveling along different paths during a lightning flash. *Journal of Geophysical Research*, *120*(19), 10267–10284. doi:10.1002/2015JD023417

Wang, X., Yuan, P., Cen, J., & Xue, S. (2016). Correlation between the spectral features and electric field changes for natural lightning return stroke followed by continuing current with M-components. *Journal of Geophysical Research*, *121*(14), 8615–8624. doi:10.1002/2016JD025314

Xue, S., Yuan, P., Cen, J., Li, Y., & Wang, X. (2015). Spectral observations of a natural bipolar cloud-to-ground lightning. *Journal of Geophysical Research*, *120*(5), 1972–1979. doi:10.1002/2014JD022598

Yuan, P., Chen, R., Su, Y., Fan, B., Wang, X. (2019) Characteristic parameters of positive cloud-to-ground lightning channel, Journal of Earth System Science, 128 (6), статья N° 164, . Doi:10.100712040-019-1196-4

KEY TERMS AND DEFINITIONS

Cloud-to-Ground Lightning Discharge: An electrical discharge during a thunderstorm that occurs between a cloud and the earth's surface. It is a natural source of forest fires.

Forest Fire: Uncontrolled aerothermochemical phenomenon characterized by step-by-step mechanism which includes following stages: inert heating, moisture evaporation, high temperature terpens evaporation, dry organic matter pyrolysis, flammable combustion and smoldering.

Forest Fuel: It can be considered like dead and live forest fuel. Main types of forest fuel which can be involved in combustion during forest fire: ground forest fuel (needles, leaves and dry grass, small branches) and crown forest fuel (needles, small branches).

Ignition: Inflammation of forest fuel caused by definite source of high temperature or energy.

Ignition Delay: Time before flame flash after forest fuel heating.

Lightning Activity: An atmospheric phenomenon characterized by discharges of the cloud-to-cloud and cloud-to-ground class.

M-Components: are observed as an increase in the brightness of the channel during continuing current. M-components may have a peak in the range of kiloamperes.

Mathematical Simulation: The production of a computer model of forest fire conditions and prerequisites, especially for the purpose of study.

Meteorological Parameters: Physical characteristics of local weather conditions in the forested area under consideration. Key parameters include ambient temperature, soil temperature, precipitation, wind speed, solar radiation, cloud cover, dew point temperature. These parameters are used for mathematical modeling of the drying of a layer of forest fuel.

Monitoring: Monitoring refers to the periodic calculation of the parameters of forest fire danger with a portion of information available in real time.

Prediction: Under the prediction of forest fires is the calculation of the parameters of forest fire danger with a certain projection in advance in order to have enough time to anticipate an emergency. The calculation in this case is carried out in a mode ahead of the real time of the development of the catastrophe - the occurrence of a forest fire.

Section 2

Mathematical Simulation of hte Deciduous Tree Ignition by the Cloud-to-Ground Lightning Discharge

Chapter 5
One-Dimensional Mathematical Models to Simulate Deciduous Tree Ignition

ABSTRACT

The results of numerical simulation of heating a coniferous tree (pine) by cloud-to-ground lightning discharge are presented. The problem is solved in a one-dimensional formulation in a cylindrical coordinate system. A four-layer structure of the coniferous trunk is considered taking into account the core, subcortical layer, bark and surface water layer. A parametric study of the effect of current-voltage characteristics typical of negative and positive cloud-to-ground lightning discharges on the process of heating the trunk wood has been carried out. The conditions for the ignition of a tree trunk in a typical range of variation of the parameters of the impact of the discharge are established.

INTRODUCTION

At present, the problem of predicting forest fire danger is highly relevant (Baranovskiy, 2020). The causes of forest fires are both anthropogenic and natural (Grishin, 2002). Of great importance is the occurrence of fires in forests from thunderstorms as a result of the action of a cloud-to-ground lightning discharge (Latham and Williams, 2001). The deterministic-probabilistic method of predicting forest fire danger (Baranovskiy, 2007; 2017) includes a subsystem for assessing the probability of forest fires from thunderstorms. However, until now the physical mechanism of wood ignition as a result of the action of a cloud-to-ground lightning discharge has not been taken into account. Cloud-to-cloud-to-ground lightning discharges can be differentiated by such parameters as their polarity, peak current and voltage, and duration (Burke and Jones, 1996). When developing a methodology for predicting forest fires, the mathematical formulation of the ignition of a tree by a cloud-to-ground lightning discharge should be used. It is difficult to experimentally conduct such a study due to the fact that such processes are characterized by high energetics. Analysis of the literature shows that the results of mathematical modeling of the ignition of a tree trunk by a cloud-to-ground lightning discharge have not yet been published.

DOI: 10.4018/978-1-7998-7250-4.ch005

Figure 1. Solution area diagram

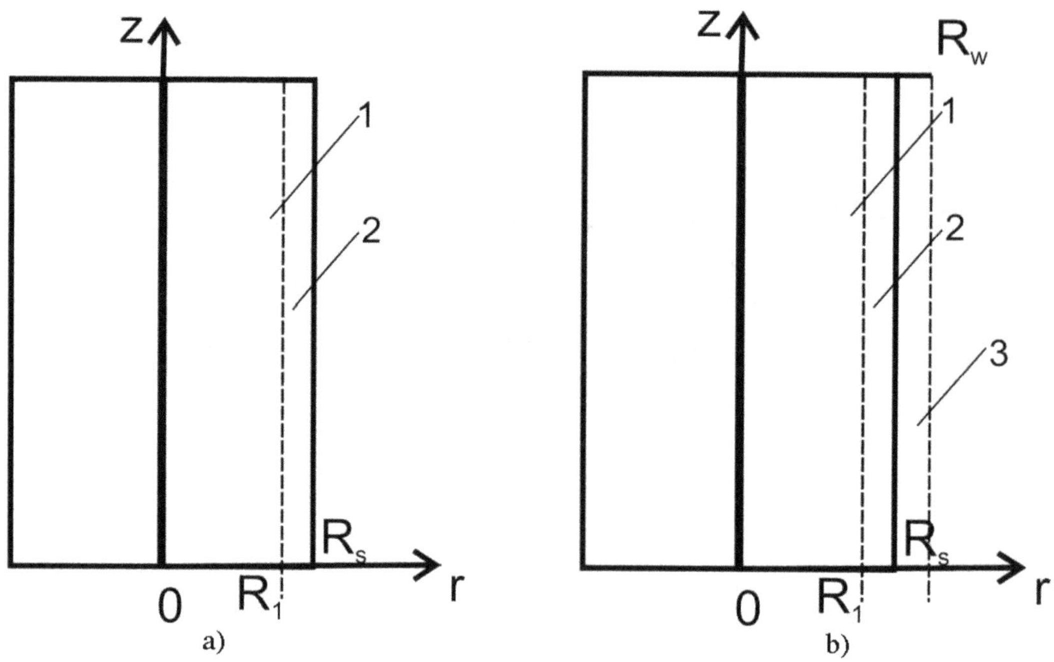

The nature of the consequences of such a high-energy external influence has not yet been established. The aim of the research is mathematical modeling of the deciduous tree ignition by a cloud-to-ground lightning discharge.

MATHEMATICAL STATEMENT

The flow of electric current in the trunk of deciduous and coniferous trees is different. This is due to the fact that in deciduous trees, moisture is transported through vessels in the massive central part (Esau, 1980). The wetter central part is in this case a conductor of electric current. The study of the tree ignition by a lightning discharge was carried out using the following physical model. A free-standing deciduous tree grows on the land surface. In order to concretize the parameters of the model, birch is considered as an object of research. A cloud-to-ground lightning discharge strikes the trunk of this tree. The discharge electric current flows along the inner part of the birch trunk and in the surface layer of water. It is assumed that the electric current has the same parameters in different sections of the trunk. The Joule heat causes the wood in the trunk to heat up. When critical values of heat fluxes to the ignition surface and its temperature are reached, the wood ignites. Moisture transfer processes are not taken into account in this model.

Geometry of the problem. A cylinder is considered that models a tree trunk. A certain section of the trunk is considered. The solution area diagram is shown in Fig. 1, where 1 is the core, 2 is the bark of the tree trunk, 3 is the surface layer of water; R_w is the boundary of the surface layer of water, R_s is the outer radius of the trunk, R_1 is the boundary between the inner region and the bark of the tree trunk.

Mathematically, the heating of tree by a cloud-to-ground lightning discharge before ignition is described by a system of non-stationary differential equations of heat conduction:

$$\rho_1 c_1 \frac{\partial T_1}{\partial t} = \frac{\lambda_1}{r}\frac{\partial}{\partial r}\left(r\frac{\partial T_1}{\partial r}\right) + JU, \tag{1}$$

$$\rho_2 c_2 \frac{\partial T_2}{\partial t} = \frac{\lambda_2}{r}\frac{\partial}{\partial r}\left(r\frac{\partial T_2}{\partial r}\right), \tag{2}$$

$$\rho_3 c_3 \frac{\partial T_3}{\partial t} = \frac{\lambda_3}{r}\frac{\partial}{\partial r}\left(r\frac{\partial T_3}{\partial r}\right) + JU, \tag{3}$$

Boundary conditions for equations (1) - (3):

$$r=0,\ \lambda_1 \frac{\partial T_1}{\partial r} = 0, \tag{4}$$

$$r=R_p,\ \lambda_1 \frac{\partial T_1}{\partial r} = \lambda_2 \frac{\partial T_2}{\partial r},\ T_1=T_2 \tag{5}$$

$$r=R_s,\ \lambda_2 \frac{\partial T_2}{\partial r} = \lambda_3 \frac{\partial T_3}{\partial r},\ T_2=T_3 \tag{6}$$

$$r=R_w,\ \lambda_3 \frac{\partial T_3}{\partial r} = \alpha(T_e - T_3), \tag{7}$$

Initial conditions for equations (1) - (3):

$$t=0,\ T_i(r) = T_{i0}(r),\ i=1,2,3 \tag{8}$$

where T_i, ρ_i, c_i, λ_i are the temperature, density, heat capacity and thermal conductivity of wood in the inner part of the trunk (i = 1), bark (i = 2) and near-surface water layer (i = 3); α is the heat transfer coefficient; J is the current of the cloud-to-ground lightning discharge; U - voltage of cloud-to-ground lightning discharge; r - coordinate; t is time. Indexes "e" and "0" correspond to the parameters of the external environment and the parameters of wood at the initial moment of time.

Initial data (birch wood): ρ=650 kg/m^3; c=2600 J/(kg•K); λ=0.35 W/(m•K). Thermophysical characteristics of the bark (Zabolotny et al., 1995): ρ=650 kg/m^3; c=1670 J/(kg•K); λ=0.29 W/(m•K). Thermophysical characteristics of the water: ρ=1000 kg/m^3; c=4125 J/(kg•K); λ=0.569 W/(m•K). Geometrical characteristics of the solution area: R_w=0.255 m; R_s=0.25 m; R_1=0.235 m. Environmental parameters: T_e=300 K, α=80 W/(m^2•K).

RESULTS AND DISCUSSION

The mathematical model with the corresponding boundary and initial conditions is solved by the finite difference method. To solve difference analogs of one-dimensional equations, the marching method was used (Samarskiy, 1983).

The parameters of cloud-to-ground lightning discharges, for example, the current, vary over a fairly wide range. Average peak current (Soriano et al., 2005): J = 23.5 kA for negative discharge and J = 35.3 kA for positive discharge. Approximately 16.5% of discharges of positive polarity have a current of less than 10 kA (Cummins et al., 1998). In order to determine the ignition conditions, a parametric study of the influence of the current-voltage characteristics of the discharge on the process of heating a tree as a result of the release of Joule heat was carried out.

It should be noted that the results of experiments to determine the kinetic mechanism (and its parameters) of the ignition of large wood blocks (including a high-energy electric discharge) are still unknown. The difficulty of implementing an experimental study of this process is obvious. However, there is an approach (Zabolotny et al., 1995), within which the conditions of ignition of deciduous wood (birch) are described by two parameters (heat flow and temperature of the ignition surface).

In fact, work (Zabolotny et al., 1995) simulates the process of gas-phase ignition of a condensed substance with high heat fluxes at relatively short times of exposure to a heating source under conditions of an excess of oxidizer. In this work, the experimental data (Zabolotny et al., 1995) on the critical temperature and heat flux to the bark surface are used as criteria for the implementation of ignition conditions. The chemical aspects of the ignition process are excluded from consideration in this article.

Birch is exposed to a negative lightning discharge of 500 ms duration, which is characterized by a peak current of 23.5 kA and a voltage of 100 kV. The distribution of temperature along the radius of the trunk of a deciduous tree without taking into account the surface layer of water at different points in time during the action of a lightning discharge is shown in Fig. 2: 1) t = 0.01 s; 2) - 0.1 s; 3) - 0.3 s; 4) - 0.5 s. The distribution of temperature along the radius of the trunk of a deciduous tree, taking into account the surface layer of water at different points in time during the action of a lightning discharge, is shown in Fig. 3: 1) t = 0.01 s; 2) - 0.1 s; 3) - 0.3 s; 4) - 0.5 s. Fig. 4 shows the dependence of the temperature of the ignition surface and the density of the heat flux to this surface on time.

In Figure 4, numbers indicate: 1 - temperature in the trunk at the border of the bark and core; 2 - temperature on the surface of the bark during heat release in the surface layer of water; 3 - heat flux to the ignition surface from the trunk core; 4 - heat flux to the ignition surface from the surface layer of water. The ignition conditions were determined from the experimental data (Zabolotny et al., 1995) (Table 1).

As a result of the action of the considered cloud-to-ground lightning discharge, the tree trunk is heated to the temperature (Fig. 2,3) of combustion of natural combustible materials (about 1000 K). Analysis

Figure 2. Temperature distribution along the radius of the tree trunk at different points in time

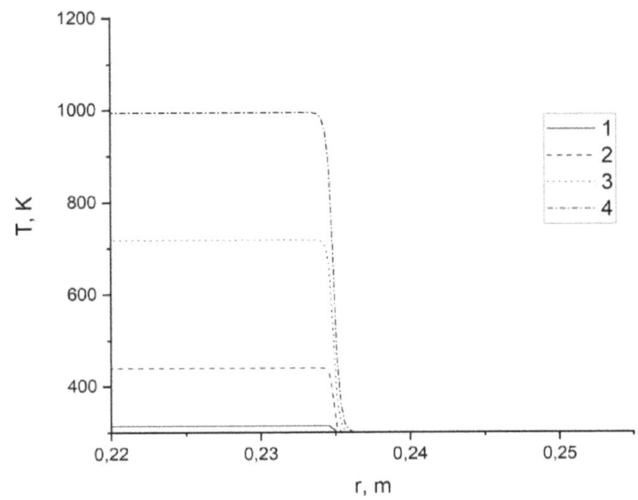

Figure 3. Distribution of temperature along the radius of the tree trunk at different times, taking into account the surface layer of water

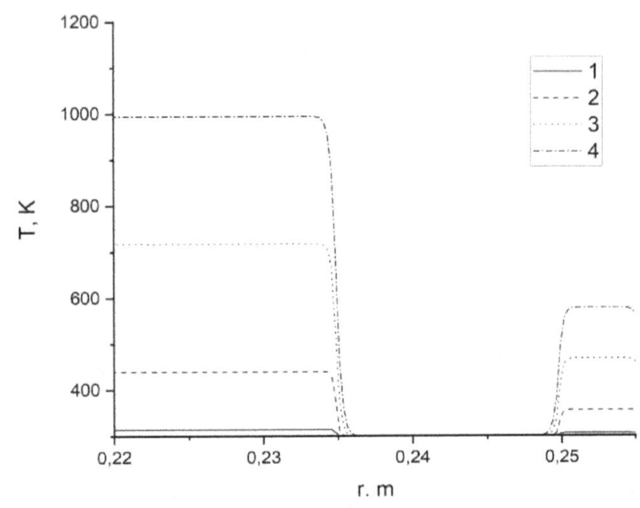

of the results allows us to conclude about the ignition of the tree trunk. This conclusion is consistent with the data of observations of thunderstorms (Plummer, 1912), where it is indicated that a lightly combustible or dry tree catches fire.

Table 2 shows the results of computational experiments to determine the ignition conditions depending on the voltage of a cloud-to-ground lightning discharge at a current of J=23.5 kA, and in table 3, depending on the current at a discharge voltage of U=100 kV.

Figure 4. Temperature of the ignition surface and heat flux to the ignition surface at different times

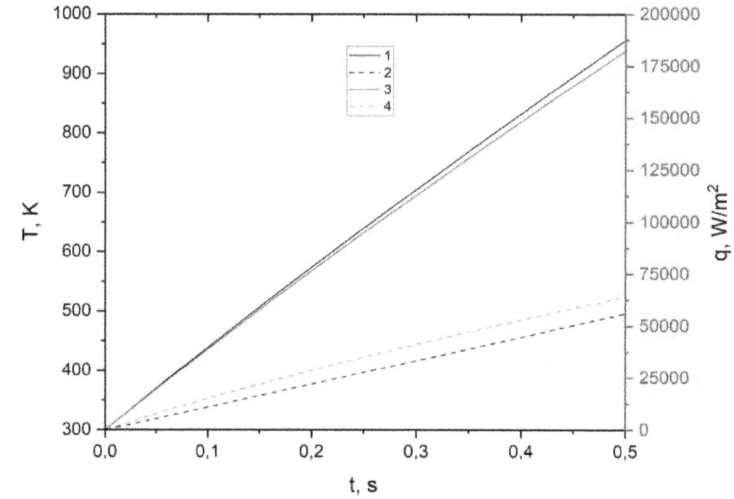

Analysis of the dependences of the magnitude of the heat flux to the ignition surface and the temperature of the interface between the inner region of the trunk and the bark (Fig. 4) shows that in terms of temperature and heat flux, the ignition conditions for the considered discharge are achieved for fairly typical parameters of a lightning discharge. It should be noted that the surface of the bark, when an electric current flows in the surface layer of water, can also theoretically ignite, but with a delay of about 10 seconds relative to the core. On the temperature dependence, two areas of increased temperature can

Table 1. Experimental data (Zabolotny et al., 1995)

Ignition delay, s	Heat flux, kW/m²	Surface temperature, K
136	15	-
61.2	21	645
17.2	42	688
1.8	125	755
0.43	210	801

Table 2 Conditions for ignition of a tree trunk from voltage at current J=23.5 kA

Voltage, *U*, kV	Ignition conditions (Zabolotny et al., 1995)	Surface temperature, K	Heat flux. kW/m²
1 – 85	No	<801	<210
90	Yes	801	226
95	Yes	801	230
100	Yes	801	232
105	Yes	801	235
110	Yes	801	238

Table 3 Conditions for the ignition of a tree trunk from the current at a voltage of U=100 kV

Current, J, kA	Ignition conditions (Zabolotny et al., 1995)	Surface temperature, K	Heat flux, kW/m^2
1 – 20	No	<801	<210
23.5	Yes	801	232
30	Yes	801	245
35	Yes	801	253

be distinguished. The first area is the core of the trunk. The second area is the surface layer of water on the trunk. As a result of a parametric study of the effect of voltage (Table 2) and current (Table 3) of the discharge, the conditions for the ignition of a tree trunk during the action of a cloud-to-ground lightning discharge were established.

Characterizing in general the results of numerical modeling of the process of heating a tree trunk by a cloud-to-ground lightning discharge, one can confidently assert that the results of this study are in agreement with the ignition conditions (Zabolotny et al., 1995).

CONCLUSION

As a result of the numerical solution of the problem of heating a tree as a result of an electric current flowing through it, the possibility of ignition of a deciduous tree by a cloud-to-ground lightning discharge is shown. The selected conditions are typical for the typical range of changes in the parameters of external influence on deciduous trees during periods of thunderstorm activity.

The results obtained are not only of theoretical importance, which consists in the development and verification of a simplified mechanism for the ignition of a deciduous tree by a cloud-to-ground lightning discharge. In addition, these results complement the theoretical basis for the further development of models for the ignition of hazardous materials.

A practical application of a real physical and mathematical model of wood ignition is possible. The existing methods of predicting and assessing forest fire danger (Perminov, 2020; Zharikova, 2020; Badmaev et al., 2020; Glagolev and Zubareva, 2020; Adab et al., 2013; Al Janabi et al., 2017; Bui et al., 2017; CWFIS, 2020; Cardille et al., 2001; Chen et al., 2017; Curt et al., 2016; Ganteaume and Guerra, 2018; Chen et al., 2020; Glagolev et al., 2020; Papagiannaki et al., 2020) can be easily improved by including in their structure a subsystem for assessing the conditions of tree ignition by a cloud-to-ground lightning discharge.

ACKNOWLEDGMENT

This work was implemented with the financial support of the Russian Foundation for Basic Research. [Project N 17-29-05093].

REFERENCES

Adab, H., Kanniah, K. D., & Solaimani, K. (2013). Modeling forest fire in the northeast of Iran using remote sensing and GIS techniques. *Natural Hazards*, *65*(3), 1723–1743. doi:10.100711069-012-0450-8

Al Janabi, S., Al Shourbaji, I., & Salman, M. A. (2017). *Assessing the suitability of soft computing approaches for forest fires prediction*. Applied Computing and Informatics. doi:10.1016/j.aci.2017.09.006

Badmaev, N. B., Bazarov, A. V., & Sychev, R. S. (2020). Forest Fire Danger Assessment Using Meteorological Trends: Case Study. In N. V. Baranovskiy (Ed.), *Predicting, Monitoring, and Assessing Forest Fire Dangers and Risks* (pp. 183–208). IGI Global. doi:10.4018/978-1-7998-1867-0.ch008

Baranovskiy, N. V. (2007). *Mathematical modeling of the most likely scenarios and conditions for forest fires* (PhD Thesis). Tomsk: Tomsk State University. (in Russian)

Baranovskiy, N. V. (2017). Forest fire danger assessment using SPMD-model of computation for massive parallel system. *International Review on Modeling and Simulations*, *10*(3), 193–201. doi:10.15866/iremos.v10i3.10570

Baranovskiy, N.V. (2020). *Predicting, Monitoring, and Assessing Forest Fire Dangers and Risks*. IGI Global. doi:10.4018/978-1-7998-1867-0

Bui, D. T., Bui, Q.-T., Nguyen, Q.-P., Pradhan, B., Nampak, H., & Trinh, P. T. (2017). A hybrid artificial intelligence using GIS-based neural-fuzzy inference system and particle swarm optimization for forest fire susceptibility modeling at a tropical area. *Agricultural and Forest Meteorology*, *233*, 32–44. doi:10.1016/j.agrformet.2016.11.002

Burke, C. P., & Jones, D. L. (1996). On the polarity and continuing current in unusually large lightning flashes deduced from ELF events. *Journal of Atmospheric and Solar-Terrestrial Physics*, *58*(5), 531–548. doi:10.1016/0021-9169(95)00054-2

Canadian Wildland Fire Information System. (n.d.). Access: https://cwfis.cfs.nrcan.gc.ca/home

Cardille, J. A., Ventura, S. J., & Turner, M. G. (2001). Environmental and Social Factors of Influencing Wildfires in the Upper Midwest, United States. *Ecological Applications*, *11*(1), 111–127. doi:10.1890/1051-0761(2001)011[0111:EASFIW]2.0.CO;2

Chen, C., Xu, T., Zhao, D., & Lei, P. (2020). Study on the Wildfires Occurring Risk Based on Fuzzy FTA. *IOP Conference Series. Earth and Environmental Science*, *440*(5), 052100. Advance online publication. doi:10.1088/1755-1315/440/5/052100

Chen, W., Xie, X., Wang, J., Pradhan, B., Hong, H., Tien Bui, D., Duan, Z., & Ma, J. (2017). A comparative study of logistic model tree, random forest and classification and regression tree models for spatial prediction of landslide susceptibility. *Catena*, *151*, 147–160. doi:10.1016/j.catena.2016.11.032

Cummins, K. L., Murphy, M. J., Bardo, E. A., Hiscox, W. L., Pyle, R. B., & Pifer, A. E. (1998). A combined TOA/MDF technology upgrade of the U.S. national lightning detection network. *Journal of Geophysical Research*, *103*(D8), 9035–9044. doi:10.1029/98JD00153

Curt, T., Frejaville, T., & Lahaye, S. (2016). Modelling the spatial patterns of ignition causes and fire regime features in southern France: Implications for the prevention policy. *International Journal of Wildland Fire*, *25*(7), 785–796. doi:10.1071/WF15205

Esau, K. (1980) Anatomy of seed plants. Moscow: Mir. (in Russian)

Ganteaume, A., & Guerra, F. (2018). Explaining the spatio-seasonal variation of fires by their causes: The case of southeastern France. *Applied Geography (Sevenoaks, England)*, *90*, 69–81. doi:10.1016/j.apgeog.2017.11.012

Glagolev, V., Bazhenov, R., Povkh, I., Chuyko, O., & Dolmatova, O. (2020). Designing and producing a geographic information system to forecast vegetation fire danger according to weather conditions. *IOP Conference Series. Materials Science and Engineering*, *862*(5), 052019. Advance online publication. doi:10.1088/1757-899X/862/5/052019

Glagolev, V. A., & Zubareva, A. M. (2020). Using Geographic Information Systems for Study of the Natural Fire Risk on the Territory of the Jewish Autonomous Region. In N. V. Baranovskiy (Ed.), *Predicting, Monitoring, and Assessing Forest Fire Dangers and Risks* (pp. 209–231). IGI Global. doi:10.4018/978-1-7998-1867-0.ch009

Grishin, A. M. (2002). Modeling and forecasting disasters. Tomsk: Publishing House of Tomsk University. (in Russian)

Latham, D., & Williams, E. (2001). *Lightning and forest fires // Forest fires: Behavior and Ecological Effects*. Elsevier.

Papagiannaki, K., Giannaros, T. M., Lykoudis, S., Kotroni, V., & Lagouvardos, K. (2020). Weather-related thresholds for wildfire danger in a Mediterranean region: The case of Greece. *Agricultural and Forest Meteorology*, *291*, 108076. Advance online publication. doi:10.1016/j.agrformet.2020.108076

Perminov, V. A. (2020). Mathematical Modeling of Steppe Fires. In N. V. Baranovskiy (Ed.), *Predicting, Monitoring, and Assessing Forest Fire Dangers and Risks* (pp. 48–62). IGI Global. doi:10.4018/978-1-7998-1867-0.ch002

Plummer, F. G. (1912). *Lightning in Relation to Forest Fires. Bulletin 111. USDA Forest Service*. Government Printing Office.

Samarskiy, A. A. (1983). *Theory of difference schemes*. Nauka. (in Russian)

Soriano, L. R., De Pablo, F., & Tomas, C. (2005). Ten-year study of cloud-to-cloud-to-ground lightning activity in the Iberian Peninsula. *Journal of Atmospheric and Solar-Terrestrial Physics*, *67*(16), 1632–1639. doi:10.1016/j.jastp.2005.08.019

Zabolotny, A.E., Zabolotnaya, M.M., Zabolotnaya, Y.A., & Timoshin, V.N. (1995). Determination of safe use zones of solid fuel generators of fire extinguishing aerosols. *Issues of Special Engineering*, (7-8), 15 - 21. (in Russian)

Zharikova, M. V. (2020). Dynamic Spatial-Distributed Fire Risk Analysis. In N. V. Baranovskiy (Ed.), *Predicting, Monitoring, and Assessing Forest Fire Dangers and Risks* (pp. 101–120). IGI Global. doi:10.4018/978-1-7998-1867-0.ch005

ADDITIONAL READING

Azadifar, M., Rubinstein, M., Rachidi, F., Rakov, V. A., Diendorfer, G., Schulz, W., & Pavanello, D. (2019). A Study of a Large Bipolar Lightning Event Observed at the Säntis Tower, IEEE Transactions on Electromagnetic Compatibility, 61 (3). *Article*, *61*(8715461), 796–806. doi:10.1109/TEMC.2019.2913220

De Paiva, A. R., Saba, M. M. F., Naccarato, K. P., Schumann, C., Jaques, R., Da Silva Ferro, M. A., & Warner, T. A. (2014) Detection of upward lightning by lightning location systems, 2014 International Conference on Lightning Protection, ICLP 2014, Article N° 6973425, pp. 1824-1826. DOI: 10.1109/ICLP.2014.6973425

Lu, G., Cummer, S. A., Tian, Y., Zhang, H., Lyu, F., Wang, T., Stanley, M. A., Yang, J., & Lyons, W. A. (2016). Sprite produced by consecutive impulse charge transfers following a negative stroke: Observation and simulation. *Journal of Geophysical Research*, *121*(8), 4082–4092. doi:10.1002/2015JD024644

Nag, A., Murphy, M. J., Schulz, W., & Cummins, K. L. (2014) Lightning locating systems: Characteristics and validation techniques, 2014 International Conference on Lightning Protection, ICLP 2014, Article N° 6973283, pp. 1070-1082. DOI: 10.1109/ICLP.2014.6973283

Nag, A., Rakov, V. A., & Murphy, M. J. (2014) Measurement of preliminary breakdown pulse trains in cloud-to-ground lightning using lightning locating systems, 2014 International Conference on Lightning Protection, ICLP 2014, статья N° 6973356, pp. 1437-1444. DOI: 10.1109/ICLP.2014.6973356

Saba, M. M. F., Shigihara, M., & Kodaira, C. Y. (2018) Observation of a lightning flashover with a high-speed video camera, 34th International Conference on Lightning Protection, ICLP 2018, Article N° 8503282. DOI: 10.1109/ICLP.2018.8503282

Stolzenburg, M., Marshall, T. C., Karunarathne, S., Karunarathna, N., & Orville, R. E. (2015). An M component with a concurrent dart leader traveling along different paths during a lightning flash. *Journal of Geophysical Research*, *120*(19), 10267–10284. doi:10.1002/2015JD023417

Wang, X., Yuan, P., Cen, J., & Xue, S. (2016). Correlation between the spectral features and electric field changes for natural lightning return stroke followed by continuing current with M-components. *Journal of Geophysical Research*, *121*(14), 8615–8624. doi:10.1002/2016JD025314

Xue, S., Yuan, P., Cen, J., Li, Y., & Wang, X. (2015). Spectral observations of a natural bipolar cloud-to-ground lightning. *Journal of Geophysical Research*, *120*(5), 1972–1979. doi:10.1002/2014JD022598

Yuan, P., Chen, R., Su, Y., Fan, B., Wang, X. (2019) Characteristic parameters of positive cloud-to-ground lightning channel, Journal of Earth System Science, 128 (6), статья N° 164, . Doi:10.100712040-019-1196-4

KEY TERMS AND DEFINITIONS

Cloud-to-Ground Lightning Discharge: An electrical discharge during a thunderstorm that occurs between a cloud and the earth's surface. It is a natural source of forest fires.

Forest Fire: Uncontrolled aerothermochemical phenomenon characterized by step-by-step mechanism which includes following stages: inert heating, moisture evaporation, high temperature terpens evaporation, dry organic matter pyrolysis, flammable combustion and smoldering.

Forest Fuel: It can be considered like dead and live forest fuel. Main types of forest fuel which can be involved in combustion during forest fire: cloud-to-ground forest fuel (needles, leaves and dry grass, small branches) and crown forest fuel (needles, small branches).

Ignition: Inflammation of forest fuel caused by definite source of high temperature or energy.

Ignition Delay: Time before flame flash after forest fuel heating.

Lightning Activity: An atmospheric phenomenon characterized by discharges of the cloud-to-cloud and cloud-to-cloud-to-ground class.

Mathematical Simulation: The production of a computer model of forest fire conditions and prerequisites, especially for the purpose of study.

Monitoring: Monitoring refers to the periodic calculation of the parameters of forest fire danger with a portion of information available in real time.

Prediction: Under the prediction of forest fires is the calculation of the parameters of forest fire danger with a certain projection in advance in order to have enough time to anticipate an emergency. The calculation in this case is carried out in a mode ahead of the real time of the development of the catastrophe - the occurrence of a forest fire.

Chapter 6
Two-Dimensional Mathematical Models to Simulate Deciduous Tree Ignition

ABSTRACT

A physical and mathematical formulation of the problem to simulate ignition of a deciduous tree in the approximation of large vessels is presented. The problem is considered in a flat formulation in polar coordinates. A parametric analysis of the current-voltage characteristics of cloud-to-ground lightning discharges on the process of heating the wood of a deciduous tree trunk is carried out. The conditions for the ignition of a tree trunk in the typical range of the parameters for discharge effect are established. Variants without and taking into account the surface layer of water are considered.

INTRODUCTION

Cloud-to-ground lightning discharge (Uman, 1969) is one of the causes of forest fires (Latham and Williams, 2001), especially in sparsely populated areas (Karanina et al., 2020) and mountainous terrain (Conedera et al., 2006). An electric current in a tree trunk flows through areas saturated with moisture (Esau, 1980). There are two main types of distribution in angiosperm wood. If the vessels have basically the same diameter and are evenly distributed in the annual ring, the wood is called diffuse vascular (Esau, 1980). Wood with vessels of unequal diameter, the largest of which are concentrated in early wood, is called annular vascular (Esau, 1980). Between these two extreme types there are also various intermediate ones. Within these large distribution types, individual vessels may be isolated from each other or occur in clusters of different sizes and shapes. Vessels located in groups are flattened along the line of their contact with each other (Esau, 1980). Studies of conductivity in different species using radioactive phosphorus and dyes show that in some species the vessels are connected only within the growth layer, while in others, there is a connection between individual growth layers (Braun, 1963).

The aim of the study is to develop a physical and mathematical model of ignition of a deciduous tree in the approximation of large vessels and to analyze the conditions of ignition.

DOI: 10.4018/978-1-7998-7250-4.ch006

MATHEMATICAL STATEMENT

A freestanding deciduous tree (referred to as angiosperms) is considered. At a fixed point in time, a cloud-to-ground lightning discharge strikes the tree trunk. An electrical current from a cloud-to-ground lightning discharge flows inside and along the surface of the tree trunk. It is assumed that the current parameters are the same in different sections of the trunk. It is assumed that 1) the vessels are located in groups, that is, they can be combined into large bundles (vessels); 2) the vessels are flattened along the line of contact; 3) vessels are connected between separate growth rings. The combination of the above assumptions is the basis for the approximation of large vessels. As a result of the release of Joule heat, the wood is heated. When critical values of heat fluxes from large vessels of the core or from the surface layer of water to the ignition surface and its temperature are reached, the deciduous tree ignites. The problem is solved in polar coordinates in a flat setting. A tree trunk cut along the z-axis is modeled.

The solution area diagram is shown in Fig. 1.a, 1.c, where 1 - core, 2 - bark, 3 - large vessels, 4 - surface layer of water; R_w is the boundary of the surface water layer, R_s is the outer radius of the trunk, R_1 is the boundary between the core and the bark. Γi, Γj, Γj.k - boundaries of the regions, which are shown in Fig. 1.b, 1.d.

Figure 1. Geometry of the solution region (a, c) and boundaries (b, d)

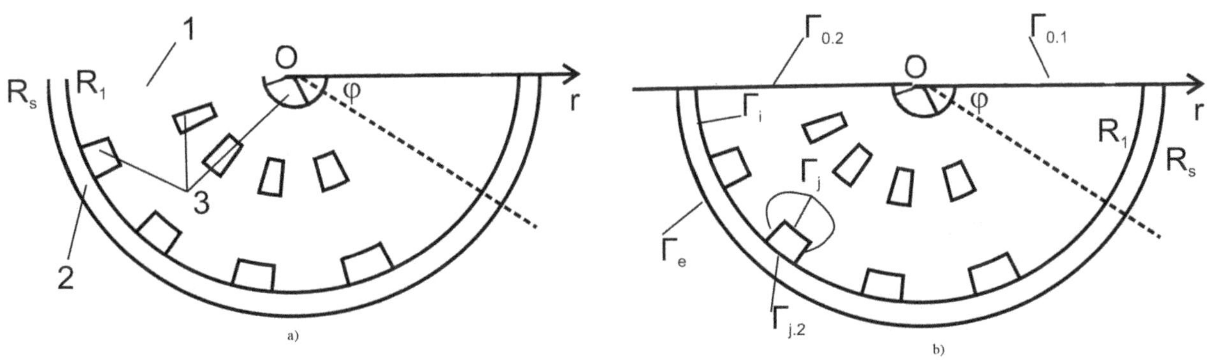

Mathematically, the process of heating a tree by a cloud-to-ground lightning discharge until the moment of ignition is described by a system of non-stationary differential equations of thermal conductivity:

$$\rho_1 c_1 \frac{\partial T_1}{\partial t} = \frac{\lambda_1}{r}\frac{\partial}{\partial r}\left(r\frac{\partial T_1}{\partial r}\right) + \frac{\lambda_1}{r^2}\frac{\partial^2 T_1}{\partial \phi^2}, \quad (1)$$

$$\rho_2 c_2 \frac{\partial T_2}{\partial t} = \frac{\lambda_2}{r}\frac{\partial}{\partial r}\left(r\frac{\partial T_2}{\partial r}\right) + \frac{\lambda_2}{r^2}\frac{\partial^2 T_2}{\partial \phi^2}, \quad (2)$$

$$\rho_3 c_3 \frac{\partial T_3}{\partial t} = \frac{\lambda_3}{r} \frac{\partial}{\partial r}\left(r \frac{\partial T_3}{\partial r} \right) + \frac{\lambda_3}{r^2} \frac{\partial^2 T_3}{\partial \phi^2} + JU, \tag{3}$$

$$\rho_4 c_4 \frac{\partial T_4}{\partial t} = \frac{\lambda_4}{r} \frac{\partial}{\partial r}\left(r \frac{\partial T_4}{\partial r} \right) + \frac{\lambda_4}{r^2} \frac{\partial^2 T_4}{\partial \phi^2} + JU, \tag{4}$$

Boundary conditions for equations (1) - (4):

$$\Gamma_{0.1}, \Gamma_{0.2} \frac{\partial T_{i1}}{\partial \phi} = 0, \tag{5}$$

$$\Gamma_i \lambda_1 \frac{\partial T_1}{\partial r} = \lambda_2 \frac{\partial T_2}{\partial r}, T_1 = T_2, \tag{6}$$

$$\Gamma_j \lambda_1 \frac{\partial T_1}{\partial r} = \lambda_3 \frac{\partial T_3}{\partial r}, T_1 = T_3, \tag{7}$$

$$\lambda_1 \frac{\partial T_1}{\partial \phi} = \lambda_3 \frac{\partial T_3}{\partial \phi}, T_1 = T_3, \tag{8}$$

$$\Gamma_{j.2} \lambda_2 \frac{\partial T_2}{\partial r} = \lambda_3 \frac{\partial T_3}{\partial r}, T_2 = T_3, \tag{9}$$

$$\Gamma_s \lambda_3 \frac{\partial T_3}{\partial r} = \lambda_4 \frac{\partial T_4}{\partial r}, T_3 = T_4, \tag{10}$$

$$\Gamma_e \lambda_2 \frac{\partial T_2}{\partial r} = \alpha (T_e - T_2), \tag{11}$$

Initial conditions for equations (1) - (4):

$$t=0,\ T_i(r,\phi) = T_{i0}(r,\phi),\ i=1,2,3,4 \tag{12}$$

where T_i, ρ_i, c_i, λ_i are the temperature, density, heat capacity and thermal conductivity, respectively, of the core (i=1), bark (i=2), large vessels (i=3) of the trunk and surface layer of water (i=4); α is the heat transfer coefficient; J - current strength; U is the voltage; r, φ - polar coordinates, t - time. Indexes "e" and "0" correspond to the parameters of the external environment and the parameters of wood at the initial moment of time.

A numerical study was carried out using the following initial data (large vessels): ρ=650 kg/m^3; c=2600 J/(kg•K); λ=0.35 W/(m•K). Thermophysical characteristics of the core (Zabolotny et al., 1995): ρ=650 kg/m^3; c=1670 J/(kg•K); λ=0.29 W/(m•K). Thermophysical characteristics of the bark: ρ=650 kg/m^3; c=1670 J/(kg•K); λ=0.12 W/(m•K). Thermophysical characteristics of water: ρ=1000 kg/m^3; c=4125 J/(kg•K); λ=0.569 W/(m•K). External exposure parameters: α=80 W/(m^2•K).

RESULTS AND DISCUSSION

The formulated mathematical model with the corresponding boundary and initial conditions was solved by a locally one-dimensional finite-difference method (Samarskiy and Vabishchevich, 2001). To solve difference analogs of differential equations, the marching method was used (Samarskiy, 1983).

Quite extensive information is known on the parameters of cloud-to-ground lightning discharges. Average peak current (Soriano et al., 2005): J=23.5 kA for negative discharge and J=35.3 kA for positive discharge. About 16.5% of positive discharges have a current of less than 10 kA (Cummins et al., 1998).

It is necessary to conduct a parametric study of the influence of the main discharge characteristics on the ignition of a deciduous tree by a cloud-to-ground lightning discharge. Conditions (Zabolotny et al., 1995) for temperature and heat flux were used as ignition criteria. The criteria for assessing the onset of ignition were the following conditions (Zabolotny et al., 1995) (Table 1).

Table 1. Experimental data (Zabolotny et al., 1995)

Ignition delay, s	Heat flux, kW/m^2	Surface temperature, K
136	15	-
61.2	21	645
17.2	42	688
1.8	125	755
0.43	210	801

Birch is exposed to a negative lightning discharge of 500 ms duration with a peak current of 23.5 kA and a voltage of 100 kV. Fig. 2 shows the temperature distribution in the horizontal section of the tree trunk at various times before and at the moment of ignition of the tree trunk by the current of a cloud-to-ground lightning discharge (the initial temperature of the tree is 300 K: a) - t = 0.01 s; b) - 0.1 s; c) - 0.3 s; d) - 0.5 s.

Figure 2. Temperature field on a horizontal cut of a tree at different times: a) 0.01 s; b) 0.1 s; c) 0.3 s; d) 0.5 s

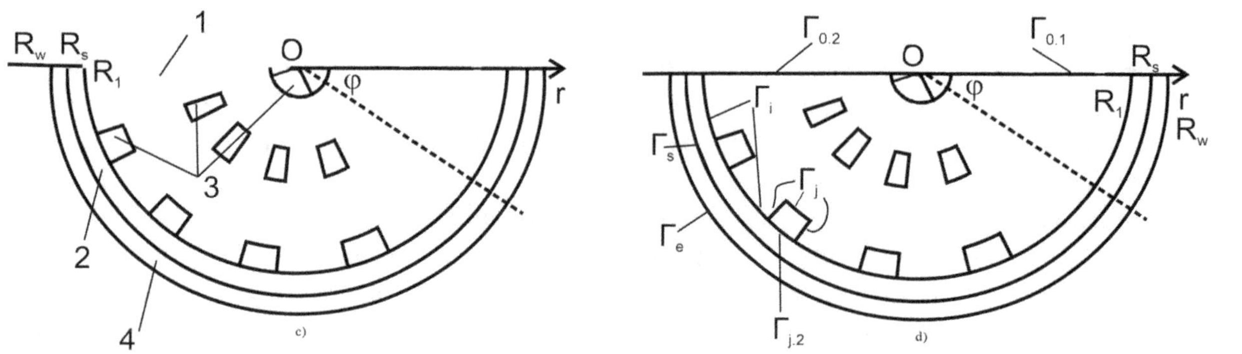

Fig. 3 shows the temperature distribution in the horizontal section of the tree trunk at different times before and at the moment of ignition of the tree trunk by a cloud-to-ground lightning discharge current, taking into account the surface layer of water (the initial temperature of the tree is 300 K: a) - t = 0.01 s; b) - 0.1 s; c) - 0.3 s; d) - 0.5 s.

Figure 3. Temperature field on a horizontal cut of a tree at different times, taking into account the surface layer of water: a) 0.01 s; b) 0.1 s; c) 0.3 s; d) 0.5 s

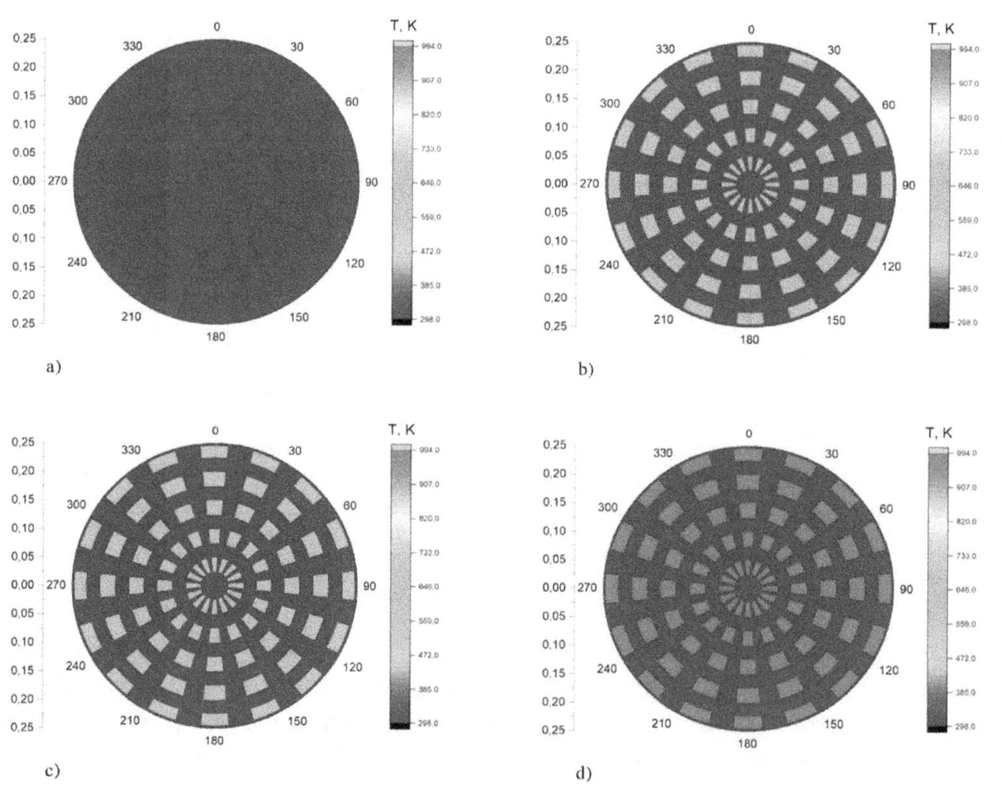

Fig. 4 shows the dependence of the heat flux to the ignition surface and its temperature on time. In Figure 4, numbers indicate: 1 - temperature in the trunk at the border of the bark and core; 2 - temperature on the surface of the bark during heat release in the surface layer of water; 3 - heat flux to the ignition surface from the core; 4 - heat flux to the ignition surface from the surface layer of water.

Figure 4. Dependence of the temperature of the ignition surface and the dependence of the heat flux to the ignition surface on time

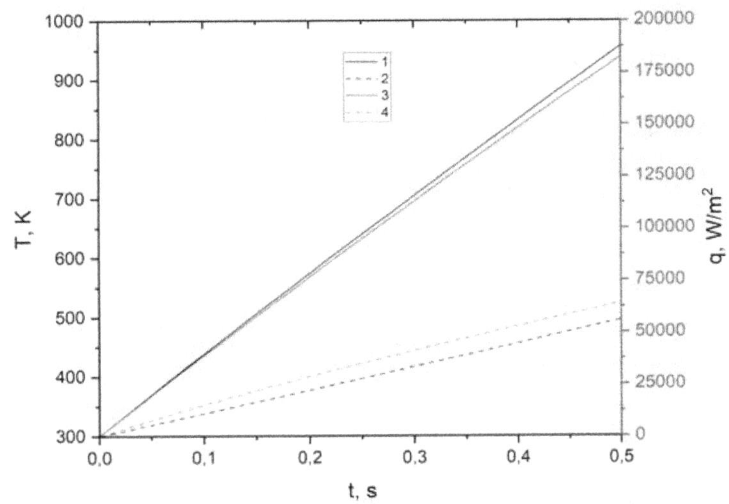

Analysis of the results presented in Fig. 2 shows that as a result of the action of the considered cloud-to-ground lightning discharge, the tree trunk heats up to the ignition temperature of the forest fuel (about 1000 K). The results suggest that the tree trunk ignites and burns. This conclusion is confirmed by the data of observations of thunderstorms (Plummer, 1912), where it is indicated that the tree ignites.

It should be noted that the internal vessels, when an electric current is passed, also release Joule heat, sufficient to ignite birch wood. However, the lack of an oxidizing agent in this zone can lead to the fact that ignition under real conditions will not occur at the boundaries of the inner vessels. This is one of the subjects of further research.

Analysis of the results in Fig. 3 shows that in the near-surface layer, the bark also warms up enough for the tree to ignite. In the zone of the surface layer of water, the temperature is approximately two times lower than in the zone of a large vessel. Figure 3 shows a ring of elevated temperature along the surface of a deciduous tree trunk. Also in Figures 2 and 3, zones of increased temperature are visible in the core of the tree trunk, where large vessels pass. It should be noted that ignition is most likely to occur in a large subsurface vessel. Internal large vessels also have an elevated temperature, but there is no access to an oxidizing agent. It is important to understand that this statement does not consider the stress-strain state of the tree trunk.

Analysis of the results, which are illustrated in Fig. 4 shows that according to the temperature criterion and the calculated value of the heat flux, the ignition conditions are achieved for a typical cloud-to-ground lightning discharge.

A parametric study of the influence of voltage and current strength of a cloud-to-ground lightning discharge showed theoretically established limits for the ignition of a deciduous tree trunk during the action of an electric discharge. It was found that when the current is less than 23.5 kA and the voltage is less than 90 kV, no ignition of the tree occurs during the period of the cloud-to-ground lightning discharge (duration is 500 ms).

Table 2 shows the results of computational experiments to determine the ignition conditions depending on the voltage of a cloud-to-ground lightning discharge at a current of J=23.5 kA, and in table 3, depending on the current at a discharge voltage of U=100 kV.

Table 2. Conditions for ignition of a tree trunk from voltage at current J=23.5 kA

Voltage, *U*, kV	Ignition conditions (Zabolotny et al., 1995)	Surface temperature, K	Heat flux. kW/m^2
1 – 85	No	<801	<210
90	Yes	801	226
95	Yes	801	230
100	Yes	801	232
105	Yes	801	235
110	Yes	801	238

Note that the approach of large vessels explains the cases when burn grooves appear on the tree after the impact of a cloud-to-ground lightning discharge. Heating and burns occur only in the area of a large surface vessel through which an electric current passes. The inter-vessel part of the wood is not heated to critical temperatures. It should be noted that as a result of the impact of a cloud-to-ground lightning discharge, heated particles (Baranovskiy et al., 2017) may form, which, falling on the layer of forest combustible material, will lead to its ignition. As a consequence, a grassland forest fire can occur (Goman, 2020).

Table 3. Conditions for the ignition of a tree trunk from the current at a voltage of U=100 kV

Current, *J*, kA	Ignition conditions (Zabolotny et al., 1995)	Surface temperature, K	Heat flux, kW/m^2
1 – 20	No	<801	<210
23.5	Yes	801	232
30	Yes	801	245
35	Yes	801	253

CONCLUSION

A physical and mathematical model to simulate the heating a deciduous tree as a result of the electric current of a lightning strike flows through it. The approximation of large vessels is used. As a result of varying the current-voltage characteristics, the conditions for the realization of the phenomenon under consideration, which are characteristic of the typical range of parameters for a cloud-to-ground lightning discharge, are identified. The possibility of ignition of a deciduous tree in conditions of a thunderstorm passing by a cloud-to-cloud-to-ground discharge is shown. The presented physical and mathematical model can become an additional module in forest fire danger assessment and prediction systems (Baranovskiy, 2020; Belikova and Glebova, 2020; Grishin, 2002; Glagolev and Zubareva, 2020; Adab et al., 2013; Al Janabi et al., 2017; Bui et al., 2017; CWFIS, 2020; Cardille et al., 2001; Chen et al., 2017; Curt et al., 2016; Ganteaume and Guerra, 2018; Chen et al., 2020; Glagolev et al., 2020; Papagiannaki et al., 2020). In addition, the results obtained have independent theoretical and fundamental significance for the theory of forest fires. They allow us to understand the physical nature of the phenomenon under study. Moreover, the developed mathematical models can be used in conjunction with atmospheric electricity models (Nagorskiy et al., 2020) and GIS systems based on remote sensing data (M D. et al., 2020).

ACKNOWLEDGMENT

This work was implemented with the financial support of the Russian Foundation for Basic Research. [Project N 17-29-05093].

REFERENCES

Adab, H., Kanniah, K. D., & Solaimani, K. (2013). Modeling forest fire in the northeast of Iran using remote sensing and GIS techniques. *Natural Hazards*, *65*(3), 1723–1743. doi:10.100711069-012-0450-8

Al Janabi, S., Al Shourbaji, I., & Salman, M. A. (2017). *Assessing the suitability of soft computing approaches for forest fires prediction*. Applied Computing and Informatics. doi:10.1016/j.aci.2017.09.006

Baranovskiy, N. V. (2020). *Predicting, Monitoring, and Assessing Forest Fire Dangers and Risks*. IGI Global. doi:10.4018/978-1-7998-1867-0

Baranovskiy, N. V., Kuznetsov, G. V., & Nemova, T. N. (2017). High temperature wood particles formation caused by the cloud-to-ground lightning discharge through the coniferous tree trunk. *Far East Journal of Mathematical Sciences*, *102*(5), 1033–1044. doi:10.17654/MS102051033

Belikova, M. Y., & Glebova, A. V. (2020). Cluster Analysis Algorithms for RS and WWLLN Data Processing. In N. V. Baranovskiy (Ed.), *Predicting, Monitoring, and Assessing Forest Fire Dangers and Risks* (pp. 151–181). IGI Global. doi:10.4018/978-1-7998-1867-0.ch007

Braun, H. J. (1963). Die organization des hydrosystems im stammholz der baume und straucher. *Deut. Bot. Gesell. Ber.*, *75*, 401–410.

Bui, D. T., Bui, Q.-T., Nguyen, Q.-P., Pradhan, B., Nampak, H., & Trinh, P. T. (2017). A hybrid artificial intelligence using GIS-based neural-fuzzy inference system and particle swarm optimization for forest fire susceptibility modeling at a tropical area. *Agricultural and Forest Meteorology*, *233*, 32–44. doi:10.1016/j.agrformet.2016.11.002

Canadian Wildland Fire Information System. (n.d.). Access: https://cwfis.cfs.nrcan.gc.ca/home

Cardille, J. A., Ventura, S. J., & Turner, M. G. (2001). Environmental and Social Factors of Influencing Wildfires in the Upper Midwest, United States. *Ecological Applications*, *11*(1), 111–127. doi:10.1890/1051-0761(2001)011[0111:EASFIW]2.0.CO;2

Chen, W., Xie, X., Wang, J., Pradhan, B., Hong, H., Tien Bui, D., Duan, Z., & Ma, J. (2017). A comparative study of logistic model tree, random forest and classification and regression tree models for spatial prediction of landslide susceptibility. *Catena*, *151*, 147–160. doi:10.1016/j.catena.2016.11.032

Chen, C., Xu, T., Zhao, D., & Lei, P. (2020). Study on the Wildfires Occurring Risk Based on Fuzzy FTA. *IOP Conference Series. Earth and Environmental Science*, *440*(5), 052100. Advance online publication. doi:10.1088/1755-1315/440/5/052100

Conedera, M., Cesti, G., Pezzatti, G. B., Zumbrunnen, T., & Spinedi, F. (2006). Lightning-induced fires in the Alpine region: An increasing problem. *Forest Ecology and Management*, *234*(Supplement 1.), S68. doi:10.1016/j.foreco.2006.08.096

Cummins, K. L., Murphy, M. J., Bardo, E. A., Hiscox, W. L., Pyle, R. B., & Pifer, A. E. (1998). A combined TOA/MDF technology upgrade of the U.S. national lightning detection network. *Journal of Geophysical Research*, *103*(D8), 9035–9044. doi:10.1029/98JD00153

Curt, T., Frejaville, T., & Lahaye, S. (2016). Modelling the spatial patterns of ignition causes and fire regime features in southern France: Implications for the prevention policy. *International Journal of Wildland Fire*, *25*(7), 785–796. doi:10.1071/WF15205

Esau, K. (1980) Anatomy of seed plants. Moscow: Mir. (in Russian)

Ganteaume, A., & Guerra, F. (2018). Explaining the spatio-seasonal variation of fires by their causes: The case of southeastern France. *Applied Geography (Sevenoaks, England)*, *90*, 69–81. doi:10.1016/j.apgeog.2017.11.012

Glagolev, V. A., & Zubareva, A. M. (2020). Using Geographic Information Systems for Study of the Natural Fire Risk on the Territory of the Jewish Autonomous Region. In N. V. Baranovskiy (Ed.), *Predicting, Monitoring, and Assessing Forest Fire Dangers and Risks* (pp. 209–231). IGI Global. doi:10.4018/978-1-7998-1867-0.ch009

Glagolev, V., Bazhenov, R., Povkh, I., Chuyko, O., & Dolmatova, O. (2020). Designing and producing a geographic information system to forecast vegetation fire danger according to weather conditions. *IOP Conference Series. Materials Science and Engineering*, *862*(5), 052019. Advance online publication. doi:10.1088/1757-899X/862/5/052019

Goman, P. N. (2020). The Dynamics of Surface Forest Fire and Forest Fuel Ignition Under the Heat Radiation From the Fire Line. In N. V. Baranovskiy (Ed.), *Predicting, Monitoring, and Assessing Forest Fire Dangers and Risks* (pp. 1–47). IGI Global. doi:10.4018/978-1-7998-1867-0.ch001

Grishin, A. M. (2002) Modeling and forecasting disasters. Tomsk: Publishing House of Tomsk University. (in Russian)

Karanina, S. Y., Kocheeva, N. A., Karanin, A. V., & Belikova, M. Y. (2020). Assessing the Territory According to the Degree of Thunderstorm Danger. In N. V. Baranovskiy (Ed.), *Predicting, Monitoring, and Assessing Forest Fire Dangers and Risks* (pp. 122–150). IGI Global. doi:10.4018/978-1-7998-1867-0.ch006

Latham, D., & Williams, E. (2001). *Lightning and forest fires. Forest fires: Behavior and Ecological Effects*. Elsevier.

M., D., Bhatt, H. P., & S. P., V. (2020). Analysis of Terrestrial Vegetation Trends and Correlation Between Vegetation Indices and Climatic Factors. In *Predicting, Monitoring, and Assessing Forest Fire Dangers and Risks* (pp. 233-257). IGI Global. doi:10.4018/978-1-7998-1867-0.ch010

Nagorskiy, P. M., Kabanov, M. V., & Pustovalov, K. N. (2020). The Influence of Smoke From Forest Fires on the Meteorological and Electrical Characteristics of the Atmosphere. In N. V. Baranovskiy (Ed.), *Predicting, Monitoring, and Assessing Forest Fire Dangers and Risks* (pp. 322–344). IGI Global. doi:10.4018/978-1-7998-1867-0.ch014

Papagiannaki, K., Giannaros, T. M., Lykoudis, S., Kotroni, V., & Lagouvardos, K. (2020). Weather-related thresholds for wildfire danger in a Mediterranean region: The case of Greece. *Agricultural and Forest Meteorology*, *291*, 108076. Advance online publication. doi:10.1016/j.agrformet.2020.108076

Plummer, F. G. (1912). *Lightning in Relation to Forest Fires. Bulletin 111. USDA Forest Service*. Government Printing Office.

Samarskiy, A. A. (1983). *Theory of difference schemes*. Nauka. (in Russian)

Samarskiy, A. A., & Vabishchevich, P. N. (2001). *Additive schemes for problems of mathematical physics*. Science. (in Russian)

Soriano, L. R., De Pablo, F., & Tomas, C. (2005). Ten-year study of cloud-to-cloud-to-ground lightning activity in the Iberian Peninsula. *Journal of Atmospheric and Solar-Terrestrial Physics*, *67*(16), 1632–1639. doi:10.1016/j.jastp.2005.08.019

Uman, M. A. (1969). *Lightning*. McGraw-Hill.

Zabolotny, A.E., Zabolotnaya, M.M., Zabolotnaya, Y.A., & Timoshin, V.N. (1995). Determination of safe use zones of solid fuel generators of fire extinguishing aerosols. *Issues of Special Engineering*, (7-8), 15 - 21. (in Russian)

ADDITIONAL READING

Azadifar, M., Rubinstein, M., Rachidi, F., Rakov, V. A., Diendorfer, G., Schulz, W., & Pavanello, D. (2019). A Study of a Large Bipolar Lightning Event Observed at the Säntis Tower, IEEE Transactions on Electromagnetic Compatibility, 61 (3). *Article*, *61*(8715461), 796–806. doi:10.1109/TEMC.2019.2913220

De Paiva, A. R., Saba, M. M. F., Naccarato, K. P., Schumann, C., Jaques, R., Da Silva Ferro, M. A., & Warner, T. A. (2014) Detection of upward lightning by lightning location systems, 2014 International Conference on Lightning Protection, ICLP 2014, Article N° 6973425, pp. 1824-1826. DOI: 10.1109/ICLP.2014.6973425

Lu, G., Cummer, S. A., Tian, Y., Zhang, H., Lyu, F., Wang, T., Stanley, M. A., Yang, J., & Lyons, W. A. (2016). Sprite produced by consecutive impulse charge transfers following a negative stroke: Observation and simulation. *Journal of Geophysical Research*, *121*(8), 4082–4092. doi:10.1002/2015JD024644

Nag, A., Rakov, V. A., & Murphy, M. J. (2014) Measurement of preliminary breakdown pulse trains in cloud-to-ground lightning using lightning locating systems, 2014 International Conference on Lightning Protection, ICLP 2014, статья N° 6973356, pp. 1437-1444. DOI: 10.1109/ICLP.2014.6973356

Nag, A., Murphy, M. J., Schulz, W., & Cummins, K. L. (2014) Lightning locating systems: Characteristics and validation techniques, 2014 International Conference on Lightning Protection, ICLP 2014, Article N° 6973283, pp. 1070-1082. DOI: 10.1109/ICLP.2014.6973283

Saba, M. M. F., Shigihara, M., & Kodaira, C. Y. (2018) Observation of a lightning flashover with a high-speed video camera, 34th International Conference on Lightning Protection, ICLP 2018, Article N° 8503282. DOI: 10.1109/ICLP.2018.8503282

Stolzenburg, M., Marshall, T. C., Karunarathne, S., Karunarathna, N., & Orville, R. E. (2015). An M component with a concurrent dart leader traveling along different paths during a lightning flash. *Journal of Geophysical Research*, *120*(19), 10267–10284. doi:10.1002/2015JD023417

Wang, X., Yuan, P., Cen, J., & Xue, S. (2016). Correlation between the spectral features and electric field changes for natural lightning return stroke followed by continuing current with M-components. *Journal of Geophysical Research*, *121*(14), 8615–8624. doi:10.1002/2016JD025314

Xue, S., Yuan, P., Cen, J., Li, Y., & Wang, X. (2015). Spectral observations of a natural bipolar cloud-to-ground lightning. *Journal of Geophysical Research*, *120*(5), 1972–1979. doi:10.1002/2014JD022598

Yuan, P., Chen, R., Su, Y., Fan, B., Wang, X. (2019) Characteristic parameters of positive cloud-to-ground lightning channel, Journal of Earth System Science, 128 (6), статья N° 164, . Doi:10.100712040-019-1196-4

KEY TERMS AND DEFINITIONS

Cloud-to-Ground Lightning Discharge: An electrical discharge during a thunderstorm that occurs between a cloud and the earth's surface. It is a natural source of forest fires.

Forest Fire: Uncontrolled aerothermochemical phenomenon characterized by step-by-step mechanism which includes following stages: inert heating, moisture evaporation, high temperature terpens evaporation, dry organic matter pyrolysis, flammable combustion and smoldering.

Forest Fuel: It can be considered like dead and live forest fuel. Main types of forest fuel which can be involved in combustion during forest fire: cloud-to-ground forest fuel (needles, leaves and dry grass, small branches) and crown forest fuel (needles, small branches).

Ignition: Inflammation of forest fuel caused by definite source of high temperature or energy.

Ignition Delay: Time before flame flash after forest fuel heating.

Lightning Activity: An atmospheric phenomenon characterized by discharges of the cloud-to-cloud and cloud-to-cloud-to-ground class.

Mathematical Simulation: The production of a computer model of forest fire conditions and prerequisites, especially for the purpose of study.

Meteorological Parameters: Physical characteristics of local weather conditions in the forested area under consideration. Key parameters include ambient temperature, soil temperature, precipitation, wind speed, solar radiation, cloud cover, dew point temperature. These parameters are used for mathematical modeling of the drying of a layer of forest fuel.

Monitoring: Monitoring refers to the periodic calculation of the parameters of forest fire danger with a portion of information available in real time.

Prediction: Under the prediction of forest fires is the calculation of the parameters of forest fire danger with a certain projection in advance in order to have enough time to anticipate an emergency. The calculation in this case is carried out in a mode ahead of the real time of the development of the catastrophe - the occurrence of a forest fire.

Chapter 7
Three-Dimensional Mathematical Models to Simulate Deciduous Tree Ignition

ABSTRACT

The results of the numerical implementation for the spatial formulation of the problem of the deciduous tree ignition by an electric current of a cloud-to-ground lightning discharge are presented. The problem was solved in a three-dimensional formulation in a cylindrical coordinate system. A symmetric formulation of the problem is considered. The localization of reactive wood, structural heterogeneity of the bark, volt-ampere characteristics of a cloud-to-ground lightning discharge, and chemical reaction in the gas phase are taken into account. The conditions of deciduous tree ignition in the typical range of the discharge effect are revealed.

INTRODUCTION

The widely known American (WFAS, 2020), Canadian (CWFIS, 2020), European (EFFIS, 2020) systems and domestic state standard GOST R 22.1.09.99, based on the Nesterov criterion (Nesterov, 1949), give an abstract numerical index. In fact, all of the above methods do not take into account the real physico-chemical processes occurring when forest fires occur.

You should focus on the deterministic-probabilistic approach to assessing forest fire danger (Baranovskiy, 2017). The mathematical content of such a system should be adequate to the physics of the process and models of forest fuels ignition should be designed.

According to the generalized mechanism of a forest fire from a thunderstorm, it is necessary to develop models for the ignition of deciduous trees by a cloud-to-ground lightning discharge. For conifers, such models are also being developed. It is of interest to develop models of deciduous tree ignition prior to spatial setting, taking into account the chemical reaction in the gas phase.

DOI: 10.4018/978-1-7998-7250-4.ch007

The aim of the study is to develop a spatial physical and mathematical model of the deciduous tree ignition by a cloud-to-ground lightning discharge, taking into account a wide range of factors.

MATHEMATICAL MODEL

According to (Esau, 1980), the electric current of a cloud-to-ground lightning discharge passes through the core of a deciduous tree trunk. The main assumptions and suggestions: 1) reactive wood is formed in the upper part of the branches (Esau, 1980); 2) the approximation of an "ideal" crack in the bark is used; 3) in the calculations, only a part of the branch growing from the trunk is considered. The rest of the continuation is not considered, since it was previously established that during the exposure to an electric current, the outer part of the branch does not have time to warm up; 4) the main product of pyrolysis is carbon monoxide (Grishin, 1997); 5) the leading chemical reaction is the oxidation of carbon monoxide to carbon dioxide (Grishin and Shipulina, 2002); 6) the tree is considered as a conductor of the resistor type, for which Ohm's and Joule-Lenz's laws are valid (Sivukhin, 1977); 7) the approximation of large vessels is used; 8) moisture evaporation is described by the Knudsen-Langmuir equation (Pankratov et al., 1975). To describe the process under study, the following physical model is adopted. A freestanding deciduous tree is considered. At a fixed moment in time, a lightning discharge of a certain polarity and duration strikes the tree trunk. It is believed that the current-voltage characteristics of the discharge are the same for different sections of the tree trunk. As a result of the flow of electric current in a large vessel, the wood heats up due to the release of Joule heat. In the process of further heating, thermal decomposition of wood occurs with the formation of gaseous pyrolysis products, which instantly enter the gas phase region and mix with the oxidizing agent. At certain temperatures and concentrations of reagents, a chemical reaction of carbon monoxide oxidation occurs. It is considered that ignition occurs if critical values of the following parameters are reached: 1) the heat flux from the chemical reaction zone exceeds the heat flux from the subcrustal zone of the tree; 2) the temperature of the gas mixture reaches a critical value. The solution area is shown in Fig. 1.

The process of ignition of a deciduous tree by a cloud-to-ground lightning discharge is described by a system of three-dimensional non-stationary nonlinear equations of heat conduction and diffusion with corresponding initial and boundary conditions. For numerical implementation, a finite-difference method was used (Samarskiy and Vabishchevich, 2001). Two-dimensional and three-dimensional difference equations were solved by a locally one-dimensional method (Samarskiy and Vabishchevich, 2001). Difference analogues of the one-dimensional equations of heat conduction and diffusion were solved by the marching method in combination with the simple iteration method (Samarskiy, 1983).

Mathematical model:

$$\rho_{ef1}c_{ef1}\frac{\partial T_1}{\partial t}=\frac{\lambda_{ef1}}{r}\frac{\partial}{\partial r}\left(r\frac{\partial T_1}{\partial r}\right)+\frac{\lambda_{ef1}}{r^2}\frac{\partial^2 T_1}{\partial \phi^2}+\lambda_{ef1}\frac{\partial^2 T_1}{\partial z^2}-QW_{ev}\phi_{13}-Q_p k_p \rho_{12}\phi_{12}\exp\left(-\frac{E_1}{RT_1}\right), \quad (1)$$

Figure 1.

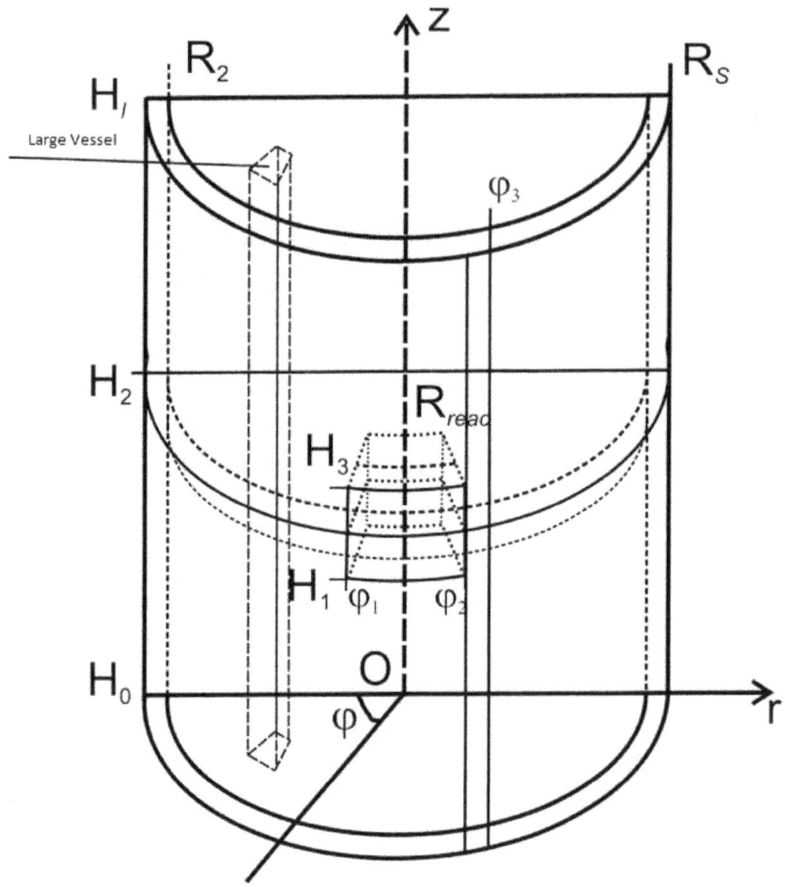

$$\rho_2 c_2 \frac{\partial T_2}{\partial t} = \frac{\lambda_2}{r}\frac{\partial}{\partial r}\left(r\frac{\partial T_2}{\partial r}\right) + \frac{\lambda_2}{r^2}\frac{\partial^2 T_2}{\partial \phi^2} + \lambda_2 \frac{\partial^2 T_2}{\partial z^2} - Q_p k_p \rho_{12}\phi_{12} \exp\left(-\frac{E_1}{RT_2}\right), \tag{2}$$

$$\rho_{ef3} c_{ef3} \frac{\partial T_3}{\partial t} = \frac{\lambda_{ef3}}{r}\frac{\partial}{\partial r}\left(r\frac{\partial T_3}{\partial r}\right) + \frac{\lambda_{ef3}}{r^2}\frac{\partial^2 T_3}{\partial \phi^2} + \lambda_{ef3} \frac{\partial^2 T_3}{\partial z^2} + JU - QW_{ev}\phi_{13} - Q_p k_p \rho_{12}\phi_{12} \exp\left(-\frac{E_1}{RT_3}\right), \tag{3}$$

$$\rho_{ef4} c_{ef4} \frac{\partial T_4}{\partial t} = \frac{\lambda_{ef4}}{r}\frac{\partial}{\partial r}\left(r\frac{\partial T_4}{\partial r}\right) + \frac{\lambda_{ef4}}{r^2}\frac{\partial^2 T_4}{\partial \phi^2} + \lambda_{ef4} \frac{\partial^2 T_4}{\partial z^2} + JU - QW_{ev}\phi_{13} - Q_p k_p \rho_{12}\phi_{12} \exp\left(-\frac{E_1}{RT_4}\right), \tag{4}$$

$$\rho_{ef5}c_{ef5}\frac{\partial T_5}{\partial t}=\frac{\lambda_{ef5}}{r}\frac{\partial}{\partial r}\left(r\frac{\partial T_5}{\partial r}\right)+\frac{\lambda_{ef5}}{r^2}\frac{\partial^2 T_5}{\partial \phi^2}+\lambda_{ef5}\frac{\partial^2 T_5}{\partial z^2}-QW_{ev}\phi_{13}-Q_p k_p \rho_{12}\phi_{12}\exp\left(-\frac{E_1}{RT_5}\right), \quad (5)$$

$$\rho_{ef6}c_{ef6}\frac{\partial T_6}{\partial t}=\frac{\lambda_{ef6}}{r}\frac{\partial}{\partial r}\left(r\frac{\partial T_6}{\partial r}\right)+\frac{\lambda_{ef6}}{r^2}\frac{\partial^2 T_6}{\partial \phi^2}+\lambda_{ef6}\frac{\partial^2 T_6}{\partial z^2}+JU-QW_{ev}\phi_{13}-Q_p k_p \rho_{12}\phi_{12}\exp\left(-\frac{E_1}{RT_6}\right), \quad (6)$$

$$\rho_{ef7}c_{ef7}\frac{\partial T_7}{\partial t}=\frac{\lambda_{ef7}}{r}\frac{\partial}{\partial r}\left(r\frac{\partial T_7}{\partial r}\right)+\frac{\lambda_{ef7}}{r^2}\frac{\partial^2 T_7}{\partial \phi^2}+\lambda_{ef7}\frac{\partial^2 T_7}{\partial z^2}-QW_{ev}\phi_{13}-Q_p k_p \rho_{12}\phi_{12}\exp\left(-\frac{E_1}{RT_7}\right), \quad (7)$$

$$\rho_g c_g\frac{\partial T_g}{\partial t}=\frac{\lambda_g}{r}\frac{\partial}{\partial r}\left(r\frac{\partial T_g}{\partial r}\right)+\frac{\lambda_g}{r^2}\frac{\partial^2 T_g}{\partial \phi^2}\lambda_g\frac{\partial^2 T_g}{\partial z^2}+Q_5\left(1-\nu_5\right)R_5, \quad (8)$$

$$\frac{\partial C_8}{\partial t}=\frac{D}{r}\frac{\partial}{\partial r}\left(r\frac{\partial C_8}{\partial r}\right)+\frac{D}{r^2}\frac{\partial^2 C_8}{\partial \phi^2}+D\frac{\partial^2 C_4}{\partial z^2}-R_5\frac{M_8}{M_9}, \quad (9)$$

$$\frac{\partial C_9}{\partial t}=\frac{D}{r}\frac{\partial}{\partial r}\left(r\frac{\partial C_9}{\partial r}\right)+\frac{D}{r^2}\frac{\partial^2 C_9}{\partial \phi^2}+D\frac{\partial C_9}{\partial z^2}-R_5, \quad (10)$$

$$\frac{\partial C_{10}}{\partial t}=\frac{D}{r}\frac{\partial}{\partial r}\left(r\frac{\partial C_{10}}{\partial r}\right)+\frac{D}{r^2}\frac{\partial^2 C_{10}}{\partial \phi^2}+D\frac{\partial C_{10}}{\partial z^2}, \quad (11)$$

$$\sum_{i=8}^{11} C_i=1. \quad (12)$$

Kinetic equations:

$$\rho_{12}\frac{\partial \phi_{12}}{\partial t}=-k_p\rho_{12}\phi_{12}\exp\left(-\frac{E_1}{RT_1}\right), \quad (13)$$

$$\rho_{12}\frac{\partial \phi_{12}}{\partial t} = -k_p \rho_{12}\phi_{12} \exp\left(-\frac{E_1}{RT_2}\right), \quad (14)$$

$$\rho_{12}\frac{\partial \phi_{12}}{\partial t} = -k_p \rho_{12}\phi_{12} \exp\left(-\frac{E_1}{RT_3}\right), \quad (15)$$

$$\rho_{12}\frac{\partial \phi_{12}}{\partial t} = -k_p \rho_{12}\phi_{12} \exp\left(-\frac{E_1}{RT_4}\right), \quad (16)$$

$$\rho_{12}\frac{\partial \phi_{12}}{\partial t} = -k_p \rho_{12}\phi_{12} \exp\left(-\frac{E_1}{RT_5}\right), \quad (17)$$

$$\rho_{12}\frac{\partial \phi_{12}}{\partial t} = -k_p \rho_{12}\phi_{12} \exp\left(-\frac{E_1}{RT_6}\right), \quad (18)$$

$$\rho_{12}\frac{\partial \phi_{12}}{\partial t} = -k_p \rho_{12}\phi_{12} \exp\left(-\frac{E_1}{RT_7}\right), \quad (19)$$

$$\rho_{13}\frac{\partial \phi_{13}}{\partial t} = -W_{ev}, \quad (20)$$

$$\sum_{i=12}^{14} \phi_i = 1, \quad (21)$$

$$W_{ev} = \frac{A(P^s - P)}{\sqrt{\frac{2\pi RT}{M}}}. \quad (22)$$

$$R_5 = k_5 M_9 T_g^{-2.25} \exp\left(-\frac{E_5}{RT_g}\right) \cdot \begin{cases} x_8^{0.25} x_9, x_8 > 0.05 \\ x_8 x_9, x_8 \le 0.05 \end{cases} \text{[377]}, \tag{23}$$

$$x_i = \frac{C_i}{\sum_{k=8}^{11} \frac{C_k}{M_k} M_i}. \tag{24}$$

Equation of state:

$$P = \frac{\rho R T_g}{M}, \frac{1}{M} = \frac{C_8}{M_8} + \frac{C_9}{M_9} + \frac{C_{10}}{M_{10}} + \frac{C_{11}}{M_{11}}, \tag{25}$$

$$\rho_{efi} = \rho_{12}\phi_{12} + \rho_{13}\phi_{13} + \rho_{14}\phi_{14},\ c_{efi} = c_{12}\phi_{12} + c_{13}\phi_{13} + c_{14}\phi_{14},\ \lambda_{efi} = \lambda_{12}\phi_{12} + \lambda_{13}\phi_{13} + \lambda_{14}\phi_{14}. \tag{26}$$

Initial conditions:

$$t=0,\ T_i(r,\phi,z) = T_{i0}(r,\phi,z),\ i=1,2,3,4,5,6,7,g \tag{27}$$

$$C_i(r,\phi,z) = C_{i0}(r,\phi,z),\ i=8,9,10,11 \tag{28}$$

$$\phi_i(r,\phi,z) = \phi_{i0}(r,\phi,z),\ i=12,13,14 \tag{29}$$

Boundary conditions:

Boundary conditions outside the branch area:

$$r=0,\ \lambda_{ef1}\frac{\partial T_1}{\partial r} = 0, \tag{30}$$

$$r=R_2,\ \lambda_{ef1}\frac{\partial T_1}{\partial r} = \lambda_2 \frac{\partial T_2}{\partial r}, \tag{31}$$

$$r=R_S,\ \lambda_2 \frac{\partial T_2}{\partial r} = \lambda_g \frac{\partial T_g}{\partial r}, \quad (32)$$

$$r=R_e,\ T_g=T_e, \quad (33)$$

$$\varphi=0,\ \frac{\partial T_i}{\partial \phi} = 0, \quad (34)$$

$$\varphi=\pi,\ \frac{\partial T_i}{\partial \phi} = 0, \quad (35)$$

$$z=H_0,\ \lambda_i \frac{\partial T_i}{\partial z} = 0, \quad (36)$$

$$z=H_l,\ \lambda_i \frac{\partial T_i}{\partial z} = 0. \quad (37)$$

Boundary conditions on the inner side of the branch:

$$\Gamma_0,\ \lambda_{ef4} \frac{\partial T_4}{\partial r} = \lambda_{ef1} \frac{\partial T_1}{\partial r}, \quad (38)$$

$$\Gamma_1,\ \lambda_{ef6} \frac{\partial T_6}{\partial r} = \lambda_{ef1} \frac{\partial T_1}{\partial r}. \quad (39)$$

Boundary conditions at the boundary of the right side of a branch and a crack:

$$\Gamma_2,\ \lambda_{ef4} \frac{\partial T_4}{\partial \phi} = \lambda_{ef1} \frac{\partial T_1}{\partial \phi}, \quad (40)$$

$$\Gamma_3,\ \lambda_{ef5} \frac{\partial T_5}{\partial \phi} = \lambda_g \frac{\partial T_g}{\partial \phi}, \quad (41)$$

$$\Gamma_4,\ \lambda_{ef6}\frac{\partial T_6}{\partial \phi} = \lambda_{ef1}\frac{\partial T_1}{\partial \phi}, \tag{42}$$

$$\Gamma_5,\ \lambda_{ef7}\frac{\partial T_7}{\partial \phi} = \lambda_g\frac{\partial T_g}{\partial \phi}. \tag{43}$$

Boundary conditions at the outer cut of the branch:

$$\Gamma_6,\ \lambda_{ef5}\frac{\partial T_5}{\partial r} = \lambda_g\frac{\partial T_g}{\partial r}, \tag{44}$$

$$\Gamma_7,\ \lambda_{ef7}\frac{\partial T_7}{\partial r} = \lambda_g\frac{\partial T_g}{\partial r}. \tag{45}$$

Boundary conditions on the lower edge of the branch:

$$\Gamma_8,\ \lambda_{ef6}\frac{\partial T_6}{\partial z} = \lambda_{ef1}\frac{\partial T_1}{\partial z}, \tag{46}$$

$$\Gamma_9,\ \lambda_{ef7}\frac{\partial T_7}{\partial z} = \lambda_{ef1}\frac{\partial T_1}{\partial z}. \tag{47}$$

Boundary conditions on the left side of the branch:

$$\Gamma_{10},\ \lambda_{ef1}\frac{\partial T_1}{\partial \phi} = \lambda_{ef4}\frac{\partial T_4}{\partial \phi}, \tag{48}$$

$$\Gamma_{11},\ \lambda_2\frac{\partial T_2}{\partial \phi} = \lambda_{ef5}\frac{\partial T_5}{\partial \phi}, \tag{49}$$

$$\Gamma_{12},\ \lambda_{ef1}\frac{\partial T_1}{\partial \phi} = \lambda_{ef6}\frac{\partial T_6}{\partial \phi}, \tag{50}$$

$$\Gamma_{13}, \ \lambda_2 \frac{\partial T_2}{\partial \phi} = \lambda_{ef7} \frac{\partial T_7}{\partial \phi}. \tag{51}$$

Boundary conditions on the upper edge of the branch:

$$\Gamma_{14}, \ \lambda_{ef4} \frac{\partial T_4}{\partial z} = \lambda_{ef1} \frac{\partial T_1}{\partial z}, \tag{52}$$

$$\Gamma_{15}, \ \lambda_{ef5} \frac{\partial T_5}{\partial z} = \lambda_{ef1} \frac{\partial T_1}{\partial z}. \tag{53}$$

Conditions at the crack boundary, excluding the right branch edge:

$$\Gamma_{16}, \ \lambda_2 \frac{\partial T_2}{\partial \phi} = \lambda_g \frac{\partial T_g}{\partial \phi}, \tag{54}$$

$$\rho D \frac{\partial C_8}{\partial \phi} = 0, \tag{55}$$

$$\rho D \frac{\partial C_9}{\partial \phi} = 0, \tag{56}$$

$$\rho D \frac{\partial C_{10}}{\partial \phi} = 0, \tag{57}$$

$$\Gamma_{17}, \ \lambda_g \frac{\partial T_g}{\partial \phi} = \lambda_2 \frac{\partial T_2}{\partial \phi}, \tag{58}$$

$$\rho D \frac{\partial C_8}{\partial \phi} = 0, \tag{59}$$

$$\rho D\frac{\partial C_9}{\partial \phi}=0, \tag{60}$$

$$\rho D\frac{\partial C_{10}}{\partial \phi}=0, \tag{61}$$

$$\Gamma_{18},\ \lambda_{ef3}\frac{\partial T_3}{\partial r}=\lambda_g\frac{\partial T_g}{\partial r}, \tag{62}$$

$$\rho D\frac{\partial C_8}{\partial r}=0, \tag{63}$$

$$\rho D\frac{\partial C_9}{\partial r}=Y_9, \tag{64}$$

$$\rho D\frac{\partial C_{10}}{\partial r}=Y_{10}, \tag{65}$$

$$\Gamma_{19},\ T_g=T_e, \tag{66}$$

$$\rho D\frac{\partial C_8}{\partial r}=0, \tag{67}$$

$$\rho D\frac{\partial C_9}{\partial r}=0, \tag{68}$$

$$\rho D\frac{\partial C_{10}}{\partial r}=0, \tag{69}$$

$$\Gamma_{20}, \lambda_g \frac{\partial T_g}{\partial z} = 0, \quad (70)$$

$$\rho D \frac{\partial C_8}{\partial z} = 0, \quad (71)$$

$$\rho D \frac{\partial C_9}{\partial z} = 0, \quad (72)$$

$$\rho D \frac{\partial C_{10}}{\partial z} = 0, \quad (73)$$

$$\Gamma_{21}, \lambda_g \frac{\partial T_g}{\partial z} = 0, \quad (74)$$

$$\rho D \frac{\partial C_8}{\partial z} = 0, \quad (75)$$

$$\rho D \frac{\partial C_9}{\partial z} = 0, \quad (76)$$

$$\rho D \frac{\partial C_{10}}{\partial z} = 0, \quad (77)$$

where T_i, ρ_{efi}, c_{efi}, λ_{efi} are the temperature, effective density, heat capacity and thermal conductivity, respectively, of the core (i=1), bark (i=2), large vessels (i=3) of the trunk, reactive wood (i=4, 5), the lower part of the branches (i=6.7); T_g, ρ_g, c_g, λ_g - temperature, density, heat capacity and thermal conductivity of the gas phase; C_i, M_i - concentration and molar mass of oxygen (i=8), carbon monoxide (i=9), water vapor (i=10) and inert components (i=11); φ_k, ρ_k, c_k, λ_k - volume fraction, density, heat capacity and thermal conductivity of organic matter (k=12), water (k=13) and gas mixture (k=14); α_e is the heat transfer coefficient; J - current; U is the voltage; Q is the thermal effect of moisture evaporation; W_{ev} is the mass rate of water evaporation, A is the accommodation coefficient, P^s is the pressure of saturated water vapor, P is the partial pressure of water vapor in the air, R is the universal gas constant;

Y_9 - mass flow (carbon monoxide); Y_{10} - mass flow (water vapor); Q_p is the thermal effect of pyrolysis; k_p - preexponential factor of the pyrolysis reaction; E_1 is the activation energy of the pyrolysis reaction; Q_5 is the heat effect of the oxidation reaction of carbon monoxide; k_5 is the preexponential factor of the oxidation reaction of carbon monoxide; E_5 is the activation energy of the oxidation reaction of carbon monoxide; $\nu 5$ - part of heat absorbed by the wood layer; x_8, x_9 - auxiliary variables; r, φ, z - cylindrical

Figure 2. Temperature distribution in the horizontal section of the tree trunk at different points in time: a) 0.01 s; b) 0.1 s; c) 0.3 s; d) 0.4 s

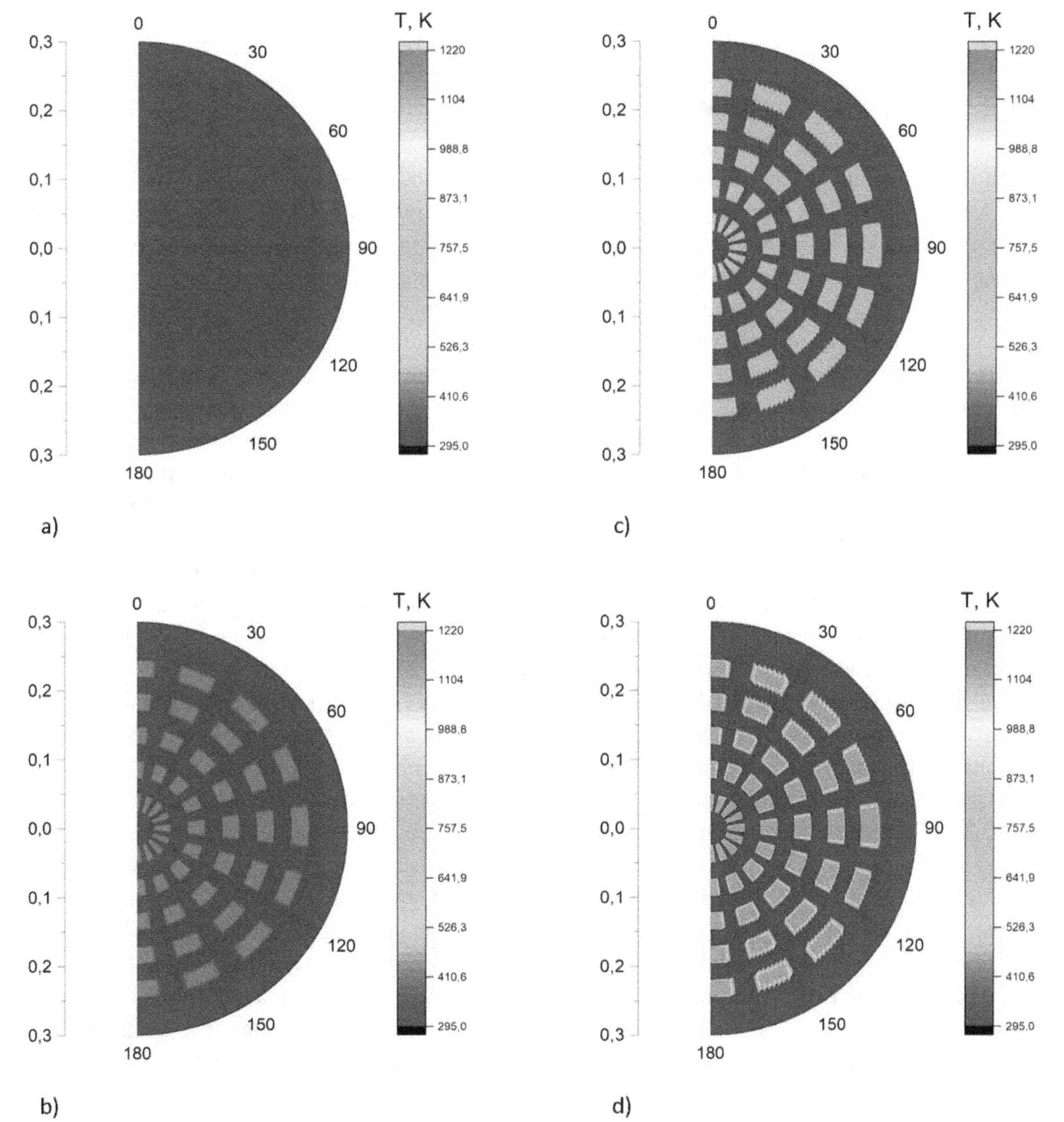

coordinates, t - time. Indices "e" and "0" correspond to the parameters of the external environment and parameters at the initial moment of time.

The numerical study was carried out using the following initial data: $\rho12=650$ kg/m³; c12=1670 J/(kg·K); $\lambda12=0.29$ W/(m·K); $\rho13=1000$ kg/m³; c13=4180 J/(kg·K); $\lambda13=0.588$ W/(m·K); $\rho14=0.598$ kg/m³; c14=2130 J/(kg·K); $\lambda14=0.024$ W/(m·K); $\rho_g=1.2$ kg/m³; $c_g=2130$ J/(kg·K); $\lambda_g=0.102$ W/(m·K). Evaporation parameters: Q=2250 J/kg; A=0.1; R=8.31 J/(mol·K); $M_6=0.010$ kg/mol. Thermokinetic parameters: $Q_p=1000$ J/kg; $k_p=3.63\times10^4$ 1/s; $E_1/R=9400$ K; $Q_5=10^7$ J/kg; $k_5=3\times10^{13}$ 1/s; $E_5/R=11500$ K; $\nu_5=0.3$; $M_8=0.032$ kg/mol; $M_9=0.028$ kg/mol; $M_{10}=0.018$ kg/mol; $M_{11}=0.044$ kg/mol. External influence parameters: $\alpha=80$ W/(m²·K), $T_e=300$ K. Geometric characteristics of the solution area: $R_s=0.25$ m; $R_2=0.245$ m.

RESULTS AND DISCUSSION

The following scheme of the process under study is considered. For example, a birch tree trunk is exposed to a 500 ms positive lightning discharge with a peak current of 23.5 kA (Soriano et al., 2005) and a voltage of 100 kV. Fig. 2 shows the temperature distribution in a horizontal section of a tree trunk at various times before and during the ignition of a cloud-to-ground lightning discharge current.

Analysis of Fig. 2 shows that as a result of the action of a typical cloud-to-ground lightning discharge, the tree trunk in the area of large vessels is heated to temperatures at which the formation of pyrolysis products occurs. The formed gaseous combustible products of pyrolysis and water vapor instantly enter the region of the gas phase. The distribution of the concentrations of the components of the gas phase along the radial coordinate is obtained. The concentration of water vapor exceeds the concentration of gaseous combustible products of pyrolysis (the maximum is observed at the bottom of the crack, where they are blown in from near-surface large vessels). However, its value is less than when an electric current passes through the trunk of a coniferous tree. Nevertheless, ignition occurs, since the gas phase in the crack area warms up to a higher temperature than occurs when a lightning discharge affects a coniferous tree (or a deciduous tree with a uniform distribution of moisture-conducting paths in the horizontal section of the trunk). The zone of localization of the chemical reaction is clearly seen in Figure 3. It is in the crack that the gaseous pyrolysis products are ignited.

It has been established that outside the crack zone, the bark layer does not allow the gas mixture to warm up to critical temperatures. The presence of structural inhomogeneities (cracks) in the bark changes the situation. An increased temperature field is formed in the zone of large vessels. It was found that if a crack partially or completely borders on a surface vessel, then it is in the crack that the gas mixture heats up to the required temperatures, and at certain concentrations of reagents, carbon monoxide is oxidized to carbon dioxide with the release of heat. At some distance from the bottom of the crack, the temperature field has a peak due to heat input from a chemical reaction. In the immediate vicinity of the trunk surface (at the bottom of the crack), the temperature of the gas phase is lower, since here pyrolysis products and water vapor are blown in, and the concentration of the oxidant becomes lower. No chemical reaction occurs under these conditions. Note that when solving the problem, an additional assumption was used that only from the layer of near-surface large vessels gaseous substances can instantly appear at the boundary of the trunk and the gas phase. It is assumed that the pyrolysis products and water vapor from the internal vessels do not have time to reach this boundary during the period of exposure to the lightning discharge. As a result of a numerical study of the model, it was found that at a current of

Figure 3. Temperature distribution in a horizontal section at the moment of ignition

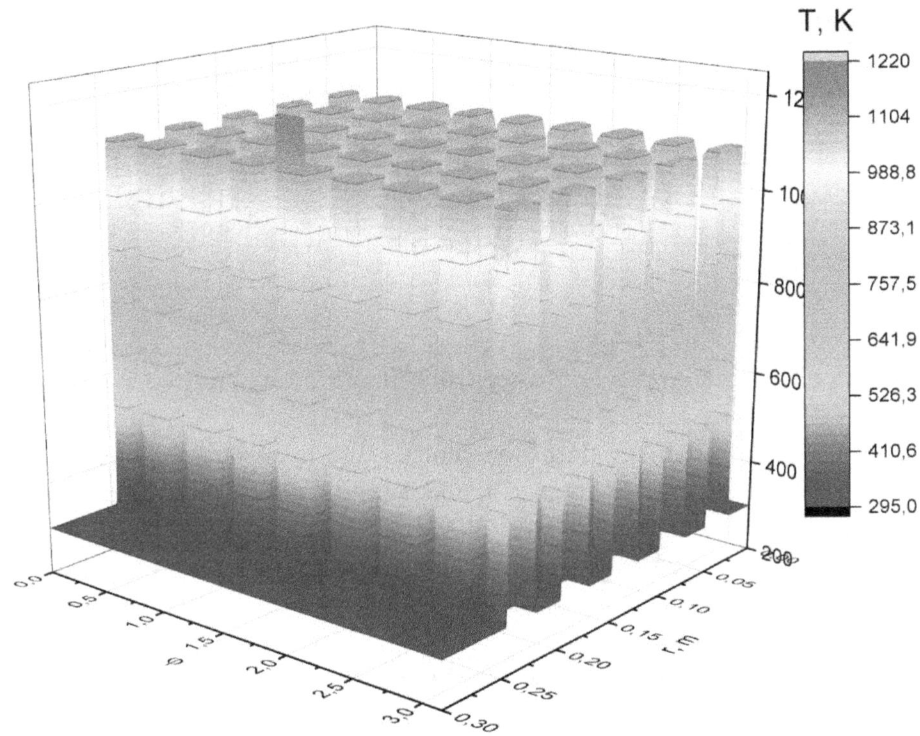

1-15 kA and a voltage of 1-60 kV, a deciduous tree does not ignite when exposed to a cloud-to-ground lightning discharge (Table 1 and 2).

As in the case of coniferous wood, the M components have no significant effect on the ignition of deciduous wood. There is no need to take them into account when modeling and creating a new forest fire danger predicting system.

Table 1 Ignition delay depending on voltage at J=23.5 kA

Voltage, *U*, kV	**Ignition delay, s**
1 – 60	No
70	0.480
80	0.445
90	0.415
100	0.400
105	0.391
110	0.380

Table 2. Ignition delay depending on the current at U=100 kV

Current, *J*, kA	Ignition delay, s
1 – 15	No
20	0.428
23.5	0.400
25	0.385
30	0.350
35	0.320

The main results correspond to those obtained from simplified statements:

1. In the zone of reactive wood, a field of low temperature is formed, and the release of pyrolysis products from this zone is carried out in a smaller amount, which should lead to a decrease in the probability of a forest fire.
2. There are two possible scenarios for wood heating: a) in the massive core area; b) in the area of large vessels. Pyrolysis products and water vapor from the near-surface layers enter the region of the gas mixture in the bark crack.
3. Ignition of deciduous wood occurs in the gas phase in the area of the crack.
4. The presence of a bark barrier in a crack even 1.5 mm thick reduces the temperature in the gas phase and ignition may not occur.
5. Differences in crack thickness under real conditions do not significantly affect the ignition delay of deciduous trees.
6. The presence of M-components of a cloud-to-ground lightning discharge practically does not affect the process of wood heating and deciduous tree ignition.

CONCLUSION

A physical and mathematical model of the gas-phase ignition of a deciduous tree as a result of the electric current of a cloud-to-ground lightning discharge flowing along its trunk is numerically implemented. The approximation of large vessels and an ideal crack is used. The possibility of gas-phase ignition of a deciduous tree under conditions of a thunderstorm passing by a cloud-to-cloud-to-ground discharge is shown.

As a result of this study, a fundamental (basic) model of gas-phase ignition of a deciduous tree by electric current of a cloud-to-ground lightning discharge has been developed. The model makes it possible to substantiate the very possibility of gas-phase ignition of a deciduous tree under the indicated conditions. However, many questions require additional consideration. For example, the process of the inflow of water vapor and gaseous pyrolysis products should be more adequately modeled taking into account their filtration, both from near-surface and from internal large vessels. It is necessary to study in more detail the effect of water vapor on the ignition conditions of deciduous trees. The process of the impact of a cloud-to-ground lightning discharge on a deciduous tree is significantly different in comparison with a coniferous tree. In some cases, damage or fragmentation of the trunk is possible under the influence of increasing pressure in its core (as a result of moisture evaporation). Thus, the presented model creates the foundation for the development on its basis of a whole set of more complex and meaningful physical and mathematical models of ignition and damage to deciduous trees, taking into account a wide range of factors of the process under study.

The presented physical and mathematical model can become an additional module in forest fire danger prediction and assessment systems (Glagolev and Zubareva, 2020; Adab et al., 2013; Al Janabi et al., 2017; Chen et al., 2017; Chen et al., 2020; Glagolev et al., 2020; Karanina et al., 2020; Papagiannaki et al., 2020). In addition, the results obtained are of independent fundamental importance for the theory of forest fires. They make it possible to explain the physical nature of the studied phenomenon of ignition of a deciduous tree by a cloud-to-ground lightning discharge.

ACKNOWLEDGMENT

This work was implemented with the financial support of the Russian Foundation for Basic Research. [Project N 17-29-05093 and 16-41-700831].

REFERENCES

Adab, H., Kanniah, K. D., & Solaimani, K. (2013). Modeling forest fire in the northeast of Iran using remote sensing and GIS techniques. *Natural Dangers*, *65*(3), 1723–1743. doi:10.100711069-012-0450-8

Al Janabi, S., Al Shourbaji, I., & Salman, M. A. (2017). *Assessing the suitability of soft computing approaches for forest fires prediction*. Applied Computing and Informatics. doi:10.1016/j.aci.2017.09.006

Baranovskiy, N. V. (2017). Forest fire danger assessment using SPMD-model of computation for massive parallel system. *International Review on Modeling and Simulations*, *10*(3), 193–201. doi:10.15866/iremos.v10i3.10570

Canadian Wildland Fire Information System. (n.d.). Access: https://cwfis.cfs.nrcan.gc.ca/home

Chen, C., Xu, T., Zhao, D., & Lei, P. (2020). Study on the Wildfires Occurring Risk Based on Fuzzy FTA. *IOP Conference Series. Earth and Environmental Science*, *440*(5), 052100. Advance online publication. doi:10.1088/1755-1315/440/5/052100

Chen, W., Xie, X., Wang, J., Pradhan, B., Hong, H., Tien Bui, D., Duan, Z., & Ma, J. (2017). A comparative study of logistic model tree, random forest and classification and regression tree models for spatial prediction of landslide susceptibility. *Catena*, *151*, 147–160. doi:10.1016/j.catena.2016.11.032

Esau, K. (1980) Anatomy of seed plants. Moscow: Mir. (in Russian)

European Forest Fire Information System. (n.d.). Access: http://effis.jrc.ec.europa.eu

Glagolev, V., Bazhenov, R., Povkh, I., Chuyko, O., & Dolmatova, O. (2020). Designing and producing a geographic information system to forecast vegetation fire danger according to weather conditions. *IOP Conference Series. Materials Science and Engineering*, *862*(5), 052019. Advance online publication. doi:10.1088/1757-899X/862/5/052019

Glagolev, V. A., & Zubareva, A. M. (2020). Using Geographic Information Systems for Study of the Natural Fire Risk on the Territory of the Jewish Autonomous Region. In N. V. Baranovskiy (Ed.), *Predicting, Monitoring, and Assessing Forest Fire Dangers and Risks* (pp. 209–231). IGI Global. doi:10.4018/978-1-7998-1867-0.ch009

Grishin, A. M. (1997). *Mathematical modeling of forest fire and new methods of fighting them. Russia.* Publishing House of the Tomsk State University.

Grishin, A. M., & Shipulina, O. V. (2002). Mathematical model for spread of crown fires in homogeneous forests and along openings. *Combustion, Explosion, and Shock Waves, 38*(6), 622–632. doi:10.1023/A:1021127924913

Karanina, S. Y., Kocheeva, N. A., Karanin, A. V., & Belikova, M. Y. (2020). Assessing the Territory According to the Degree of Thunderstorm Danger. In N. V. Baranovskiy (Ed.), *Predicting, Monitoring, and Assessing Forest Fire Dangers and Risks* (pp. 122–150). IGI Global. doi:10.4018/978-1-7998-1867-0.ch006

Nesterov, V. G. (1949). *Combustibility of the forest and methods of its determination.* Goslesbumizdat.

Pankratov, B. M., Polezhaev, Y. V., & Rudko, A. K. (1975). Interaction of Materials with Gas Streams. Moscow: Mechanical Engineering. (in Russian)

Papagiannaki, K., Giannaros, T. M., Lykoudis, S., Kotroni, V., & Lagouvardos, K. (2020). Weather-related thresholds for wildfire danger in a Mediterranean region: The case of Greece. *Agricultural and Forest Meteorology, 291*, 108076. Advance online publication. doi:10.1016/j.agrformet.2020.108076

Samarskiy, A. A. (1983). *Theory of difference schemes.* Nauka. (in Russian)

Samarskiy, A. A., & Vabishchevich, P. N. (2001). *Additive schemes for problems of mathematical physics.* Science. (in Russian)

Sivukhin, D. V. (1977) General course of physics. Moscow: Nauka. (in Russian)

Soriano, L. R., De Pablo, F., & Tomas, C. (2005). Ten-year study of cloud-to-cloud-to-cloud-to-ground lightning activity in the Iberian Peninsula. *Journal of Atmospheric and Solar-Terrestrial Physics, 67*(16), 1632–1639. doi:10.1016/j.jastp.2005.08.019

WFAS Wildland Fire Assessment System. (n.d.). Access: http://www.wfas.net

ADDITIONAL READING

Azadifar, M., Rubinstein, M., Rachidi, F., Rakov, V. A., Diendorfer, G., Schulz, W., & Pavanello, D. (2019). A Study of a Large Bipolar Lightning Event Observed at the Säntis Tower, IEEE Transactions on Electromagnetic Compatibility, 61 (3). *Article, 61*(8715461), 796–806. doi:10.1109/TEMC.2019.2913220

De Paiva, A. R., Saba, M. M. F., Naccarato, K. P., Schumann, C., Jaques, R., Da Silva Ferro, M. A., & Warner, T. A. (2014) Detection of upward lightning by lightning location systems, 2014 International Conference on Lightning Protection, ICLP 2014, Article N° 6973425, pp. 1824-1826. DOI: 10.1109/ICLP.2014.6973425

Lu, G., Cummer, S. A., Tian, Y., Zhang, H., Lyu, F., Wang, T., Stanley, M. A., Yang, J., & Lyons, W. A. (2016). Sprite produced by consecutive impulse charge transfers following a negative stroke: Observation and simulation. *Journal of Geophysical Research, 121*(8), 4082–4092. doi:10.1002/2015JD024644

Nag, A., Murphy, M. J., Schulz, W., & Cummins, K. L. (2014) Lightning locating systems: Characteristics and validation techniques, 2014 International Conference on Lightning Protection, ICLP 2014, Article N° 6973283, pp. 1070-1082. DOI: 10.1109/ICLP.2014.6973283

Nag, A., Rakov, V. A., & Murphy, M. J. (2014) Measurement of preliminary breakdown pulse trains in cloud-to-ground lightning using lightning locating systems, 2014 International Conference on Lightning Protection, ICLP 2014, статья N° 6973356, pp. 1437-1444. DOI: 10.1109/ICLP.2014.6973356

Saba, M. M. F., Shigihara, M., & Kodaira, C. Y. (2018) Observation of a lightning flashover with a high-speed video camera, 34th International Conference on Lightning Protection, ICLP 2018, Article N° 8503282. DOI: 10.1109/ICLP.2018.8503282

Stolzenburg, M., Marshall, T. C., Karunarathne, S., Karunarathna, N., & Orville, R. E. (2015). An M component with a concurrent dart leader traveling along different paths during a lightning flash. *Journal of Geophysical Research, 120*(19), 10267–10284. doi:10.1002/2015JD023417

Wang, X., Yuan, P., Cen, J., & Xue, S. (2016). Correlation between the spectral features and electric field changes for natural lightning return stroke followed by continuing current with M-components. *Journal of Geophysical Research, 121*(14), 8615–8624. doi:10.1002/2016JD025314

Xue, S., Yuan, P., Cen, J., Li, Y., & Wang, X. (2015). Spectral observations of a natural bipolar cloud-to-ground lightning. *Journal of Geophysical Research, 120*(5), 1972–1979. doi:10.1002/2014JD022598

Yuan, P., Chen, R., Su, Y., Fan, B., Wang, X. (2019) Characteristic parameters of positive cloud-to-ground lightning channel, Journal of Earth System Science, 128 (6), статья N° 164, . Doi:10.100712040-019-1196-4

KEY TERMS AND DEFINITIONS

Cloud-to-Ground Lightning Discharge: An electrical discharge during a thunderstorm that occurs between a cloud and the earth's surface. It is a natural source of forest fires.

Forest Fire: Uncontrolled aerothermochemical phenomenon characterized by step-by-step mechanism which includes following stages: inert heating, moisture evaporation, high temperature terpens evaporation, dry organic matter pyrolysis, flammable combustion and smoldering.

Forest Fuel: It can be considered like dead and live forest fuel. Main types of forest fuel which can be involved in combustion during forest fire: cloud-to-cloud-to-ground forest fuel (needles, leaves and dry grass, small branches) and crown forest fuel (needles, small branches).

Ignition: Inflammation of forest fuel caused by definite source of high temperature or energy.

Ignition Delay: Time before flame flash after forest fuel heating.

Lightning Activity: An atmospheric phenomenon characterized by discharges of the cloud-to-cloud and cloud-to-cloud-to-cloud-to-ground class.

Mathematical Simulation: The production of a computer model of forest fire conditions and prerequisites, especially for the purpose of study.

Meteorological Parameters: Physical characteristics of local weather conditions in the forested area under consideration. Key parameters include ambient temperature, soil temperature, precipitation, wind speed, solar radiation, cloud cover, dew point temperature. These parameters are used for mathematical modeling of the drying of a layer of forest fuel.

Monitoring: Monitoring refers to the periodic calculation of the parameters of forest fire danger with a portion of information available in real time.

Prediction: Under the prediction of forest fires is the calculation of the parameters of forest fire danger with a certain projection in advance in order to have enough time to anticipate an emergency. The calculation in this case is carried out in a mode ahead of the real time of the development of the catastrophe - the occurrence of a forest fire.

Chapter 8
M-Components Mathematical Modeling for Deciduous Tree Ignition

ABSTRACT

Based on a one-dimensional two-layer physical and mathematical model of ignition of a deciduous tree (birch), the influence of the m-components of a cloud-to-ground lightning discharge is estimated. The problem is solved in a cylindrical coordinate system. Typical cloud-to-ground lightning discharges are considered. An assessment of the influence of m-components was carried out for a typical range of changes in their characteristics. Surface temperature of tree trunk and heat flux to this surface are obtained during simulation.

INTRODUCTION

The main cause of forest fires in the Republic of Buryatia (Russian Federation) is cloud-to-ground lightning discharges (Karanina et al., 2020; Belikova and Glebova, 2020). Thunderstorm activity is characterized by a number of parameters, including rainfall (Pineda and Rigo, 2017). According to (Nieto et al., 2012), rainfall plays an important role in the occurrence of forest fires. Precipitation is expected to affect the moisture content of forest fuels. Many researchers have found a positive correlation between precipitation and thunderstorm activity (Marshall and Radhakant, 1978; Sheridan et al., 1997; Petersen and Montanya, 1998; Pineda et al., 2007). Despite the fact that most lightning discharges occur during heavy rains, there are a small number of so-called dry thunderstorms (Hall, 2008). It is these discharges that can lead to the appearance of active forest fires. In this paper, the scenario of dry thunderstorms is considered.

Lightning direction finding systems exist, for example, European Cooperation for Lightning Detection (EUCLID) (EUCLID, 2020; Moris et al., 2020). Such systems make it possible to record a number of parameters of lightning discharges, including location, discharge current, and duration of discharge (Moris et al., 2020). This information can be used for mathematical modeling of ignition of deciduous

DOI: 10.4018/978-1-7998-7250-4.ch008

tree by cloud-to-ground lightning discharge. It should be noted that there is another characteristic of a ground lightning discharge, namely, the M-components (Campos et al., 2007), which can affect the occurrence of forest fires.

M-components, first described in (Malan and Collens, 1937), are observed as an increase in the channel brightness during a continuing current. M-components may have a peak in the range of kiloamperes (Campos et al., 2007). It is necessary to study the influence of M-components on the ignition process of a deciduous tree trunk in order to assess the feasibility of supplementing the existing empirical (Kurbatskiy and Kostyrina, 1977; Larjavaara et al., 2005) and deterministic-probabilistic (Baranovskiy, 2009a; 2009b) methods for predicting forest fire danger with subsystems for accounting for the presence of M-components of a lightning discharge. As a physical and mathematical model of ignition of a deciduous tree by a cloud-to-ground lightning discharge, a one-dimensional two-layer model was used (Kuznetsov and Baranovskiy, 2009).

The purpose of the study is to assess the influence of M-components on the ignition of a deciduous tree by a cloud-to-ground lightning discharge, as well as to determine the ignition conditions depending on the discharge parameters.

PHYSICAL AND MATHEMATICAL STATEMENT

The core of the deciduous tree is more saturated with moisture and, as a result, the electric current of the discharge passes in this area (Esau, 1980). The following physical model is used. At a certain point in time, a lightning discharge of a given polarity and duration of action strikes the trunk of a deciduous tree. It is assumed that current–voltage characteristics of the discharge are the same for different sections of the tree trunk. The discharge current has M components. Fig. 1 shows a typical discharge with M-components. The time is plotted on the abscissa, and the channel brightness in arbitrary units on the ordinate.

The warming of the trunk wood occurs due to the Joule heat released in the core of the tree trunk. As a result of the flow of electric current, the wood heats up and when critical heat fluxes from the core of the trunk to the ignition surface and its temperature are reached, the wood ignites. The solution domain is shown in Fig. 2, where the numbers indicate the zones: 1 - the core of the tree trunk; 2 - tree bark.

Mathematically, the process of heating a tree trunk before ignition with a ground lightning discharge is described by a system of non-stationary differential equations (Kuznetsov and Baranovskiy, 2009):

$$\rho_{ef} c_{ef} \frac{\partial T_1}{\partial t} = \frac{\lambda_{ef}}{r} \frac{\partial}{\partial r}\left(r \frac{\partial T_1}{\partial r} \right) + JU - QW\phi_2, \tag{1}$$

$$\rho_2 c_2 \frac{\partial T_2}{\partial t} = \frac{\lambda_2}{r} \frac{\partial}{\partial r}\left(r \frac{\partial T_2}{\partial r} \right), \tag{2}$$

Figure 1. Discharge with M-components (Campos et al., 2007)

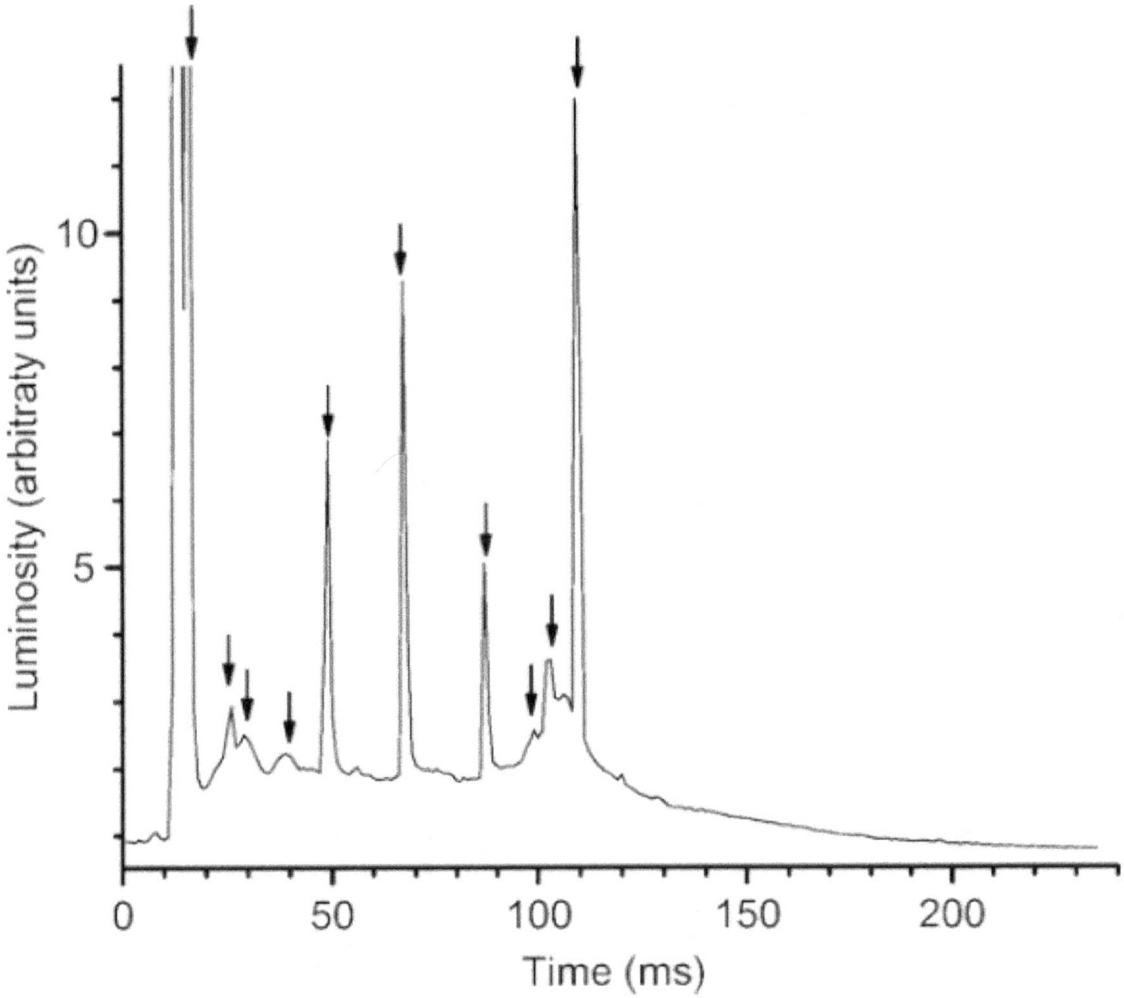

$$\rho_3 \frac{\partial \phi_1}{\partial t} = 0, \tag{3}$$

$$\rho_4 \frac{\partial \phi_2}{\partial t} = -W, \tag{4}$$

$$\sum_{i=3}^{5} \phi_i = 1, \tag{5}$$

Figure 2. Scheme of the solution area

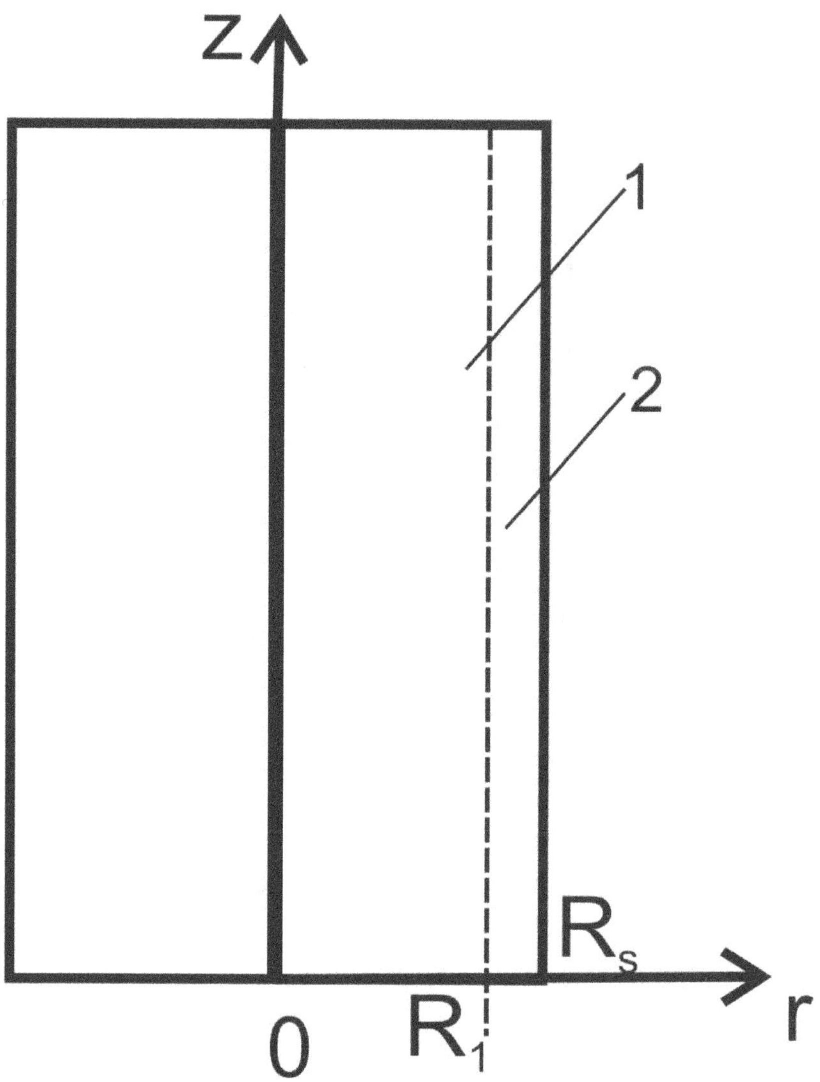

$$W = \frac{A(P^s - P)}{\sqrt{\frac{2\pi RT}{M}}}, \quad (6)$$

$$\rho_{ef} = \rho_3\phi_3 + \rho_4\phi_4 + \rho_5\phi_5,\ c_{ef} = c_3\phi_3 + c_4\phi_4 + c_5\phi_5,\ \lambda_{ef} = \lambda_3\phi_3 + \lambda_4\phi_4 + \lambda_5\phi_5, \quad (7)$$

The boundary conditions for equations (1) - (2):

$$r=0,\ \lambda_{\bullet\bullet} \frac{\partial T_1}{\partial r} = 0, \quad (8)$$

$$r=R_1,\ \lambda_{\bullet\bullet} \frac{\partial T_1}{\partial r} = \lambda_2 \frac{\partial T_2}{\partial r},\ T_1=T_2 \quad (9)$$

$$r=R,\ \lambda_2 \frac{\partial T_2}{\partial r} = \alpha(T_e - T_{Rs}), \quad (10)$$

The initial conditions for equations (1) - (5):

$$t=0,\ T_i(r) = T_{i0}(r) \quad (11)$$

$$\varphi 3_(0)=0.715, \quad (12)$$

$$\varphi 4_(0)=0.285, \quad (13)$$

where T_i is the temperature of the inner part of the trunk (i = 1) and the bark (i = 2); φ_i is the volume fraction of: organic matter (i= 3), water (i= 4) and water vapor (i= 5); ρ_i, c_i, λ_i — density, heat capacity and thermal conductivity of the crust (i=2), organic matter (i=3), water (i=4) and water vapor (i=5); ρ_{ef}, c_{ef}, λ_{ef} - effective density, heat capacity and thermal conductivity of wood of the inner part of the trunk; α - heat transfer coefficient; J is the current strength; U is the voltage; Q is the thermal effect of moisture evaporation; r is the coordinate; t is time. W is the mass rate of water evaporation, A is the accommodation coefficient, P^s is the pressure of saturated water vapor, P is the partial pressure of water vapor in the air, R is the universal gas constant, M is the molecular weight of water. The indices "Rs", "e" and "0" correspond to the parameters on the outer boundary of the tree trunk, the external environment and the parameters at the initial time. A numerical study was carried out using the following initial data: ρ_3=650 kg/m^3; c_3=1670 J/(kg·K); λ_3=0.29 W/(m·K) (Zabolotny et al., 1995); ρ_4=1000 kg/m^3; c_4=4180 J/(kg·K); λ_4=0.588 W/(m·K); ρ_5=0.598 kg/m^3; c_5=2130 J/(kg·K); λ_5=0.024 W/(m·K). Evaporation parameters: Q=2250 J/kg; A=0.1; R=8.31 J/(mol·K); M = 0.010 kg/mol. External exposure parameters: α=80 W/(m^2·K).

RESULTS AND DISCUSSION

The mathematical model (1) - (7) with boundary and initial conditions (8) - (13) is implemented by the finite-difference method (Samarskiy, 1983). To solve the difference analogues of one-dimensional equations, the marching method (Samarskiy, 1983) was used. As a result of an experimental study (Zabolotny et al., 1995), the criteria for ignition are determined by temperature and heat flux to the surface of the ignition. A scenario is considered when a negative polarity discharge strikes a pine trunk with a duration of 300–500 ms with an average current of 23.5 kA and 35 kA and a voltage of 100 kV (Soriano et al., 2005;Cummins et al., 1998). The current has 5 M components.

Figure 3 shows the time dependence of the current of a cloud-to-ground lightning discharge and M-components. In Fig. 3.a, curves 1 and 2 correspond to discharges with a duration of 500 and 300 ms, respectively. Figure 3b shows the dependence of the current and the M component for a discharge with a large average current (J = 35 kA).

Figure 4a shows the dependence of the temperature of the ignition surface at different instants of time for a discharge lasting 500 ms. Figure 4b shows the time dependence of the heat flux to the ignition surface for a discharge with a duration of 500 ms. Curves 1 and 2 correspond to discharges with and without M-components. The ignition conditions were evaluated according to the following criteria (table. 1):

Table 1. Experimentally determined ignition conditions (Zabolotny et al., 1995)

Ignition delay, s	Heat flux, kW/m^2	Surface temperature, K
136	15	-
61.2	21	645
17.2	42	688
1.8	125	755
0.43	210	801

An analysis of the results (Fig. 4) shows that the ignition conditions are satisfied for a typical cloud-to-ground discharge (with a duration of about 500 ms). Studies have shown that a short-term discharge (lasting less than 500 ms) with average volt-ampere characteristics does not ignite a deciduous tree trunk. That is, a clear dependence is established on the time of exposure of the tree trunk by the current of a cloud-to-ground lightning discharge and the process of its ignition as a result of the release of Joule heat.

An analysis of the influence of the M-components of a cloud-to-ground lightning discharge revealed a weak dependence of the ignition process on their presence. The M-components of typical cloud-to-ground lightning discharges show only a negligible effect on the ignition of deciduous trees (Fig. 4). The influence of the M-components of a ground lightning discharge with a high average current is even less noticeable. The curves in this case lie almost one in one.

In this work, the effect of the presence of M-components was assessed on the ignition of a deciduous tree by cloud-to-ground lightning discharge. The ignition conditions that are characteristic of a typical range of lightning discharge parameters are revealed. It was found that a lightning discharge lasting about 500 ms leads to ignition of the trunk of a deciduous tree, regardless of the presence or absence of M-components. Moreover, the dependence on the presence of M-components is weak and decreases with an increase in the average current of a cloud-to-ground lightning discharge. An explicit dependence on the time of exposure of the tree trunk by the current of cloud-to-ground lightning discharge has also been established. A short-term (less than 300 ms) discharge with typical volt-ampere characteristics cannot lead to ignition of the material of a deciduous tree trunk.

Figure 3. Current and M-components adopted in computational experiments

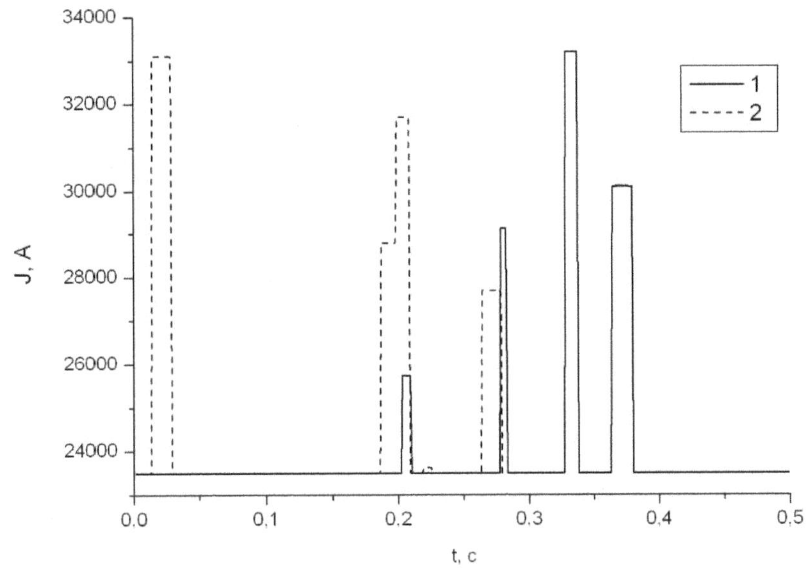

a)

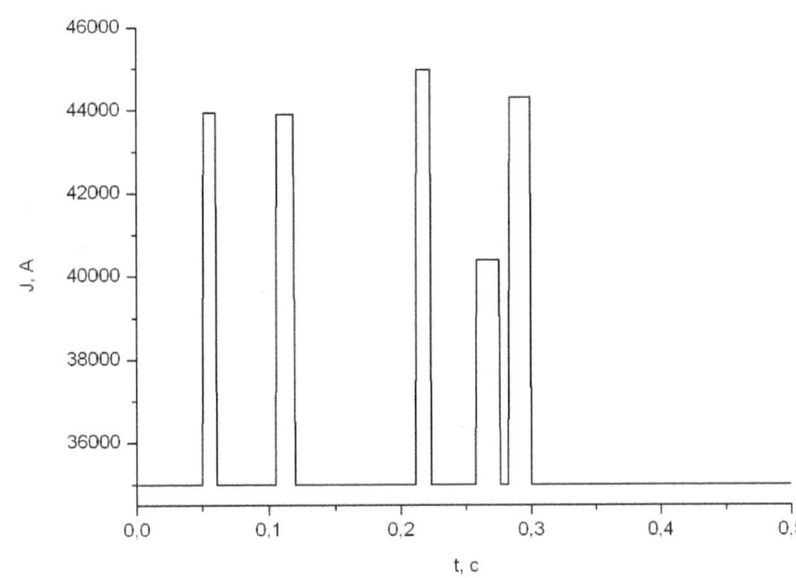

b)

Figure 4. The temperature of the ignition surface (a) and the heat flux to it (b) at various points in time (prolonged ground lightning discharge)

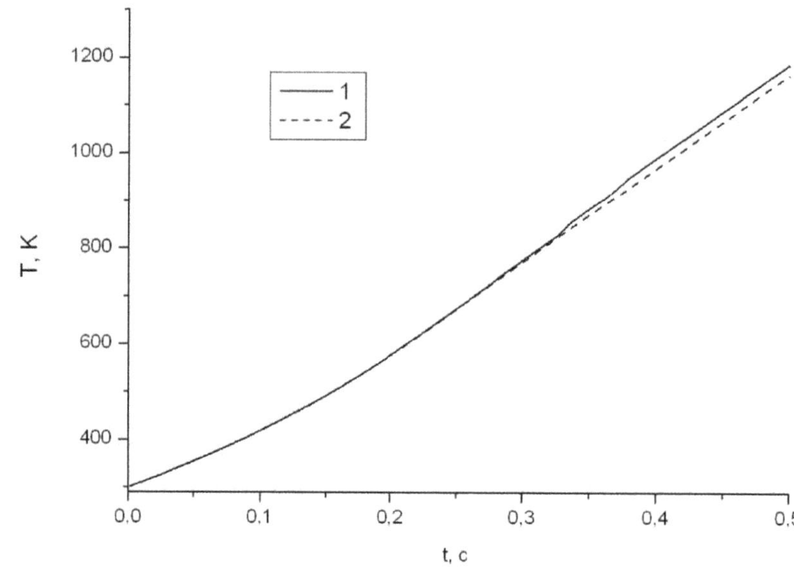

a)

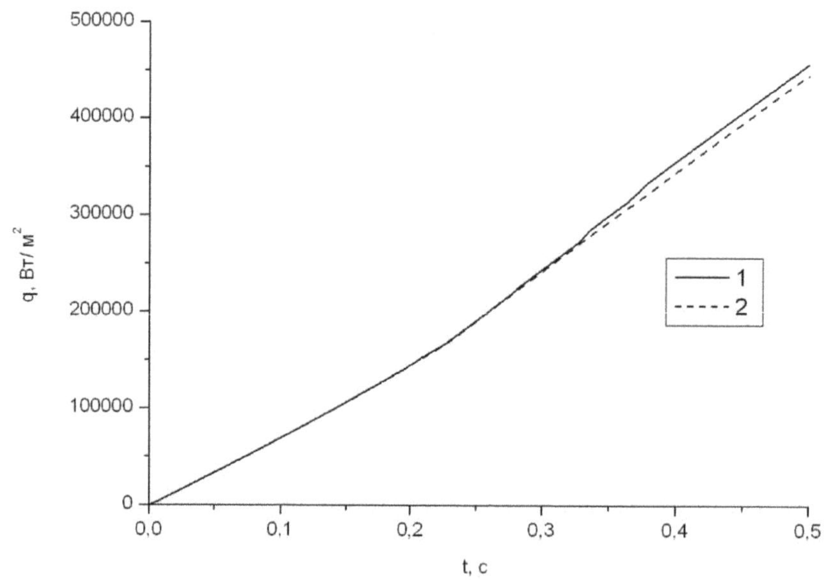

b)

CONCLUSION

In this work, the effect of the presence of M-components was assessed on the ignition of a deciduous tree by cloud-to-ground lightning discharge. The ignition conditions that are characteristic of a typical range of lightning discharge parameters are revealed. It was found that a lightning discharge lasting about 500 ms leads to ignition of the trunk of a deciduous tree, regardless of the presence or absence of M-components. Moreover, the dependence on the presence of M-components is weak and decreases with an increase in the average current of a cloud-to-ground lightning discharge. An explicit dependence on the time of exposure of the tree trunk by the current of cloud-to-ground lightning discharge has also been established. A short-term (less than 300 ms) discharge with typical volt-ampere characteristics cannot lead to ignition of the material of a deciduous tree trunk.

Developing software for forest fire danger predicting, it is not necessary to take into account the presence of M-components. This fact allows us to operate with simpler physical and mathematical models when constructing information-forecasting systems for the needs of forest protection from fires. In addition, the database of such a system may not contain information on the M-components of a cloud-to-ground lightning discharge.

On the other hand, in the present work, the ignition of a deciduous tree by cloud-to-ground lightning discharge was simulated in a more complete physical formulation than in (Kuznetsov and Baranovskiy, 2009). The results obtained are of independent importance for the development of the theory of forest fires (Chen et al., 2020; Dupuy et al., 2020; Eskandari et al., 2020; Glagolev et al., 2020; Papagiannaki et al., 2020; Yankovich and Yankovich, 2020). In the present work, the location and duration of the M-components of the discharge were determined randomly from the range of their variation determined statistically (Campos et al., 2007). This is programmatically implemented using a random number generator. It is known that continuing current has various waveforms (Campos et al., 2007) depending on the location and duration of the M-components. In subsequent works, while conducting basic research, this issue can be worked out in more detail.

ACKNOWLEDGMENT

This work was implemented with the financial support of the Russian Foundation for Basic Research. [Project N 17-29-05093 and Project N 16-41-700831].

REFERENCES

Baranovskiy, N. V. (2009). Deterministic-probabilistic forecast of forest fire hazard based on experimental data on the ignition of forest combustible material. [in Russian]. *High-tech*, (6), 66–70.

Baranovskiy, N. V. (2009). Estimation of the probability of forest fires taking into account weather conditions, anthropogenic load and thunderstorm activity. [in Russian]. *Fire Safety*, (1), 93–99.

Belikova, M. Y., & Glebova, A. V. (2020). Cluster Analysis Algorithms for RS and WWLLN Data Processing. In N. V. Baranovskiy (Ed.), *Predicting, Monitoring, and Assessing Forest Fire Dangers and Risks* (pp. 151–181). IGI Global. doi:10.4018/978-1-7998-1867-0.ch007

Campos, L. Z. S., Saba, M. M. F., Pinto, O. Jr, & Ballarotti, M. G. (2007). Waveshapes of continuing currents and properties of M-components in natural negative cloud-to-ground lightning from high-speed video observations. *Atmospheric Research*, *84*(9), 302–310. doi:10.1016/j.atmosres.2006.09.002

Chen, C., Xu, T., Zhao, D., & Lei, P. (2020). Study on the Wildfires Occurring Risk Based on Fuzzy FTA. *IOP Conference Series. Earth and Environmental Science*, *440*(5), 052100. Advance online publication. doi:10.1088/1755-1315/440/5/052100

Cummins, K. L., Murphy, M. J., Bardo, E. A., Hiscox, W. L., Pyle, R. B., & Pifer, A. E. (1998). A combined TOA/MDF technology upgrade of the U.S. national lightning detection network. *Journal of Geophysical Research*, *103*(D8), 9035–9044. doi:10.1029/98JD00153

Dupuy, J.-L., Fargeon, H., Martin-StPaul, N., Pimont, F., Ruffault, J., Guijarro, M., Hernando, C., Madrigal, J., & Fernandes, P. (2020). Climate change impact on future wildfire danger and activity in southern Europe: A review. *Annals of Forest Science*, *77*(2), 35. Advance online publication. doi:10.100713595-020-00933-5

Esau, K. (1980). Anatomy of seed plants. Moscow: Mir. (in Russian)

Eskandari, S., Pourghasemi, H. R., & Tiefenbacher, J. P. (2020). Relations of land cover, topography, and climate to fire occurrence in natural regions of Iran: Applying new data mining techniques for modeling and mapping fire danger. *Forest Ecology and Management*, *473*, 118338. Advance online publication. doi:10.1016/j.foreco.2020.118338

European Cooperation for Lightning Detection (EUCLID). (n.d.). Available online: http://www.euclid.org

Glagolev, V., Bazhenov, R., Povkh, I., Chuyko, O., & Dolmatova, O. (2020). Designing and producing a geographic information system to forecast vegetation fire danger according to weather conditions. *IOP Conference Series. Materials Science and Engineering*, *862*(5), 052019. Advance online publication. doi:10.1088/1757-899X/862/5/052019

Hall, B. L. (2008). Fire ignitions related to radar reflectivity patterns in Arizona and New Mexico. *International Journal of Wildland Fire*, *17*(3), 317–327. doi:10.1071/WF06110

Karanina, S. Y., Kocheeva, N. A., Karanin, A. V., & Belikova, M. Y. (2020). Assessing the Territory According to the Degree of Thunderstorm Danger. In N. V. Baranovskiy (Ed.), *Predicting, Monitoring, and Assessing Forest Fire Dangers and Risks* (pp. 122–150). IGI Global. doi:10.4018/978-1-7998-1867-0.ch006

Kurbatskiy, N. P., & Kostyrina, T. V. (1977). US National Fire Hazard Calculation System. In *Detection and analysis of forest fires* (pp. 38–90). IliD SB AS USSR. (in Russian)

Kuznetsov, G.V., & Baranovskiy, N.V. (2009). Mathematical modeling of ignition of hardwood by ground lightning discharge. *Fire and Explosion Safety, 4*, 19-22. (in Russian)

Larjavaara, M., Kuuluvainen, T., & Rita, H. (2005). Spatial distribution of lightning-ignited fires in Finland. *Forest Ecology and Management*, *208*(1-3), 177–188. doi:10.1016/j.foreco.2004.12.005

Malan, D. J., & Collens, H. (1937). Progressive lightning, III, the fine structure of return lightning strokes. *Proceedings of the Royal Society of London. Series A, Mathematical and Physical Sciences*, *162*(909), 175–203. doi:10.1098/rspa.1937.0175

Marshall, J. S., & Radhakant, S. (1978). Radar precipitation maps as lightning indicators. *Journal of Applied Meteorology*, *17*(2), 206–212. doi:10.1175/1520-0450(1978)017<0206:RPMALI>2.0.CO;2

Moris, J. V., Conedera, M., Nisi, L., Bernardi, M., Cesti, G., & Pezzatti, G. B. (2020). Lightning-caused fires in the Alps: Identifying the igniting strokes. *Agricultural and Forest Meteorology*, *290*, 107990. doi:10.1016/j.agrformet.2020.107990

Nieto, H., Aguado, I., Garcia, M., & Chuvieco, E. (2012). Lightning-caused fires in Central Spain: Development of a probability model of occurrence for two Spanish regions. *Agricultural and Forest Meteorology*, *162-163*, 35–43. doi:10.1016/j.agrformet.2012.04.002

Papagiannaki, K., Giannaros, T. M., Lykoudis, S., Kotroni, V., & Lagouvardos, K. (2020). Weather-related thresholds for wildfire danger in a Mediterranean region: The case of Greece. *Agricultural and Forest Meteorology*, *291*, 108076. Advance online publication. doi:10.1016/j.agrformet.2020.108076

Petersen, W. A., & Montanya, J. (1998). On the relationship between cloud-to-ground lightning and convective rainfall. *Journal of Geophysical Research*, *103*(D12), 14025–14040. doi:10.1029/97JD02064

Pineda, N., & Rigo, T. (2017). The rainfall factor in lightning-ignited wildfires in Catalonia. *Agricultural and Forest Meteorology*, *239*, 249–263. doi:10.1016/j.agrformet.2017.03.016

Pineda, N., Rigo, T., Bech, J., & Soler, X. (2007). Lightning and precipitation relationship in summer thunderstorms: Case studies in the north western Mediterranean region. *Atmospheric Research*, *85*(2), 159–170. doi:10.1016/j.atmosres.2006.12.004

Samarskiy, A. A. (1983). *Theory of difference schemes*. Nauka. (in Russian)

Sheridan, S. C., Griffiths, J. F., & Orville, R. E. (1997). Warm season cloud-to-ground lightning-precipitation relationships in the south-central United States. *Weather and Forecasting*, *12*(3), 449–458. doi:10.1175/1520-0434(1997)012<0449:WSCTGL>2.0.CO;2

Soriano, L. R., De Pablo, F., & Tomas, C. (2005). Ten-year study of cloud-to-ground lightning activity in the Iberian Peninsula. *Journal of Atmospheric and Solar-Terrestrial Physics*, *67*(16), 1632–1639. doi:10.1016/j.jastp.2005.08.019

Yankovich, E. P., & Yankovich, K. S. (2020). Classification of Territory on Forest Fire Danger Level Using GIS and Remote Sensing. In N. V. Baranovskiy (Ed.), *Predicting, Monitoring, and Assessing Forest Fire Dangers and Risks* (pp. 258–267). IGI Global. doi:10.4018/978-1-7998-1867-0.ch011

Zabolotny, A.E., Zabolotnaya, M.M., Zabolotnaya, Y.A., & Timoshin, V.N. (1995). Determination of safe use zones of solid fuel generators of fire extinguishing aerosols. *Issues of Special Engineering*, (7-8), 15-21. (in Russian)

ADDITIONAL READING

Azadifar, M., Rubinstein, M., Rachidi, F., Rakov, V. A., Diendorfer, G., Schulz, W., & Pavanello, D. (2019). A Study of a Large Bipolar Lightning Event Observed at the Säntis Tower, IEEE Transactions on Electromagnetic Compatibility, 61 (3). *Article*, *61*(8715461), 796–806. doi:10.1109/TEMC.2019.2913220

De Paiva, A. R., Saba, M. M. F., Naccarato, K. P., Schumann, C., Jaques, R., Da Silva Ferro, M. A., & Warner, T. A. (2014) Detection of upward lightning by lightning location systems, 2014 International Conference on Lightning Protection, ICLP 2014, Article N° 6973425, pp. 1824-1826. DOI: 10.1109/ICLP.2014.6973425

Lu, G., Cummer, S. A., Tian, Y., Zhang, H., Lyu, F., Wang, T., Stanley, M. A., Yang, J., & Lyons, W. A. (2016). Sprite produced by consecutive impulse charge transfers following a negative stroke: Observation and simulation. *Journal of Geophysical Research*, *121*(8), 4082–4092. doi:10.1002/2015JD024644

Nag, A., Rakov, V. A., & Murphy, M. J. (2014) Measurement of preliminary breakdown pulse trains in cloud-to-ground lightning using lightning locating systems, 2014 International Conference on Lightning Protection, ICLP 2014, статья N° 6973356, pp. 1437-1444. DOI: 10.1109/ICLP.2014.6973356

Nag, A., Murphy, M. J., Schulz, W., & Cummins, K. L. (2014) Lightning locating systems: Characteristics and validation techniques, 2014 International Conference on Lightning Protection, ICLP 2014, Article N° 6973283, pp. 1070-1082. DOI: 10.1109/ICLP.2014.6973283

Saba, M. M. F., Shigihara, M., & Kodaira, C. Y. (2018) Observation of a lightning flashover with a high-speed video camera, 34th International Conference on Lightning Protection, ICLP 2018, Article N° 8503282. DOI: 10.1109/ICLP.2018.8503282

Stolzenburg, M., Marshall, T. C., Karunarathne, S., Karunarathna, N., & Orville, R. E. (2015). An M component with a concurrent dart leader traveling along different paths during a lightning flash. *Journal of Geophysical Research*, *120*(19), 10267–10284. doi:10.1002/2015JD023417

Wang, X., Yuan, P., Cen, J., & Xue, S. (2016). Correlation between the spectral features and electric field changes for natural lightning return stroke followed by continuing current with M-components. *Journal of Geophysical Research*, *121*(14), 8615–8624. doi:10.1002/2016JD025314

Xue, S., Yuan, P., Cen, J., Li, Y., & Wang, X. (2015). Spectral observations of a natural bipolar cloud-to-ground lightning. *Journal of Geophysical Research*, *120*(5), 1972–1979. doi:10.1002/2014JD022598

Yuan, P., Chen, R., Su, Y., Fan, B., Wang, X. (2019) Characteristic parameters of positive cloud-to-ground lightning channel, Journal of Earth System Science, 128 (6), статья N° 164, . Doi:10.100712040-019-1196-4

KEY TERMS AND DEFINITIONS

Cloud-to-Ground Lightning Discharge: An electrical discharge during a thunderstorm that occurs between a cloud and the earth's surface. It is a natural source of forest fires.

Forest Fire: Uncontrolled aerothermochemical phenomenon characterized by step-by-step mechanism which includes following stages: inert heating, moisture evaporation, high temperature terpens evaporation, dry organic matter pyrolysis, flammable combustion and smoldering.

Forest Fuel: It can be considered like dead and live forest fuel. Main types of forest fuel which can be involved in combustion during forest fire: ground forest fuel (needles, leaves and dry grass, small branches) and crown forest fuel (needles, small branches).

Ignition: Inflammation of forest fuel caused by definite source of high temperature or energy.

Ignition Delay: Time before flame flash after forest fuel heating.

Lightning Activity: An atmospheric phenomenon characterized by discharges of the cloud-to-cloud and cloud-to-ground class.

M-Components: Are observed as an increase in the brightness of the channel during continuing current. M-components may have a peak in the range of kiloamperes.

Mathematical Simulation: The production of a computer model of forest fire conditions and prerequisites, especially for the purpose of study.

Meteorological Parameters: Physical characteristics of local weather conditions in the forested area under consideration. Key parameters include ambient temperature, soil temperature, precipitation, wind speed, solar radiation, cloud cover, dew point temperature. These parameters are used for mathematical modeling of the drying of a layer of forest fuel.

Monitoring: Monitoring refers to the periodic calculation of the parameters of forest fire danger with a portion of information available in real time.

Prediction: Under the prediction of forest fires is the calculation of the parameters of forest fire danger with a certain projection in advance in order to have enough time to anticipate an emergency. The calculation in this case is carried out in a mode ahead of the real time of the development of the catastrophe - the occurrence of a forest fire.

Section 3

Forest Fuel Ignition Physical Modeling and Mathematical Simulation

Chapter 9
Physical Modeling of Forest Fuel Ignition by the Heated Up to High Temperatures Particle

ABSTRACT

This chapter discusses a comparative analysis of the results of physical modeling of ignition of various forest fuels by a particle of metals and nonmetals heated to high temperatures. Typical forest fuels are considered: pine litter, birch leaves, and grass. The results of physical modeling of ignition of spruce branches with needles are discussed separately. Conclusions are drawn about the lower limit of ignition and fire danger of particles heated to high temperatures. The physical mechanism of ignition of forest fuels by a particle heated to high temperatures is presented.

INTRODUCTION

It is possible to ignite forest fuels with a single particle heated to high temperatures. Real particles, as a rule, have the shape of irregular polyhedrons and are in a solid state during deposition. In order to predict the forest fire danger, it is impossible to take into account the real configuration of particles falling onto the surface of a fuel. For experiments, particles similar to parallelepipeds and cylinders were selected.

The results of experimental and theoretical studies of the ignition of natural fuels by particles heated to high temperatures have been published. It is known (Hadden et al., 2011) that burning particles (firebrands, coals) are the main mechanism for the propagation of forest fires and fires passing to the village (WUI Fires - Wildland Urban Interface Fires). Often, such sources of local heating are formed during the burning of forest fuels and are transferred by the wind or as a result of the plume (convective column) from the fire to the forest fuel layer untouched by the fire and lead to its ignition. For example, an analysis of the 1994 fire statistics in Sydney showed that 75% of houses caught fire as a result of heated particles and only 25% caught fire from particles and radiation from a flame (Ramsey and McArthur, 1995). In addition, molten or burning particles can be formed as a result of collision of power line wires at high wind speeds (Hadden et al., 2011). The processes of forest fuel ignition by heated particles have not been

DOI: 10.4018/978-1-7998-7250-4.ch009

sufficiently studied, with the exception of some field studies (Stokes, 1990; Rowntree and Stokes, 1994; Manzello et al., 2006a,b; Pitts, 2007; Manzello et al., 2008; Caine et al., 2009). In (Hadden et al., 2011), the results of an experimental and theoretical study of the forest fuel ignition by heated particles are presented. Steel spheres that were inert to chemical interaction were considered as a source of ignition. At the same time, cellulose in the form of a powder was used as a fuel, since it formed homogeneous layers with known thermophysical properties. The following range of sizes of hot particles is known, which are formed as a result of burning vegetation or the interaction of power transmission line wires: fragments with characteristic sizes of 200 to 10 mm in diameter are formed from Douglas spruce wood (Manzello et al., 2008a,b; Yoshioka et al., 2004). The objectives of the research (Hadden et al., 2011) was to evaluate the ignition conditions; smoldering; decay, turning into a flame; lack of ignition. These conditions depend on the size and state of the particle (temperature, heat capacity, smoldering or burning in a flame mode), as well as on the properties of the fuel. It was established (Manzello et al., 2008a) that when using single hot fragments from Douglas spruce (5 mm and 10 mm in diameter, 51 mm and 76 mm in length, respectively) at a blown stream speed of 0.5 m / s or 1 m / s, paper scraps ignited in smoldering mode, but did not occur in pine needles and hardwood. For burning particles, ignition was recorded in all experiments, with the exception of "mulch" of deciduous trees. In (Caine et al., 2009), hot springs obtained as a result of electric heating of a spiral were used. No explicit relationship was found between the diameter of the heating source and the temperature necessary for ignition. As a result of theoretical studies, the Joule heat release from the spiral was evaluated as an ignition criterion by analogy with the minimum ignition energy for the concept of gas ignition (Stokes, 1990; Rowntree and Stokes, 1994). However, this approach was evaluated critically by the authors of (Babrauskas, 2003), since particles of different sizes with the same energy did not always initiate ignition. In (Babrauskas, 2003), it was concluded that the ratios of its size and initial temperature are important for ignition by a particle. The first works on ignition of condensed matter were published in the 60-70s of the last century (Goldschleger et al., 1973; Zinn, 1962; Boddington, 1963; Friedman, 1963; Thomas, 1965) and later applied to the study of the ignition of natural fuels (Jones, 1993; 1994; 1995). Later, a mathematical simulation of ignition was carried out when a fuel was heated by sources of constant temperature and large sizes (Zvyagilskaya and Subbotin, 1996; Grishin et al., 1998; Lautenberger and Fernandez-Pello, 2008).

The purpose of this study is a comparative analysis of the results of experimental studies of the ignition of forest fuels by a metal or non-metal particle heated to high temperatures.

STUDY AREA

The ecological system of forests of the Russian Federation occupies 1.2 billion hectares of territory and contains about 25% of the forest resources of the entire planet. Russian forests are not only an economic, but also an important environmental resource, since the Russian Federation provides an annual carbon storage of 29 billion tons. The global processes of regulating the state of the environment, biodiversity, climate, river flows are significantly affected by the forests of the Russian Federation (Kuznetsov et al., 2005).

Tomsk Region

Tomsk region, especially its northern part, is a fairly typical forest-covered territory of the boreal forest zone. On her example, a fairly general description of the conditions for the occurrence of fires is possible. The region has large forest resources. Forest fund occupy 90.5% of its entire territory. An area of 17 million hectares is covered with tree species, including 9.9 million hectares - coniferous (Panevin, 2006). The main relief types within the Tomsk Region are the watershed plains and river valleys along with the hollows of the ancient runoff. Dividing plains are represented by positive and negative morphostructures.

Forests of the region are located in the river basin (The Ob) are on exclusively flat territory with excess moisture and are of great environmental importance (Panevin, 2006). Relatively harsh climatic conditions determine a fairly limited species composition of forests. The most common types of forest formers are common pine, Siberian cedar, Siberian spruce, Siberian fir, saggy and fluffy birch, Siberian aspen and larch (Panevin, 2006).

The fire danger of the forests of the Tomsk Region is determined by the presence of a significant proportion of coniferous forests, developed by the burning ground cover and hot dry summers. The climate of the Tomsk region is sharply continental of the boreal type. In territories with a continental climate, conditions especially favorable for the occurrence of forest fires are created (Kurbatskiy, 1964). Depending on weather conditions, all three peaks of seasonal incidence are expressed in the forests of the region: spring wave of fires, summer steady fires and autumn fires (Panevin, 2006).

A feature of the forests of the Tomsk region is the presence of fuel in all stands. Mostly surface fires develop in the region (98.5%), 1.1% of incidents and 12.5% of the burned area account for the crown fires, and underground fires occur even less frequently (Panevin, 2006). The share of fires for anthropogenic reasons is quite stable over the years, and fires from a lightning discharge are cyclical in nature. Periods with massive thunderstorms give way to calmer ones. The firewood of the region's forests also varies significantly over the months of the fire danger season. The most "burning" months are June and July. The duration of the fire danger season according to the weather is from 137 to 161 days (Panevin, 2006). According to Rosleskhoz, the statistics of forest fires in the Tomsk Region suggests that approximately 200 forest fires are caused by anthropogenic factors and about 75 forest fires arise as a result of thunderstorm activity (according to 2016 data). Part of the fires arises as a result of the transition of agricultural bollards to forested areas.

Annually field observations and collection of forest fuel samples for experimental studies is carried out in the Timiryazevskiy forestry of the Tomsk region. The specified forestry of Tomsk Forest Management is located between two large rivers (Ob and Tom) in the territory of three administrative regions of Tomsk Region: Tomskiy, Shegarskiy and Kozhevnikovskiy. The length of the territory from North to South is 64 km, from West to East is 50 km. Forests are mainly represented by a single forests, except for the isolated near-village pine forests of the settlements of Zorkaltsevo, Nizhne-Sechenovo and Gubino (Matsenko et al., 1999).

According to the forest vegetation zoning of Western Siberia, the territory of the Timiryazev forestry of the Tomsk region belongs to the southern taiga zone (the Ob-Tomsk cedar-pine forest district). According to the agroclimatic zoning of the Tomsk Region adopted by the Tomsk branch of Sibgiprozem, the forestry area is classified as a moderately humid area. The growing season is 120 days. The predominant main breed is pine (39.6%); Aspen (26.2%) and birch (21.2%); cedar, larch, spruce and fir account for 13% (Matsenko et al., 1999). For several years, the former Kaltai forestry as the Kaltai precinct forestry has been part of the Timiryazevskiy forestry.

Republic of Buryatia

The study was conducted in the Gilbirinskiy forestland (Ivolginskiy district of the Republic of Buryatia) in a protected natural area. The study area is located between latitudes 51⁰35'50'' N and 51⁰46'40'' N and longitudes 106⁰42' E and 107⁰02' E (Roslesinforg, 2020). This area is 55 km southeast of Lake Baikal, and belongs to the northern tip of the Selenga middle mountain. The area of the forestland is about 270 km^2 (Reglamenty, 2020). The forestland was divided into 126 forest areas. The forest quarter is a part of the forestland and it has permanent boundaries. The forest quarter is the main accounting unit of the forest fund in the Russian Federation. The study region, like the whole of Buryatia, has a sharply continental climate, with cold winters and hot summers. The average temperature in summer is about +18.5 ⁰C, in winter is about -22 ⁰C, and the average annual temperature is about -1.6 ⁰C. The average annual rainfall is 244 mm. A significant feature of the climate is the long duration of sunshine for 1900–2200 hours per year (Ivolginskiy Rayon, 2020). About 90% of the territory is occupied by natural plantations. There are types of glades, slopes, pebbles, forest cultures, arable land, pastures, etc. The main forest-forming species are larch, pine, cedar, birch, and aspen (Forest resources, 2020; Yankovich et al., 2019).

MATERIALS AND METHODS

Experiments were conducted with various forest fuels.

Pine litter. The first object of study was the model layers of a typical forest fuel (pine needles), which were formed in a glass refractory cuvette by means of a random laying of needles in a uniform layer. It should be noted that the packing density corresponded to the density of coniferous litter in a real forest with a porosity of more than 0.9). Characteristics of needles: a) on the appearance of brownish; b) the needles of last year's gathering partially decomposed; c) pre-dried; d) the main fraction was composed of individual needles with a size of 7-8 cm in the longitudinal and 0.7-1.2 mm in the transverse direction; e) the fraction differing from the main part of the needles was about 5% (probably the presence of such needles in the litter can be explained by some tree diseases and, accordingly, their lower mechanical strength (Esau, 1980).

Leaves of birch. The second series of experiments was performed with birch leaves dried and purified from foreign inclusions. Immediately before the experiments, the leaves were also dried in an oven for 2-3 hours. The experiments were carried out with weighed portions of the studied forest fuel, which included 5-6 leaves. Laying leaves was uniform.

Birch leaves are biologically transformed. The object of the study was hinges prepared from a typical forest fuel — litter of hardwood (birch leaves) with a long residence time in the ground layer. Real layers of leaves were modeled by their random laying on the substrate. The packing density corresponded to the density of deciduous hardwood in mixed forests. The experiments were carried out with leaves, in which, after partial decomposition under external influences (precipitation, solar radiation, periodic temperature changes), the content (fraction) of combustible components can change. Decomposition processes can affect the flammable properties of brown foliage.

Spruce needles with branches. The object of the study was the model layers of a typical forest fuel (spruce needles with branches), which were formed in a refractory cuvette by means of a random arrangement of branches with needles in a uniform layer. Characteristics of forest fuel: a) the needles in appearance are green with a slight gray-brown tint; b) needles and branches of the current collection are practically not decomposed; c) pre-dried material; d) the main fraction consisted of needles with a size of (1.5-2) cm in the longitudinal and (0.7-1.3) mm in the transverse direction; e) the fraction of twigs different from the main part of the needles was about 25%.

Grass. The object of research was samples of grass using stacking densities corresponding to real environmental conditions. The material for the experiment was collected along transport and railway lines at the borders with forests (Tomsk Region). Immediately before carrying out experiments on ignition, the samples of grass were dried in an oven until the moisture completely evaporated from the material. The grass has a complex structure, including dead morphological parts of grassy vegetation, such as a stem, leaf plate, and sometimes cereals. When forming the sample, all elements from the totality of morphological parts were used with a predominance of leaf blades in the total mass. Such a composition is characteristic of grass located in real natural conditions.

The following sources of elevated temperature were used:

Particle of metal. The choice of steel for the manufacture of particles used as a local heating source is due to several reasons (Zakharevich et al., 2012). Firstly, in real practice, the sources of ignition in forested areas are mainly carbon particles that are formed when the wind blows unheated fires. But there are frequent cases of fires when particles that form during the welding or cutting of metals in forest-covered areas hit the surface of a forest fuel. Secondly, earlier attempts were made to conduct experiments with solid wood combustion products heated to high temperatures. It was found that in this case it is very difficult to obtain particles that differ slightly in configuration and size. But the shape of the particle and its characteristic dimensions strongly affect the ignition conditions of the forest fuel. The temperature of a larger particle, separated from a large sample of charred wood, as a rule, differs significantly from the temperature of a small particle. In addition, their structure (porosity, crack sizes, pore surface configuration) also differ significantly. Therefore, it is extremely difficult to ensure stability and repeatability of conditions in individual experiments. For these reasons, metal particles with fixed constant sizes and a structure that does not change during the experiment are significantly more preferable. Thirdly, also in the process of preliminary stage experiments, it was found that the main parameter determining the conditions and ignition delay of forest fuel is the particle temperature. Deviations of 50 - 100 K lead to significant changes in the conditions of ignition of forest fuel. At the same time, carbon particles formed during the burning of, for example, wood when heated to high temperatures (more than 1000 K) are destroyed, as a rule, as a result of physicochemical transformations and the action of thermal stresses. For these reasons, steel particles were chosen as a local heating source. Stage experiments carried out during the development of experimental research methods showed that particles weighing less than 2.5 g (and, correspondingly, with smaller characteristic sizes), even when heated to high temperatures, do not initiate ignition of the studied forest fuel. Therefore, experiments were carried out only with sufficiently large particles (characteristic longitudinal and transverse sizes of at least 13 mm and 6 mm, respectively).

Carbon particle. The source of elevated temperature was modeled by parallelepiped-shaped graphite particles. It should be noted that the process of ignition by a carbon particle is different from a similar process with a steel particle as a source of ignition. Single carbon particles at high temperatures are characterized by the gasification process, which occurs during the intense penetration of gas reagents through the porous structure of the particle (Vilensky and Khzmalyan, 1978; Golovina, 1983; Morell et al., 1990). That is, diffusion processes prove to be important (it was shown (Samuylov et al., 2004) that diffusion phenomena have a significant influence on the gasification process for large particles) and the transformation of components inside the particle, which leads to a change in the porous structure itself (Jones et al., 1999). The form of the global gasification reaction is as follows (Samuylov et al., 2004): C+CO2®2CO. According to (Laurendau, 1978), this process includes a chain of reactions on the surface of pores:

C_f+CO_2«$C(O)_L+CO$,
$C(O)_L$®C_f+CO,
$C(O)_L$«$C(O)_S$,
$C(O)_S$®C_f+CO,

Where C_f, $C(O)_L$, $C(O)_S$ – respectively, free active carbon centers, an oxygen atom connected to a carbon atom by a mobile ionic bond, and an oxygen atom forming a fixed carbonyl bond with a carbon atom. To date, new approaches to studying the mechanism and laws of heterogeneous combustion and carbon gasification reactions have developed (Golovina, 2002). If in the framework of the diffusion-kinetic theory of heterogeneous combustion and gasification of carbon the laws of the process were judged by the behavior of only the gas phase, then now, along with the gas phase, changes in the solid phase are also taken into account. For this, the concept of active surface centers (ASA) or, more broadly, reactive active centers (RSA) is introduced. The reactive surface is determined by the concentration of active carbon atoms, on which a carbon-oxygen complex is formed, which gives a gaseous product upon decomposition (Golovina, 2002; Lizzo et al., 1990).

According to the results of test experiments, particles in the form of a parallelepiped made of a graphite bar with characteristic dimensions were selected for the experiments: height h = 14 mm; width x1 = 8 mm; thickness x2 = 8mm; m = 1.3 g. Their preliminary placement in an induction furnace showed that in the temperature range 1113 - 1273 K a graphite particle burns in a flame mode. The precipitation of such a particle on the forest fuel layer also unambiguously leads to ignition. It is likely that the initial heating of a graphite particle is accompanied by the release of any volatile compounds that burn in the gas phase. Subsequent heating of the particle is not accompanied by the appearance of a flame around the particle. A series of preliminary experiments showed that a carbon particle is characterized by its burnout over time. This can be explained by the processes of gasification of a carbon particle considered above, as well as by the heterogeneous oxidation of carbon itself.

For the experiments, an experimental setup was made, the circuit diagram of which is shown in Fig. 1.

The experiments were carried out according to the classical plan with randomization due to the fact that a mathematical model has not yet been determined that describes the relationship between the ignition delay of the forest fuel and the initial temperature of a local heating source. At a constant value of T_0, 5–7 experiments were performed, the standard deviation and confidence intervals for determining t_{ign} were calculated with a confidence probability of P=0.95 (Gorban, 2017). The normal distribution of the measured quantity (ignition delay) was assumed.

PHYSICAL AND CHEMICAL MECHANISM OF IGNITION

This section provides a description of the physicochemical mechanisms of ignition of various forest fuels with a metal and carbon particle.

Figure 1. Experimental installation to study the ignition of forest fuel by the single particle heated to high temperatures: 1 - platform, 2 - heater, 3 - camera, 4 - movable holder with terminal at end, 5 - cuvette with sample of forest fuel, 6 - high-temperature particle (Baranovskiy and Zakharevich, 2018)

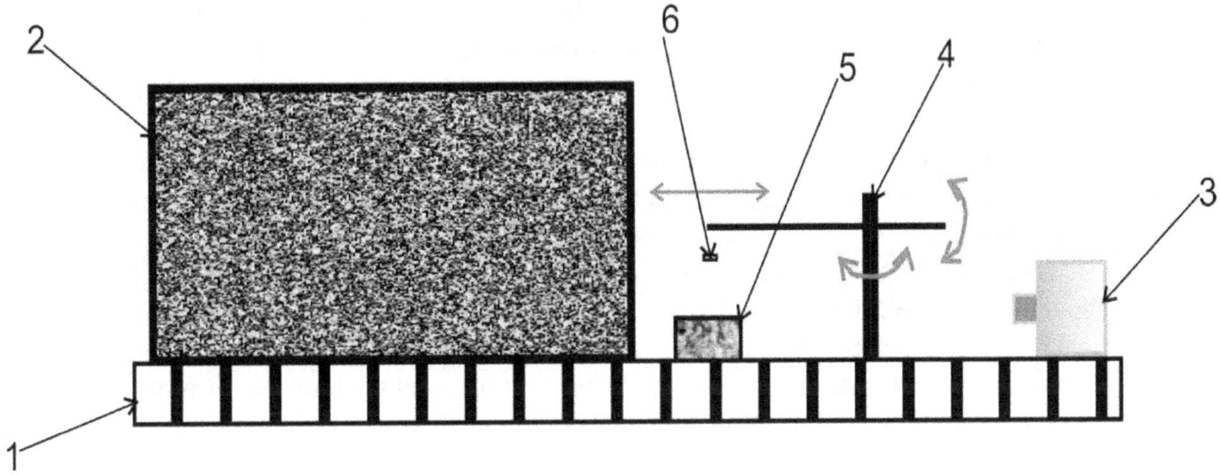

Pine litter. For a metal particle, the following regularities of the process are established. After a short period of inert heating of the forest fuel layer, thermal decomposition of the material begins with the release of gaseous pyrolysis products. The forest fuel in the surface layer decomposes almost completely with a small coke residue, which falls on the substrate. In the porous medium of forest fuel, gaseous pyrolysis products are filtered to the heated surface of the layer and mixed with an oxidizing agent, and the gas mixture is heated, followed by ignition. Then the flame appears around the perimeter of the particle.

The following regularities of the process under study are established for a carbon particle. There are two options for implementing the ignition conditions (Zakharevich et al., 2012b):

Initially, the mechanism of ignition resulting from the action of a burning graphite particle was investigated. A carbon particle, accompanied by a flame torch of combustion of volatile compounds, falls on a layer of forest fuel, which is heated as a result of the action of three heat transfer mechanisms: conduction, convection and radiation (most likely convective and radiant transfer are the main ones in this ignition mechanism). Individual needles warm up and begin to thermally decompose with the release of gaseous pyrolysis products. There is injection of gaseous combustible products and ignition of forest fuel in the gas phase. For the moment of ignition, the appearance of a second flame torch is characteristic. The first is formed as a result of combustion of volatile products released by the particle. Then there is a union of torches and the subsequent spread of flame over a layer of forest fuel.

Leaves of birch. Visual observations and video frame analysis of the processes of ignition of litter from birch leaves allow us to describe the following mechanism of the ignition of forest fuel as a result of the action of a metal particle heated to high temperatures (Zakharevich et al., 2012a). The first short stage is the inert heating of the leave by a local source of elevated temperature. Then the process can go in two ways (depending on the orientation of the leave relative to the precipitated particles).

Particle precipitation on the back side of the leave: after inert heating, the thermal decomposition of the forest fuel begins with the formation of gaseous pyrolysis products (first, the surface layer of the leave undergoes thermal decomposition). Filtration occurs in the microporous structure of the forest fuel of the pyrolysis products to the "leave-particle" contact surface, heating of the fuel and oxidizer mixture and ignition in the gas phase.

If the particle falls on the front side of the leave: after inert heating, high-temperature evaporation of essential oils and analogues of terpene compounds begins. These vapors are mixed with atmospheric oxygen and subsequent explosive ignition of the mixture as a result of local heating. Visually, the picture of the process as a whole corresponds to the picture of ignition of liquid fuels (Zakharevich et al., 2008). In a number of cases, ignition occurs even before the significant thermal decomposition of the forest fuel, which is the next stage. The gaseous products of pyrolysis are blown into the heated region and enhance the burning area. It should be noted that in almost all experiments the formation of a flame plume was observed along the entire perimeter of the particle. In some cases, in a fraction of a second, a volumetric flame torch was formed over the entire surface of the ignition source (around the perimeter and above the particle). It is known (Kansa et al., 1977; Grishin, 1992) that gaseous, liquid, and solid thermal decomposition products are formed during the pyrolysis of hardwood. In the conducted experiments, a noticeable formation of liquid products of pyrolysis of the birch leaf was also observed.

In the case of a carbon particle, after the initial stage of a short period of inert heating of the forest fuel layer, thermal decomposition of the material begins with the release of gaseous pyrolysis products. In the near-surface layer, forest fuel decomposes almost completely with a small coke residue, which falls on the substrate. In the porous medium of forest fuel, gaseous pyrolysis products are filtered to the heated surface of the layer and mixed with an oxidizing agent, and the gas mixture is heated, followed by ignition. Then the flame appears around the perimeter of the particle. It is established that after a particle falls out on the front side of the leave, an inert heating of the material occurs for 0.1-0.2 s. The next stage of the process is the gasification of ether compounds and gas-phase ignition of a mixture of fuel with air. If the particle falls on the back of the leave, then the ignition mechanism is somewhat different. A separate leave warms up and begins to thermally decompose with the release of gaseous pyrolysis products and their subsequent blowing into the region of elevated temperature and ignition of forest fuel in the gas phase. For the ignition conditions of this forest fuel, the appearance of a second flame torch is characteristic (the first is formed as a result of burning of volatile products released by the particle). Then there is a union of torches and the subsequent spread of flame over a layer of forest fuel. Visual observations of the processes of ignition of litter from birch leaves and viewing of video frames allow us to formulate a physical model of the occurrence of forest fuel ignition when heated by a small particle (flameless heating mode). The first short stage is the inert heating of the leave by a local source of elevated temperature. In the future, two options for the development of the process are possible.

If the particle is on the back side of the leave after inert heating, thermal decomposition of the forest fuel begins with the formation of gaseous pyrolysis products. The pyrolysis products are filtered in the microporous structure of the forest fuel to the "leave-particle" contact surface, the mixture of fuel and oxidizer is heated and ignition occurs at some distance from the particle surface.

If the particle is on the front side of the leave after inert heating, high-temperature gasification of essential oils and analogues of terpene compounds begins. Diffusion of these vapors in air leads to explosive ignition of a mixture of fuel and an oxidizing agent as a result of local heating (visually the picture of the process as a whole corresponds to the picture of ignition of liquid fuels). In some cases, ignition begins before significant thermal decomposition of the forest fuel. As a result, the gaseous products of pyrolysis are blown into the heated region and increase the burning area. In most experiments, a flame torch was formed around the entire perimeter of the particle. Sometimes, in a split second, a volume flame torch arose over the entire surface of the particle.

Grass. It should be noted that there are two possible variants of the interaction of the heated particle with the forest fuel layer: the heating source falls on the surface of the leave plate or falls into the depth of the grass rag layer. Penetration of a particle into a layer is characteristic of steel sources due to their greater mass compared to carbon particles. Therefore, we consider only the ignition mechanism of grass rags as a result of exposure to the surface of the leave plate. For a short period, an inert heating of the forest fuel layer occurs, followed by thermal decomposition of the material with the release of gaseous pyrolysis products. It should be noted that the isolation of the products of pyrolysis is more intense than that of the litter of pine needles. The combustible forest material in the near-surface layer decomposes almost completely with a small amount of coke residue that falls on the substrate. In a microporous forest fuel medium, gaseous pyrolysis products are transported to the heated surface of the layer and mixed with an oxidizing agent. Then the gas mixture is heated, followed by the ignition stage. After a fraction of a second, a flame appears around the perimeter of the particle.

Birch leaves are biologically transformed. Any forest fuel in the process of exposure to external conditions and meso-, microfauna of the soil is subject to biological transformation (Pinaeva, 2009). These processes begin from the moment the forest fuel element falls onto the ground cover and lasts several years (Belyakova, 2001). A significant amount of litter decomposes during the first year. Moreover, the decomposition coefficient is maximum in tropical forests and minimal in boreal forests. Possible impact of both biological and environmental factors (leaching, thermal destruction, etc.) (Chastukhin and Nikolaevskaya, 1969). The cycle of organic matter in the tundra, forest tundra, coniferous taiga and coniferous-deciduous forests is greatly inhibited. The rate of accumulation of dead plant debris prevails over their decomposition in broad-leaved forests, where the organic cycle is also inhibited (Belyakova, 2001). To study the destruction of litter, weighed portions of plant material are used, for which the change in weight over time is determined (Belyakova, 2001; Microbiological…, 1987). This method allows you to develop mathematical models of the cycle of substances in biogeocenosis. So the models (Titlyanova, 1971) describe biogeocenosis as a system of blocks with a certain stock of biological substance, which passes from one block to another. Almost throughout the study of transformation processes, the biological factor was considered as determining. However, in some conditions, the influence of ultraviolet radiation on plant residues may be a decisive factor (Kokovina, 1967). Biological researchers agree that climatic conditions — the mutual influence of moisture, temperature, solar radiation and their distribution over the seasons of the year — are most significant in decomposition processes. Observations showed that with the same composition of forest-forming species under different climate conditions, decomposition proceeds in different ways. Most researchers studying the decomposition of plant residues agree (Chastukhin and Nikolaevskaya, 1969) that it is impossible to accurately determine which of the factors (abiotic or biotic) is leading in ensuring the conversion of organic substances and their mineralization. As a result of the deposition of a particle heated to high temperatures, inert heating begins in a certain local region of the forest fuel element (leave). Then the process of thermal decomposition of the fuel occurs with the formation of gaseous and solid pyrolysis products. In contrast to the thermal decomposition of dry green birch leaf, the appearance of a liquid fraction of pyrolysis products is not observed. Gaseous combustible components diffuse into the region of the gas mixture, where they are mixed with atmospheric oxygen. At certain concentrations of reagents, a chemical reaction occurs of the oxidation of gaseous products of pyrolysis with atmospheric oxygen. It must be assumed that, as in the case of softwood litter, the main component of the pyrolysis products is carbon monoxide. The flame microflare appears on the side of the local heating source, which indicates the filtration in the microporous structure of the leave of thermal decomposition products to its surface. An analysis of the frames of the video showed that the pattern of the development of the flame structure is typical, in which part of the side surface of the particle is also covered by a flame torch. It is possible that the gaseous pyrolysis products not only mix with atmospheric oxygen, but also penetrate into the microporous surface layers of the carbon particle, in the pores of which complex physical and chemical processes occur (the interaction of the pyrolysis products with the material of the particle and its thermal decomposition products, heterogeneous reactions on the surface of micropores, oxidation of gaseous pyrolysis products). Based on the results obtained, it can be concluded that the source of heating itself has a significant effect on the thermokinetic processes that occur during combustion. The carbon particle not only maintains heat inflow into the gasification zone of the forest fuel and heats the gaseous products of pyrolysis of forest fuels, but also directly participates in chemical interaction with the gaseous products of pyrolysis.

Spruce needles with branches. The following patterns of the process under study are established. There are two options for implementing the ignition conditions (Baranovskiy and Zakharevich, 2020):

a) Initially, the mechanism of ignition resulting from the action of a burning graphite particle was investigated. A carbon particle, accompanied by a flame torch of combustion of volatile compounds, falls on a layer of forest fuel, which is heated as a result of the action of three heat transfer mechanisms: conduction, convection and radiation (most likely convective and radiant transfer are the main ones in this ignition mechanism). Individual needles warm up and begin to thermally decompose with the release of gaseous pyrolysis products. There is injection of gaseous combustible products and ignition of forest fuel in the gas phase. For the moment of ignition, the appearance of a second flame is characteristic (the first is formed as a result of the combustion of volatile products released by the particle). Then there is a combination of torches and the subsequent spread of flame along the forest fuel layer.

b) Flameless mode. After the initial stage of a short period of inert heating of the forest fuel layer, thermal decomposition of the material begins with the release of gaseous pyrolysis products. In the contact zone, the needles from the heterogeneous layer of forest fuel decompose almost completely with a small coke residue that falls on the substrate. Thin branches are thermally decomposed in a thin surface layer. In the porous medium of forest fuel, gaseous pyrolysis products are filtered to the heated surface of the layer and mixed with an oxidizing agent, and the gas mixture is heated, followed by ignition. In most experiments, a flame torch formed over a heated particle.

Unlike pine litter, the test sample was characterized by an ordered distribution of individual needles, thin branches and often a fixed distance between them. The structure of the sample was characterized by porosity due to the ordered structure of the needles, as well as large pore space due to the morphology of spruce branches, which led to rather high values of the standard deviations of the measurement results t_{ign} from the average values. Due to this, in each particular experiment from a series of experiments at the same initial temperature, the heat exchange conditions between the ignition source and the layer of forest fuel were also different.

COMPARATIVE RESULTS

Table 1 presents the comparative results on the average ignition delay time of various forest fuels of a particle heated to high temperatures.

Table 2 presents comparative results on confidence intervals of the averaged delay time of ignition of various forest fuels of a particle heated to high temperatures.

Table 1. Ignition delay for different forest fuels subjected to hot particle

T,K	Pine, steel [*]	Pine, carbon [**]	Leaf, steel [3*]	Leaf, carbon [4*]	Grass, steel [5*]	Grass, carbon [6*]	Leaf, transformed [7*]	Spruce and branches [8*]
1073						0,296		
1113		0,432	0,308	0,272	0,29	0,272	0,208	0,256
1153	0,32	0,304	0,224	0,2	0,25	0,24	0,152	0,2
1193	0,32	0,272	0,208	0,168	0,224	0,22	0,136	0,264
1233	0,236	0,168	0,136	0,12	0,184	0,192	0,096	0,188
1273	0,24	0,168	0,096	0,08	0,176	0,168	0,076	0,176

An analysis of the results presented in Table 1 shows that the smallest lower ignition limit for temperature is characteristic for grass. Moreover, this happens in the case of ignition of the grass with a carbon particle. This is because a carbon particle, in contrast to a steel particle, is chemically active. Heterogeneous chemical oxidation reactions occur on the surface of the carbon particle with the release of additional heat, which in reality leads to an increase in the real surface temperature of the precipitating particle on the forest fuel layer.

Table 2. Confidence intervals of ignition delay for different forest fuels subjected to hot particle

T,K	Pine, steel [*]	Pine, carbon [**]	Leaf, steel [3*]	Leaf, carbon [4*]	Grass, steel [5*]	Grass, carbon [6*]	Leaf, transformed [7*]	Spruce and branches [8*]
1073						0,027		
1113		0,174	0,054	0,041	0,08	0,089	0,081	0,097
1153	0,061	0,152	0,083	0,035	0,061	0,035	0,022	0,061
1193	0,093	0,108	0,022	0,022	0,027	0,025	0,027	0,027
1233	0,09	0,065	0,027	0,025	0,027	0,022	0,027	0,048
1273	0,137	0,108	0,021	0,005	0,087	0,022	0,007	0,027

* (Baranovskiy et al., 2015); ** (Baranovskiy and Zakharevich, 2016a); 3* (Zakharevich et al., 2012a); 4* (Zakharevich et al., 2012c); 5* (Baranovskiy and Zakharevich, 2016b); 6* (Baranovskiy and Zakharevich, 2015); 7* (Zakharevich et al., 2012d); 8* (Baranovskiy and Zakharevich, 2020)

An exception is the case of ignition of needles with a steel particle. In this case, the highest ignition limit in temperature. This is explained by the developed porous space and the small contact area of the particle and individual needles. In other cases, the lower limit of ignition in temperature is 1113 K. Moreover, the same lower limit is also characteristic for spruce needles with the inclusion of small branches.

CONCLUSION

The generalized results of experimental studies on the ignition of typical forest fuels by a metal and non-metal particle heated to high temperatures are presented. Several types of forest fuel were considered: pine litter, birch leaves, grass and spruce branches. The smallest lower ignition limit for temperature is characteristic of grass. Moreover, this happens in the case of ignition of the grass with a carbon particle. An exception is the case of ignition of needles with a steel particle. In this case, the highest ignition limit in temperature. In other cases, the lower limit of ignition in temperature is 1113 K. Moreover, the same lower limit is also characteristic of spruce needles with the inclusion of small branches.

ACKNOWLEDGMENT

This work was implemented with the financial support of the Russian Foundation for Basic Research. [Project N 17-29-05093 and Project N 16-41-700831].

REFERENCES

Babrauskas, V. (2003). *Ignition handbook: principles and applications to fire safety engineering, fire investigation, risk management, and forensic science*. Fire Science Publishers.

Baranovskii, N. V., & Zakharevich, A. V. (2016b). Conditions and Characteristics of the Ignition of a Typical Vegetable Fuel by a Local Energy Source. *Journal of Engineering Physics and Thermophysics, 89*(6), 1538–1544. doi:10.100710891-016-1524-3

Baranovskiy, N. V., & Zakharevich, A. V. (2015). *Experimental research of grassy rags ignition by heated up to high temperatures carbon particle*. Multifunctional Materials and Modeling.

Baranovskiy, N. V., & Zakharevich, A. V. (2016a). Experimental investigation of processes of typical forest fuel ignition by a high-temperature steel particle. *Heat Transfer Research, 47*(7), 681–689. doi:10.1615/HeatTransRes.2016010609

Baranovskiy, N. V., & Zakharevich, A. V. (2018). Mechanical, thermophysical and aerothermochemical processes at forest fuel ignition caused by the heated to high temperatures single metal particle. *International Review of Mechanical Engineering, 12*(6), 522–532. doi:10.15866/ireme.v12i6.15065

Baranovskiy, N. V., & Zakharevich, A. V. (2020). *Experimental modeling of spruce needles ignition by a carbonaceous particle heated to high temperatures* (Vol. 3). Forest Science Issue.

Baranovskiy, N. V., Zakharevich, A. V., & Osotova, D. S. (2015). Experimental study of pine forest fuel layer ignition by the steel heated particle. *EPJ Web of Conferences, 82*. 10.1051/epjconf/20158201020

Belyakova, O. I. (2001). *Long-term dynamics of decay of plant litter in the main ecosystems of the Central forest-steppe* (PhD dissertation). Voronezh: Central Black Earth State Biosphere Reserve named after V.V. Alekhine.

Boddington, T. (1963). The growth and decay of hot spots and the relation between structure and stability. *Proceedings of the Combustion Institute*, *9*(1), 287–293. doi:10.1016/S0082-0784(63)80036-3

Caine, P., Puttick, S. J., Brindley, J., McIntosh, A. C., & Griffiths, J. F. (2009). Ignition of bulk solid materials by a localised hotspot. Dangers XXI, Symposium Series, 155, 191–200.

Chastukhin, V. Ya., & Nikolaevskaya, M. A. (1969). *Biological decay and resynthesis of organic substances in nature*. Nauka.

Esau, K. (1980) Anatomy of seed plants. Moscow: Mir.

Friedman, M. H. (1963). A correlation of impact sensitivities by means of the hot spot model. *Proceedings of the Combustion Institute*, *9*(1), 294–302. doi:10.1016/S0082-0784(63)80037-5

Geoportal FGBU. (n.d.). *Roslesinforg*. Available online: http://geoportal.roslesinforg.ru:8080/

Goldschleger, U. I., Pribytkova, K. V., & Barzykin, V. V. (1973). Ignition of condensed explosives by a heated body of finite dimensions. *Combustion and Explosion Physics.*, *9*(1), 119–123.

Golovina, E. S. (1983). *High temperature combustion and gasification of carbon*. Energoatomizdat.

Golovina, E. S. (2002). The study of heterogeneous combustion and gasification of carbon and solid fuels (Review). Combustion and Explosion Physics, 38(4), 25 - 34.

Gorban, I. I. (2017). *Fundamentals of the Mathematical Statistics of Probability Theory // Randomness and Hyper-randomness*. Springer.

Grishin, A. M. (1992). *Mathematical modeling of forest fires and new methods fighting them*. Nauka.

Grishin, A. M., Dolgov, A. A., Zima, V. P., Kryuchkov, D. A., Reynaud, V. V., Subbotin, A. N., & Tsvyk, R. Sh. (1998). Investigation of the ignition of a layer of forest combustible materials. *Combustion and Explosion Physics.*, *34*(6), 14–22.

Hadden, R. M., Scott, S., Lautenberger, Ch., & Fernandez-Pello, C. (2011). Ignition of Combustible Fuel Beds by Hot Particles: An Experimental and Theoretical Study. *Fire Technology*, *47*(2), 341–355. doi:10.100710694-010-0181-x

Jones, J. C. (1993). Predictive calculations of the effect of an accidental heat source on a bed of forest litter. *Journal of Fire Sciences*, *11*(1), 80–86. doi:10.1177/073490419301100104

Jones, J. C. (1994). Further calculations concerning the accidental supply of heat to a bed of forest material. *Journal of Fire Sciences*, *12*(6), 502–505. doi:10.1177/073490419401200602

Jones, J. C. (1995). Improved calculations concerning the ignition of forest litter by hot particle ingress. *Journal of Fire Sciences*, *13*(5), 350–356. doi:10.1177/073490419501300502

Jones, J. M., Pourkashanian, M., Rena, C. D., & Williams, A. (1999). Modelling the relationship of coal structure to chair porosity. *Fuel*, *78*(14), 1737–1744. doi:10.1016/S0016-2361(99)00122-2

Kansa, S. E., Perles, H. E., & Chaiken, R. F. (1977). Mathematical model of wood pyrolysis. *Combustion and Flame*, *29*, 311–324. doi:10.1016/0010-2180(77)90121-3

Kokovina, T. P. (1967). On the issue of ash exchange in oak forests on powerful chernozem. *Transactions of Tsentr.-Chernozem. state Nature reserve*, (10), 133–154.

Kurbatskiy, N. P. (1964). *The problem of forest fires. The occurrence of forest fires*. Nauka.

Kuznetsov, V. I., Kozlov, N. I., & Khomyakov, P. M. (2005). *Mathematical modeling of forest evolution for forest management purposes*. LENAND.

Laurendau, N. M. (1978). Heterogeneous kinetics of coal char gasification and combustion. *Progress in Energy and Combustion Science*, *4*(4), 221–270. doi:10.1016/0360-1285(78)90008-4

Lautenberger, C., & Fernandez-Pello, A. C. (2008). Modeling ignition of combustible fuel beds by embers and heated particles. *Proceedings of International Conference "Forest fires 2008"*.

Lizzo, A., Jiang, H., & Radovic, R. (1990). On the kinetics of carbon (char) gasification: Reconciling model with experiments. *Carbon*, *28*(1), 7–19. doi:10.1016/0008-6223(90)90087-F

Manzello, S. L., Cleary, T. G., Shields, J. R., Maranghides, A., Mell, W., & Yang, J. C. (2008a). Experimental investigation of firebrands: Generation and ignition of fuel beds. *Fire Safety Journal*, *43*(3), 226–233. doi:10.1016/j.firesaf.2006.06.010

Manzello, S. L., Cleary, T. G., Shields, J. R., & Yang, J. C. (2006a). On the ignition of fuel beds by firebrands. *Fire and Materials*, *30*(1), 77–87. doi:10.1002/fam.901

Manzello, S. L., Cleary, T. G., Shields, J. R., & Yang, J. C. (2006b). Ignition of mulch and grasses by firebrands in wildland-urban interface fires. *International Journal of Wildland Fire*, *15*(3), 427–431. doi:10.1071/WF06031

Manzello, S. L., Shields, J. R., Cleary, T. G., Maranghides, A., Mell, W. E., Yang, J. C., Hayashi, Y., Nii, D., & Kurita, T. (2008b). On the development and characterization of a firebrand generator. *Fire Safety Journal*, *43*(4), 258–268. doi:10.1016/j.firesaf.2007.10.001

Matsenko, V. V., Sokolov, A. Ya., Kalinin, S. I., Andriyanova, F. I., Andreeva, T. A., Ananyin, S. V., Krylov, M. N., & Kazantseva, L. V. (1999). Master Plan for the fire prevention of forests. Volume 1. Explanatory note. 5-99.14-17-PM. Barnaul: State Design and Research Institute "Rosgiproles", Altai Branch.

Microbiological destruction of organic residues in a biogeocenosis. (1987). Scientific Center for Biological Research, USSR Academy of Sciences.

Morell, J. I., Amundson, N. R., & Park, S. K. (1990). Dynamics of a single particle during char gasification. *Chemical Engineering Science*, *45*(2), 387–401. doi:10.1016/0009-2509(90)87024-M

Official Site, M. O. "Ivolginskiy Rayon". (n.d.). Available online: https://egov-buryatia.ru/ivolga/o-munitsipalnomobrazovanii/ob-organizatsii/

Panevin, V. S. (2006). *Forests and forestry of the Tomsk region*. Publishing House of Tomsk University.

Pinaeva, O. N. (2009). *Transformation of organic matter of deciduous litter depending on the structure of the community of soil micro- and mesofauna* (PhD dissertation). Tolyatti: RGUTiS.

Pitts, W. (2007). Ignition of cellulosic fuels by heated and radiative surfaces. NIST technical note 1481.

Ramsey, G. C., & McArthur, N. A. (1995). Building in the urban interface: lessons from the January 1994 Sydney bushfires. In Bushfires 1995, Hobart, Australia.

ReglamentyiL. (n.d.). Online: https://egov-buryatia.ru/ralh/activities/documents/lesokhozyaystvennye-reglamenty/

Resources, F. (n.d.). online: https://egov-buryatia.ru/about_republic/nature-resources/lesnye-resursy-/.

Rowntree, G. W. G., & Stokes, A. D. (1994). Fire ignition by aluminum particles of controlled size. *Journal of Electrical and Electronics Engineering (Oradea)*, *14*, 117–123.

Samuylov, E. V., Faminskaya, M. V., & Golovina, E. S. (2004). Model and calculation of the gasification process of a single carbon particle. *Combustion and Explosion Physics.*, *40*(1), 86–94.

Stokes, A. D. (1990). Fire ignition by copper particles of controlled size. *Journal of Electrical and Electronics Engineering (Oradea)*, *10*, 188–194.

Thomas, P. H. (1965). A comparison of some hot spot theories. *Proceedings of the Combustion Institute*, *10*, 369–372.

Titlyanova, A. A. (1971) The study of the biological cycle in biogeocenosis. Novosibirsk: Science, Sib.

Vilensky, T. V., & Khzmalyan, D. M. (1978). *The dynamics of the combustion of pulverized fuel*. Energy.

Yankovich, K. S., Yankovich, E. P., & Baranovskiy, N. V. (2019). Classification of Vegetation to Estimate Forest Fire Danger Using Landsat 8 Images: Case Study. *Mathematical Problems in Engineering*, *2019*, 6296417. doi:10.1155/2019/6296417

Yoshioka, H., Hayashi, Y., Masuda, H., & Noguchi, T. (2004). Real-scale fire wind tunnel experiment on generation of firebrands from a house on fire. *Fire Science & Technology*, *23*(2), 142–150. doi:10.3210/fst.23.142

Zakharevich, A. V., Baranovsky, N. V., & Maksimov, V. I. (2012a). Ignition of forest combustible materials by single particles heated to high temperatures. Fire and Explosion Safety, 21, 13 - 16.

Zakharevich, A. V., Baranovsky, N. V., & Maksimov, V. I. (2012b). Ignition of forest combustible material by a carbon particle heated to high temperatures. *Butlerov Communications.*, *29*(2), 102–108.

Zakharevich, A. V., Baranovsky, N. V., & Maksimov, V. I. (2012c). An experimental study of the ignition processes of litter of broad-leaved tree species as a source of limited energy intensity. *Ecological systems and devices*, (7), 18–23.

Zakharevich, A. V., Baranovsky, N. V., & Strizhak, P. A. (2012d). An experimental study of the processes of ignition by a heated particle of a biologically transformed litter of hardwood. Butlerov Communications, 32, 95 - 99.

Zakharevich, A. V., Kuznetsov, G. V., & Maksimov, V. I. (2008). The mechanism of ignition of gasoline by a single metal particle heated to high temperatures. Fire and Explosion Safety, 17(5), 39 - 42.

Zinn, J. (1962). Initiation of explosions by hot spot. *The Journal of Chemical Physics*, *36*(7), 7. doi:10.1063/1.1701302

Zvyagilskaya, A. I., & Subbotin, A. N. (1996). The effect of moisture content and heat and mass transfer with the environment on the critical conditions for the occurrence of a bottom fire center. *Combustion and Explosion Physics.*, *32*(5), 99–106.

ADDITIONAL READING

Antonov, D. V., Volkov, R. S., Voitkov, I. S., Zhdanova, A. O., & Kuznetsov, G. V. (2018). Influence of Special Additives in a Water Aerosol on the Suppression of a Forest Fire with it. *Journal of Engineering Physics and Thermophysics.*, *91*(5), 1250–1259. doi:10.100710891-018-1855-3

Baranovskiy, N. V., Sherstnyov, V. S., Yankovich, E. P., Engel, M. V., & Belov, V. V. (2016) Focused sunlight factor of forest fire danger assessment using Web-GIS and RS technologies. Proceedings of SPIE - The International Society for Optical Engineering, 9688, article 968823, 6 P. 10.1117/12.2240378

Chen, T., Zhang, J., & Wu, J. (2016). Kinetic and energy production analysis of pyrolysis of lignocellulosic biomass using a three-parallel Gaussian reaction model. *Bioresource Technology*, *211*, 502–508. doi:10.1016/j.biortech.2016.03.091 PMID:27035484

Goman, P. N. (2020). The Dynamics of Surface Forest Fire and Forest Fuel Ignition Under the Heat Radiation From the Fire Line. In N. Baranovskiy (Ed.), *Predicting, Monitoring, and Assessing Forest Fire Dangers and Risks* (pp. 1–47). IGI Global. doi:10.4018/978-1-7998-1867-0.ch001

Hameed, S., Sharma, A., Pareek, V., Wu, H., & Yu, Y. (2019). A review on biomass pyrolysis models: Kinetic, network and mechanistic models. *Biomass and Bioenergy*, *123*, 104–122. doi:10.1016/j.biombioe.2019.02.008

Liu, H., Wang, C., Zhao, W., Yang, S., & Hou, X. (2018). Pyrolysis characteristics and kinetic modeling of Artemisia apiacea by thermogravimetric analysis. *Journal of Thermal Analysis and Calorimetry*, *131*(2), 1783–1792. doi:10.100710973-017-6599-3

Thomas, J. C., Hadden, R. M., & Simeoni, A. (2017). Experimental investigation of the impact of oxygen flux on the burning dynamics of forest fuel bed. *Fire Safety Journal*, *91*, 855–863. doi:10.1016/j.firesaf.2017.03.086

KEY TERMS AND DEFINITIONS

Anthropogenic Load: Different human activities on forested territories lead to forest fire occurrence and characterized by presence of fire sources.

Forest Fire: Uncontrolled aerothermochemical phenomenon characterized by step-by-step mechanism which includes following stages: inert heating, moisture evaporation, high temperature terpens evaporation, dry organic matter pyrolysis, flammable combustion. and smoldering.

Forest Fuel: It can be considered like dead and live forest fuel. Main types of forest fuel which can be involved in combustion during forest fire: ground forest fuel (needles, leaves and dry grass, small branches) and crown forest fuel (needles, small branches).

Ignition: Inflammation of forest fuel caused by definite source of high temperature or energy.

Ignition Delay: Time before flame flash after forest fuel heating.

Prediction: Under the prediction of forest fires is the calculation of the parameters of forest fire danger with a certain projection in advance in order to have enough time to anticipate an emergency. The calculation in this case is carried out in a mode ahead of the real time of the development of the catastrophe - the occurrence of a forest fire.

Chapter 10
Physical Modeling of Forest Fuel Ignition by the Molten Metal Particles

ABSTRACT

Forests in the Siberian region have suffered from frequent and intense forest fires in recent years. Measures are required to restore damaged and dead forest stands. One option for preserving forest stands from catastrophic forest fires is prescribed low-intensity burning. The purpose of the study is to study the ignition mechanism of a layer of forest fuels by crystallizing small-sized metal particles (molten metal drop). The generation of such droplets is carried out using a standard electric welding machine. The physical mechanism of forest fuel layer ignition by a group of particles of a rather small size is revealed. It is proposed to use this mechanism for burning forest fuels in order to preserve coniferous and mixed stands from intense forest fires.

INTRODUCTION

The Russian Federation has significant territories covered by boreal forests (Baranovskiy and Kuznetsov, 2017)]. In such forests, a proportion of pine and birch is noticeable. It should be noted that the areas of pine and birch include almost the entire territory of the Russian Federation (Devisilov, 2010). This allows us to consider Siberian forests as the most typical territories, on the example of which issues of forest fire danger can be considered. Frequent fires of recent years in the forests of the Siberian region have led to a significant destruction of the forest fund. The occurrence of fires in the forest is due to a wide range of natural and man-made causes (Baranovskiy and Kuznetsov, 2017). It must be said that the anthropogenic load on a number of forested areas is growing in connection with their economic development. The probability of a forest fire is affected by the type of source of elevated temperature and the specific ignition mechanism that is realized when the forest fuel is ignited (Babrauskas, 2003).

The literature presents data on experimental and theoretical studies of the ignition of fuels by particles heated to high temperatures (Hadden et al., 2011). Such particles can be formed both as a result of destruc-

DOI: 10.4018/978-1-7998-7250-4.ch010

tion of wood of trees or morphological parts of plants (Tohidi et al., 2015; Manzello et al., 2006), and as a result of destruction of building wood-glued materials in urban areas (Suzuki and Manzello, 2016). Moreover, they can have different geometry and shape. It is known that burning particles (firebrands, coals) are the main mechanism for the spread of forest fires and fires passing to the village (Manzello et al., 2006). In the case of the natural mechanism of particle formation, they undergo the following path: a) the formation and separation of the heated particle from a tree or plant; b) the rise (vertical transfer) of a particle due to the action of a convective column over a fire; c) horizontal transport by wind; d) sedimentation of a particle on a layer of forest fuel and the occurrence of a fire. For particles of natural origin, a number of models have been developed that describe their trajectory in the surface layer of the atmosphere (Matvienko et al., 2016; Kortas et al., 2009; Wadhwani et al., 2017). In addition, a number of properties of such particles (Zakharevich and Zygin, 2016) and the occurrence of fires when they are exposed to a layer of forest fuel (Baranovskiy and Zakharevich, 2016) were studied. For an experimental study of such particles, a generator has been developed that allows physical modeling of these processes (Manzello et al., 2008).

In addition, molten or burning particles can form as a result of collision of wires of high-voltage power lines with large winds (Hadden et al., 2011; Fernandez-Pello, 2017; Pleasance and Hart, 1977). It should be noted that the processes of ignition of forest fuels by such heated particles were studied only in the course of single field studies (Stokes, 1990; Rowntree and Stokes, 1994; Manzello et al., 2006b).

One of the stages of ignition of fuels is their thermal decomposition (pyrolysis). It should be noted that the problem of gasification of solids is widely described in the literature, for example, in the review (Di Blasi, 1993). The simplest approach to this problem implies that a solid decomposes with the release of volatile compounds directly at a critical temperature (often denoted by Tp). Accordingly, the critical temperature is a parameter of the problem. This approach (often called the ablation model) is mathematically similar to the Stefan problem (Carlslaw and Jaeger, 1984) and its variation with respect to the melting process is described in the classical Landau article and later implemented in a number of works. The second approach includes a kinetic mechanism for the decomposition process, which is usually established as a result of thermogravimetric analysis (Liu and Wang, 2018).

Thus, the analysis of scientific periodicals shows that the formation, transport, and the effect of wood particles or building materials on forest fuels have been studied quite well. However, only a few works have been published on the ignition of forest fuel by crystallizing metal particles, which are formed as a result of the collision of the wires of power lines (Stokes, 1990; Rowntree and Stokes, 1994; Manzello et al., 2006b). The present work is devoted to the development of an experimental setup and the physical modeling of the ignition of forest fuel by crystallized metal particles that are formed as a result of electric welding in urbanized or forested areas.

The purpose of the work is to study the ignition mechanism of a layer of forest fuels by crystallizing small-sized metal particles (molten metal drop). The purpose of the study can be achieved by solving the following problems: a) development of an experimental setup for the physical modeling of ignition of forest fuels by a crystallizing metal particle; b) conducting experiments on the ignition of forest fuel by a group of crystallizing small metal particles, typical for carrying out electric welding works; c) development of recommendations for the design of a model of the mathematical description of the studied processes and mechanisms.

STUDY AREA AND METHODOLOGY

The study area is represented by typical forest formations characteristic of many Siberian and Far Eastern forests (Tomsk Region and Republic of Buryatia). Typical forest fuel material was collected in these territories in the fall of 2019. The litter of pine, birch leaves and grass was collected, followed by removal of foreign inclusions, namely, pine cones, thin branches, small grasses and particles of the cortical layer, in accordance with the methodology (Ganteaume et al., 2014).

Before conducting semi-natural experiments on ignition, samples of forest fuel were dried in an oven to completely evaporate moisture (Bulba and Malinovsky, 2015; Grishin and Baranovskiy, 2003). An electric welding machine typical of technical work was used as a generator of molten metal particles. The experimental setup is shown in Figure 1. In Figure 1, the numbers indicate: 1 - bed, 2 - concrete block, 3 - a sample of a layer of forest fuel, 4 - a tripod, 5 - a video camera, 6 - a holder, 7 - a metal bar, 8 - welding machine, 9 - wires, 10 - electrode, 11 - group of dropping out particles.

A layer of forest fuel is laid on the concrete surface, the packing density of which corresponded to the laying density in the forest. Three types of forest fuel were considered; a) litter of needles, b) birch leaves, c) grass. Forest fuel is represented by a set of individual needles, partially decomposed and browned, combined into a layer. The leaves were arranged randomly, forming a porous layer of forest fuel. Blades of grass were also laid with high porosity. From the standpoint of the reacting media mechanics, the layer is a highly porous reactive material (Nigmatulin, 1987). A video camera was placed on the line of a sample of forest fuel, which recorded the process of precipitation of particles of molten metal and the subsequent ignition of a layer of forest fuel. In the immediate vicinity of the sample was an electric welding machine. A metal bar was attached over a sample of forest fuel. The contact electrode of the apparatus acted on this bar.

Figure 1. The experimental setup

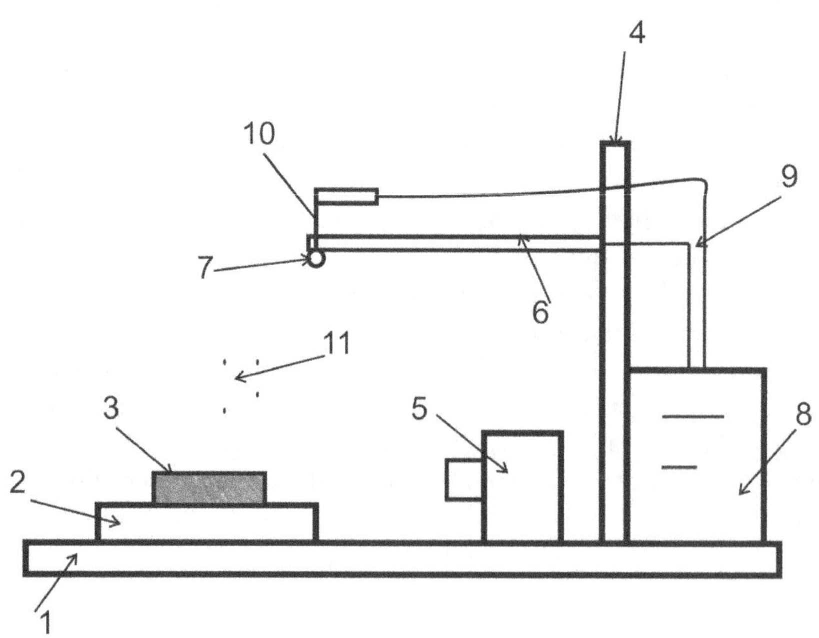

During the experiment, a video was recorded of the generation, deposition of particles, and ignition of forest fuel by these particles in accordance with the methodology (Zakharevich et al., 2008). A series of experiments was carried out with identical electrodes and metal rods (iron). As a result of processing the videograms, the ignition delays for the first and second microflares and the induction period of the propagation of the combustion front over the layer of forest fuel were established. The average values of the parameters were determined and confidence intervals were calculated from a series of seven experiments with a confidence probability of P = 0.95. The mistakes of these parameters were rejected as a result of statistical processing with a reliability of P = 0.95 (Pakshirajan, 2013).

RESULTS AND DISCUSSION

Pine litter. Visual observations were made. Figure 2 presents typical frames of a videogram demonstrating the process of the impact of a group of crystallizing particles on a layer of forest fuel (pine needles): a) the initial form of a layer of forest fuel; b) pyrolysis of forest fuel; c) the occurrence of a microflare of flame; d) the beginning of the spread of the combustion front over the layer of forest fuel.

Figure 2. The stages of ignition in pine litter when exposed to crystallizing particles

One can imagine the physical mechanism of ignition of forest fuel by a group of relatively small particles (droplets) of molten metal. The initial effect of a single drop of molten metal on a layer of forest fuel should be described. Initially, the generation and separation of a heated particle occurs in the zone of contact of the electrode with a metal bar. Then, under the influence of gravity, a particle settles on a layer of forest fuel. It should be noted that the characteristic particle (droplet) size is smaller than the pores of the forest fuel layer.

The molten particle falls onto a layer of forest fuel, and crystallization of the metal occurs and additional energy is released. It should be noted that part of the particles sediments through the highly porous material onto the substrate, without igniting it. Ignition can only result from particles that come into contact with the skeleton of a layer of forest fuel consisting of individual needles. As a result of a phase transition and a higher temperature, the heat flux into the frame in the case of a molten particle is higher than when exposed to an inertly heated metal particle (Baranovskiy and Zakharevich, 2016). First, an inert heating of the dry organic matter occurs, of which a layer of forest fuel consists. Then, the evolution of gaseous pyrolysis products is observed, which, as a result of convective-diffusive mass transfer, mixes with atmospheric oxygen in the near-wall region of the particle. At certain concentrations of the reacting components and the temperature of the gas mixture, a layer of forest fuel may ignite. Individual particles with a probability close to 1 do not lead to ignition of a layer of forest fuel. For the ignition of a layer of forest fuel to occur, a group effect of small molten particles on the framework of the forest fuel in a certain local area is necessary. This effect can be simultaneous, or it can be sequential with a very small interval between the deposition of individual particles.

Leaves of birch. Visual observations were made. Figure 3 presents typical frames of a videogram that demonstrates the effect of a group of crystallizing particles on a layer of forest fuel (birch leaves):

Figure 3. Stages of ignition in birch leaves when exposed to crystallizing particles

a) the initial form of a layer of forest fuel; b) pyrolysis of forest fuel; c) the occurrence of a microflare of flame; d) the beginning of the spread of the combustion front along the layer of forest fuel.

The molten particle also falls onto a layer of forest fuel and crystallizes with the release of additional energy. In the case of leaves, all particles fall onto the surface of a single leave, but not all lead to its ignition. Ignition, as a rule, occurs only when a relatively large particle is exposed (3-4 mm in diameter). A large number of small particles have virtually no effect on birch leaves. It should be noted that in this case, the heat sink to the birch leaf is quite noticeable. A birch leaf can be considered as a plate. The lack of ignition is explained by two situations. Firstly, the heat sink from the falling particle is large in the leave plate and it cools noticeably. The residual heat reserve is not enough to ignite a birch leaf, since a sufficient amount of gaseous pyrolysis products is not formed and the temperature of the mixture of fuel and oxidizer is low. On the other hand, relatively large particles can instantly burn a birch leaf without lingering on the surface of the leaf, and then fall onto the substrate. In this situation, a fresh portion of the gaseous pyrolysis products does not have time to form and there is no sustainable combustion of fuels.

To summarize, we can formulate three modes of exposure of a crystallizing particle to a layer of birch leaves:

a) the sedimentation of a relatively large particle on several leaves of birch, which are under each other. In this situation, the first leave is initially heated and burned out without the formation of a flame, since the amount of gaseous combustible pyrolysis products is insufficient. Then the particle moves down and, as the leaves are introduced into the material, a sufficient amount of gaseous combustible pyrolysis products is formed and mixed with an oxidizing agent (atmospheric oxygen). With sufficient concentrations of reagents and reaching a critical temperature of the mixture, they are ignited in the gas phase. Then there is the formation of a source of ignition and its subsequent distribution over a layer of forest fuel.

b) the second mode is also characterized by the action of a relatively large particle. However, in this case, the particle falls onto a single leave, which instantly burns out without the formation of a microflare of flame. The resulting amount of gaseous combustible pyrolysis products is not enough and they are distributed over a large volume under the influence of diffusion-convective mass transfer. As a result, fire dangerous concentrations of gaseous combustible pyrolysis products are not formed.

c) the third mode is characteristic for the precipitation of an individual or group of relatively small particles (1-2 mm in diameter). Such particles do not, as a rule, lead to ignition. The particle falls onto the surface of a single leave and noticeably loses its temperature as a result of heat removal into the leave material. The same result is characteristic of several particles, if they fall out in different places of the leave. Rarely enough is the regime realized when a repeated hit of subsequent or subsequent particles of relatively small size is observed in the same place with a certain small time interval. In this situation, the first particle causes an inert heating of the layer and cools. As a result of sedimentation of subsequent particles, a local increase in temperature occurs and the decomposition of dry organic matter of the leave occurs. Gaseous combustible pyrolysis products are formed. Under certain conditions, the subsequent particle ignites the forest fuel in the gas phase. The formation of a stable combustion front, which spreads over a layer of forest fuel, takes place. The following are quantitative estimates for the ignition of a layer of birch leaves as a result of the loss of a group of particles formed during electrical welding.

Grass. Visual observations were made. Figure 3 presents typical frames of a videogram demonstrating the process of the influence of a group of crystallizing particles on a layer of forest fuel (grass): a) the initial form of a layer of forest fuel; b) pyrolysis of forest fuel; c) the occurrence of a microflare of flame; d) the beginning of the spread of the combustion front along the layer of forest fuel.

Table 1. Ignition of birch leaves by the large or group of small particles

Ignition by the relatively large particle	40%
No ignition by the relatively large particle	15%
Ignition by the group of relatively small particles	10%
No ignition by the group of relatively small particles	35%
Total	100%

The ignition of grass rags is similar to the ignition of birch leaves in the sense that particles fall on the blade of grass and exhibit a similar ignition mechanism. However, in the event of a sustained fire, the further process is similar to the spread of the flame front along the set of pine needles. It has been found that in the vast majority of cases, grass rags are subject to ignition. Below are quantitative estimates for the ignition of grass rags as a result of the loss of a group of particles formed during electrical welding.

Figure 4. Stages of ignition in grass when exposed to crystallizing particles

A quantitative analysis of the results of an experiment on the ignition of forest fuels by a group of crystallizing metal particles is carried out. The ignition delays and the corresponding confidence intervals are presented in table 3.

Statistical estimates of the average value and confidence intervals of the following parameters were made: the ignition delay time of forest fuel with the appearance of the first and second microflares, as well as the beginning of the growth of the burning focus. It should be noted that in some cases, a simultaneous or sequential formation of up to five microflares of flame was observed in the studied sample of forest fuel. An analysis of the data shows that an average of 6 seconds is enough for a sustainable fire to appear. The first microflare appears on average after 6 seconds, and the second after 7-10 seconds. However,

Table 2. Ignition of grass by the large or group of small particles

Ignition by the relatively large particle	50%
No ignition by the relatively large particle	5%
Ignition by the group of relatively small particles	35%
No ignition by the group of relatively small particles	10%
Total	100%

the confidence intervals for ignition delay for different forest fuels are different. A sufficiently large spread in the ignition delay in experiments with grass indicates the probabilistic nature of this process.

There is no doubt that the implementation of technological work on forested areas can cause an anthropogenic fire. This factor must be taken into account in systems for assessing, predicting and monitoring forest fires, especially in the vicinity of industrial facilities located in forested areas. The present work shows the possibility of igniting forest fuels by a group of relatively small sources of local heating. On the other hand, this mechanism can ensure the burning of forest fuels. International experts recommend such burning out in order to prevent the accumulation of a critical stock of forest fuels in a controlled forested area. In addition, periodic fires of low intensity contribute to the enrichment of the surface soil layer in the forest with various substances and microelements necessary for the development of undergrowth of coniferous stands (Baranovskiy and Kuznetsov, 2017). Such events can ensure the development of undergrowth and the subsequent growth and restoration of coniferous stands, including pine formations. It should be noted that coniferous stands are of great value both as suppliers of industrial timber, and as participants in the global ecological cycle.

One of the practical applications of the results is the development of recommendations for creating a mathematical model of generation, transport, and the impact of a group of crystallizing particles on a layer of forest fuel. Such models can be used in new generation forest fire danger prediction systems (Baranovskiy, 2018). For these purposes, either stochastic approaches (Tohidi and Kaye, 2017) or combined deterministic-probabilistic approaches can be used. Consider the description of the latter option. A probabilistic approach is used to describe the process of generating individual particles, when the quantity, size, temperature, and initial motion vector of such particles are set using the pseudorandom number generator. Then, the motion of each particle until it comes into contact with a layer of forest fuel is described in the framework of a deterministic mathematical model. Then a group of problems should be solved on ignition of a layer of forest fuel by a single particle heated to high temperatures, taking into account the metal crystallization mechanism. In addition to inert heating, in the general case, the processes of moisture evaporation from the layer and the pyrolysis of dry organic matter based on kinetic mechanisms should be described. It is sufficient to describe the transport of gaseous pyrolysis products

Table 3. Ignition delay of forest fuels by the molten particles

Forest fuel	t_{ign}, s	Confidential Interval, s
Pine litter	6	±1.97
Birch leaves	5.6	±1.89
Grass	7	±5.87

in the region above the layer of forest fuel using the diffusion equations, since during the ignition period the thermal and diffusion relaxation lengths are several orders of magnitude longer than the convective (Vilyunov, 1984). However, the processes of convective heat and substance transfer using the general theory of heat and mass transfer (Kuznetsov and Sheremet, 2006) and processes in the gas phase based on the general mathematical model of forest fires (Grishin, 1997) can be taken into account.

CONCLUSION

Thus, there is no doubt that the implementation of technological work on forested areas can cause an anthropogenic fire. This factor must be taken into account in systems for assessing, predicting and monitoring forest fires, especially in the vicinity of industrial facilities located in forested areas. The present work shows the possibility of igniting forest fuels by a group of relatively small sources of local heating. As a practical application, it is proposed to use such an ignition mechanism during the prescribed burning of forest fuels in order to clean coniferous and mixed stands, including partially damaged ones. It is known that local damage to the stands lead to their fragmentation and subsequent death.

An experimental setup has been developed to simulate the processes of ignition of forest fuel by crystallizing metal particles that are formed during electrical welding. In the scientific periodicals no direct analogues of this installation were found. There are works on the creation of generators of heated and / or burning particles formed during the destruction of wood or building materials. It should be noted that in the vicinity of industrial facilities, the use of devices developed on the basis of this method is safer compared to devices based on liquid or gaseous fuels. The physical mechanism of ignition of forest fuel during the fallout of a group of relatively small crystallizable particles is revealed, which is of a probabilistic nature and provides recommendations on the mathematical modeling of the processes of generation, transport and the impact of such particles on a layer of forest fuel. The results can be used to verify future mathematical models. In turn, such mathematical models can be used in new generation forest fire danger prediction systems (Baranovskiy, 2020; CWFIS, 2020; WFAS, 2020; EFFIS, 2020; ISDM-Rosleskhoz, 2020).

ACKNOWLEDGMENT

This work was implemented with the financial support of the Russian Foundation for Basic Research. [Project N 17-29-05093].

REFERENCES

Babrauskas, V. (2003). *Ignition handbook: principles and applications to fire safety engineering, fire investigation, risk management, and forensic science*. Fire Science Publishers.

Baranovskii, N. V., & Zakharevich, A. V. (2016). Conditions and Characteristics of the Ignition of a Typical Vegetable Combustible Material by a Local Energy Source. *Journal of Engineering Physics and Thermophysics.*, *89*(6), 1538–1544. doi:10.100710891-016-1524-3

Baranovskiy, N. V. (2018). Mathematical foundations of the system of assimilation of data on the level of anthropogenic load in forecasting forest fires. In *Materials and reports of the VI All-Russian Conference "Safety and Monitoring of Technogenic and Natural Systems"*. Krasnoyarsk: SFU.

Baranovskiy, N. V. (2020). Mathematical Simulation of Anthropogenic Load on Forested Territories for Point Source. In N. Baranovskiy (Ed.), *Predicting, Monitoring, and Assessing Forest Fire Dangers and Risks* (pp. 64–88). IGI Global. doi:10.4018/978-1-7998-1867-0.ch003

Baranovskiy, N. V., & Kuznetsov, G. V. (2017). *Forest fire occurrences and ecological impact prediction*. Publishing House of the Siberian Branch of the Russian Academy of Science. doi:10.15372/FOREST2017BNV

Bulba, E. E., & Malinovsky, A. A. (2015). Experimental Research of Moisture Evaporation Process from Biomass in a Drying Chamber. *MATEC Web of Conferences, 23*. 10.1051/matecconf/20152301053

Canadian Wildland Fire Information System. (n.d.). Access: https://cwfis.cfs.nrcan.gc.ca/home

Carlslaw, S., & Jaeger, J. C. (1984). *Conduction of heat in solids*. Oxford University Press.

Devisilov, V. A. (2010). Russian forest is asking the protection! [in Russian]. *Safety in Technosphere*, (6), 3–7.

Di Blasi, C. (1993). Modelling and simulation of combustion processes of charring and non-charring solid fuels. *Progress in Energy and Combustion Science, 19*(1), 71–104. doi:10.1016/0360-1285(93)90022-7

European Forest Fire Information System. (n.d.). Access: http://effis.jrc.ec.europa.eu

Fernandez-Pello, C. A. (2017). Wildland fire spot ignition by sparks and firebrands. *Fire Safety Journal, 91*, 2–10. doi:10.1016/j.firesaf.2017.04.040

Ganteaume, A., Jappiot, M., Curt, T., Lampin, C., & Borgniet, L. (2014). Flammability of litter sampled according to two different methods: Comparison of results in laboratory experiments. *International Journal of Wildland Fire, 23*(8), 1061–1075. doi:10.1071/WF13045

Grishin, A. M. (1997). *Mathematical modeling of forest fire and new methods of fighting them*. Publishing House of the Tomsk State University.

Grishin, A. M., & Baranovskij, N. V. (2003). Comparative analysis of simple models of drying of the layer of forest combustibles, including the data of experiments and natural observations. *Inzhenerno-Fizicheskii Zhurnal., 76*(5), 166–169. doi:10.1023/B:JOEP.0000003234.63878.7b

Hadden, R. M., Scott, S., Lautenberger, Ch., & Fernandez-Pello, C. (2011). Ignition of Combustible Fuel Beds by Hot Particles: An Experimental and Theoretical Study. *Fire Technology, 47*(2), 341–355. doi:10.100710694-010-0181-x

Kortas, S., Mindykowski, P., Consalvi, J. L., Mhiri, H., & Porterie, B. (2009). Experimental validation of a numerical model for the transport of firebrands. *Fire Safety Journal, 44*(8), 1095–1102. doi:10.1016/j.firesaf.2009.08.001

Kuznetsov, G. V., & Sheremet, M. A. (2006). Conjugate heat transfer in a closed domain with a locally lumped heat-release source. *Journal of Engineering Physics and Thermophysics.*, *79*(1), 57–64. doi:10.100710891-006-0066-5

Liu, H., Wang, C., Zhao, W., Yang, S., & Hou, X. (2018). Pyrolysis characteristics and kinetic modeling of Artemisia apiacea by thermogravimetric analysis. *Journal of Thermal Analysis and Calorimetry*, *131*(2), 1783–1792. doi:10.100710973-017-6599-3

Manzello, S. L., Cleary, T. G., Shields, J. R., & Yang, J. C. (2006a). Ignition of mulch and grasses by firebrands in wildland-urban interface fires. *International Journal of Wildland Fire*, *15*(3), 427–431. doi:10.1071/WF06031

Manzello, S. L., Cleary, T. G., Shields, J. R., & Yang, J. C. (2006b). On the ignition of fuel beds by firebrands. *Fire and Materials*, *30*(1), 77–87. doi:10.1002/fam.901

Manzello, S. L., Shields, J. R., Cleary, T. G., Maranghides, A., Mell, W. E., Yang, J. C., Hayashi, Y., Nii, D., & Kurita, T. (2008). On the development and characterization of a firebrand generator. *Fire Safety Journal*, *43*(4), 258–268. doi:10.1016/j.firesaf.2007.10.001

Matvienko, O. V., Fil'kov, A. I., & Grishin, A. M. (2016). Computational Investigation of the Transport of Burning Particles. *Journal of Engineering Physics and Thermophysics.*, *89*(5), 1315–1324. doi:10.100710891-016-1494-5

Nigmatulin, R. I. (1987). *The dynamics of multiphase media. Part I.* Nauka. (in Russian)

Pakshirajan, R. P. (2013) Probability Theory. Springer. doi:10.1007/978-93-86279-54-5

Pleasance, G. E., & Hart, J. A. (1977). *An examination of particles from conductors clashings possible source of bushfire ignition*. State Electricity Commission of Victoria (SEC), Victoria, Australia, Research and Development Department, Report FM-1.

Podolskaya, A.S., Ershov, D.V., & Shulyak, P.P. (2011). Application of the method for assessing the probability of occurrence of forest fires in ISDM-Rosleskhoz. *Modern Problems of Remote Sensing of the Earth From Space, 8*, 118 - 126. (in Russian)

Rowntree, G. W. G., & Stokes, A. D. (1994). Fire ignition by aluminum particles of controlled size. *Journal of Electrical and Electronics Engineering, Australia, 14*, 117–123.

Stokes, A. D. (1990). Fire ignition by copper particles of controlled size. *Journal of Electrical and Electronics Engineering, Australia, 10*, 188–194.

Suzuki, S., & Manzello, S. L. (2016). Firebrand production from building components fitted with siding treatments. *Fire Safety Journal*, *80*, 64–70. doi:10.1016/j.firesaf.2016.01.004 PMID:27114643

Tohidi, A., Kaye, N., & Bridges, W. (2015). Statistical description of firebrand size and shape distribution from coniferous trees for use in Metropolis Monte Carlo simulations of firebrand flight distance. *Fire Safety Journal*, *77*, 21–35. doi:10.1016/j.firesaf.2015.07.008

Tohidi, A., & Kaye, N. B. (2017). Stochastic modeling of firebrand shower scenarios. *Fire Safety Journal*, *91*, 91–102. doi:10.1016/j.firesaf.2017.04.039

Vilyunov, V. N. (1984). *Theory of ignition of condensed matter*. Science, Sib. Department. (in Russian)

Wadhwani, R., Sutherland, D., Ooi, A., Moinuddin, K., & Thorpe, G. (2017). Verification of a Lagrangian particle model for short-range firebrand. *Fire Safety Journal, 91*, 776–783. doi:10.1016/j.firesaf.2017.03.019

WFAS Wildland Fire Assessment System. (n.d.). Access: http://www.wfas.net

Zakharevich, A. V., Kuznetsov, V. T., Kuznetsov, G. V., & Maksimov, V. I. (2008). Ignition of model composite propellants by a single particle heated to high temperatures. *Combustion, Explosion, and Shock Waves*, *44*(5), 543–546. doi:10.100710573-008-0083-0

Zakharevich, A. V., & Zygin, M. S. (2016). Experimental Study of the Influence of the Structure of Local Energy Sources on the Conditions and Characteristics of the Ignition of Liquid Fuels. *Journal of Engineering Physics and Thermophysics.*, *89*(6), 1522–1526. doi:10.100710891-016-1522-5

ADDITIONAL READING

Antonov, D. V., Volkov, R. S., Voitkov, I. S., Zhdanova, A. O., & Kuznetsov, G. V. (2018). Influence of Special Additives in a Water Aerosol on the Suppression of a Forest Fire with it. *Journal of Engineering Physics and Thermophysics.*, *91*(5), 1250–1259. doi:10.100710891-018-1855-3

Baranovskiy, N. V., Sherstnyov, V. S., Yankovich, E. P., Engel, M. V., & Belov, V. V. (2016) Focused sunlight factor of forest fire danger assessment using Web-GIS and RS technologies. Proceedings of SPIE - The International Society for Optical Engineering, 9688, article 968823, 6 P. 10.1117/12.2240378

Baranovskiy, N. V., & Zakharevich, A. V. (2015). *Experimental research of grassy rags ignition by heated up to high temperatures carbon particle*. Multifunctional Materials and Modeling., doi:10.1201/b18552

Chen, T., Zhang, J., & Wu, J. (2016). Kinetic and energy production analysis of pyrolysis of lignocellulosic biomass using a three-parallel Gaussian reaction model. *Bioresource Technology*, *211*, 502–508. doi:10.1016/j.biortech.2016.03.091 PMID:27035484

Goman, P. N. (2020). The Dynamics of Surface Forest Fire and Forest Fuel Ignition Under the Heat Radiation From the Fire Line. In N. Baranovskiy (Ed.), *Predicting, Monitoring, and Assessing Forest Fire Dangers and Risks* (pp. 1–47). IGI Global., doi:10.4018/978-1-7998-1867-0.ch001

Grishin, A. M., & Shipulina, O. V. (2002). Mathematical model for spread of crown fires in homogeneous forests and along openings. *Combustion, Explosion, and Shock Waves*, *38*(6), 622–632. doi:10.1023/A:1021127924913

Hameed, S., Sharma, A., Pareek, V., Wu, H., & Yu, Y. (2019). A review on biomass pyrolysis models: Kinetic, network and mechanistic models. *Biomass and Bioenergy*, *123*, 104–122. doi:10.1016/j.biombioe.2019.02.008

Kansa, E., Perlee, H., & Chaiken, R. (1977). Mathematical model of wood pyrolysis. *Combustion and Flame*, *29*, 311–324. doi:10.1016/0010-2180(77)90121-3

Kuznetsov, G. V., & Rudzinskii, V. P. (2000). High-temperature heat and mass transfer in a layer of coke of heat-shielding materials. *High Temperature*, *38*(4), 629–635. doi:10.1007/BF02755811

Kuznetsov, G. V., & Strizhak, P. A. (2009a). Numerical Solution of the Problem of Ignition of a Combustible Liquid by a Single Hot Particle. *Combustion, Explosion, and Shock Waves*, *45*(5), 543–550. doi:10.100710573-009-0066-9

Kuznetsov, G. V., & Strizhak, P. A. (2009b). Simulation of the ignition of a liquid fuel with a hot particle. *Russian Journal of Physical Chemistry B*, *3*(3), 441–447. doi:10.1134/S199079310903018X

Thomas, J. C., Hadden, R. M., & Simeoni, A. (2017). Experimental investigation of the impact of oxygen flux on the burning dynamics of forest fuel bed. *Fire Safety Journal*, *91*, 855–863. doi:10.1016/j.firesaf.2017.03.086

KEY TERMS AND DEFINITIONS

Anthropogenic Load: Different human activities on forested territories lead to forest fire occurrence and characterized by presence of fire sources.

Forest Fire: Uncontrolled aerothermochemical phenomenon characterized by step-by-step mechanism which includes following stages: inert heating, moisture evaporation, high temperature terpens evaporation, dry organic matter pyrolysis, flammable combustion, and smoldering.

Forest Fuel: It can be considered like dead and live forest fuel. Main types of forest fuel which can be involved in combustion during forest fire: ground forest fuel (needles, leaves and dry grass, small branches) and crown forest fuel (needles, small branches).

Ignition: Inflammation of forest fuel caused by definite source of high temperature or energy.

Ignition Delay: Time before flame flash after forest fuel heating.

Prediction: Under the prediction of forest fires is the calculation of the parameters of forest fire danger with a certain projection in advance in order to have enough time to anticipate an emergency. The calculation in this case is carried out in a mode ahead of the real time of the development of the catastrophe - the occurrence of a forest fire.

Chapter 11
Physical Modeling of Forest Fuel Ignition by the Focused Sunlight

ABSTRACT

This chapter discusses the issue of experimental modeling of the forest fuel layer ignition by the focused solar radiation. The main types of glass containers that can be found in forested areas are considered. Two experimental facilities are presented for simulating smoldering and ignition of forest fuels under the influence of focused solar radiation. Samples of forest fuels typical for the Siberian regions (Tomsk region and Republic of Buryatia) are investigated. Data on the physical mechanism and ignition delay of forest fuel are presented.

INTRODUCTION

At present, both natural (California Department of Forestry and Fire Protection, 2020; Country Fire Authority, 2020) and anthropogenic (Devisilov, 2010; Baranovskiy and Kuznetsov, 2017)] causes of fires in natural landscapes are known. Mathematical models have been developed to describe the ignition of forest fuels under the influence of single heated particles and an electric discharge under thunderstorm conditions, as well as deterministic-probabilistic models for taking these factors into account when predicting a forest fire danger (Baranovskiy, 2017). At the same time, the probability of a forest fire as a result of the action of focused solar radiation is quite high (Babrauskas, 2003). The capacity of the energy concentrator can be either containers with liquid or glass fragments, or large resin drops of coniferous trees. According to fire safety rules, it is forbidden to scatter glass containers in forests, as these objects can become sources of ignition (Fire Safety Rules, 2007). But so far, the results of theoretical or experimental studies have not been published justifying the conditions for the occurrence of ignitions of forest fuels when exposed to focused solar radiation. Of interest is the experimental study of the forest fuel ignition by the focused solar radiation.

One can distinguish several ignition modes of combustible materials of plant origin (Babrauskas, 2007), which differ from similar ignition modes of high-energy materials (Kuznetsov et al., 2004a; 2004b) and substances (Kuznetsov and Strizhak, 2009a; 2009b) used in special-purpose power plants.

DOI: 10.4018/978-1-7998-7250-4.ch011

The most typical mode of ignition with the formation of a flame in literature is denoted by the term ignition. According to (Vilyunov, 1984), under such a regime, a leading exothermic chemical reaction occurring in the gas phase can be distinguished. The second type of ignition is characteristic of materials when they turn into a red-hot state (indicated in literature by the keyword glowing ignition). In this state, for example, ignition of cellulose-containing materials (wood) by a radiation heat flux can occur (Babrauskas, 2002). When igniting porous materials, a third ignition mode can be realized, accompanied by smoldering (in the literature, the term of smoldering ignition corresponds to this type) (Hadden et al. 2011). Both the second and the third ignition modes are characterized by the localization of the leading exothermic reaction in the structure of the combustible material. For these modes, heterogeneous reactions are typical, both on the surface and in the interstitial space of the material (Vilyunov, 1984; Samuilov et al., 2004).

Correspondingly, the material is characterized by a developed pore space (Moskalev and Shitov, 2007) or an initially monolithic structure (then the formation of a porous structure occurs similar to the mechanisms that are realized in surface layers of heat-protective materials heated to high temperatures (Kuznetsov and Rudzinskii, 2000; Strakhov et al., 2001). Layers of forest fuels, as a rule, are developed porous structures in which energy transfer is possible not only due to thermal conductivity, but also due to radiation and convection (Kuznetsov and Rudzinskii, 2000; Strakhov et al., 2001). A theoretical analysis of such processes under conditions of time-varying porosity of the heated layer is too complicated a task. Therefore, it is of interest to experimentally study the ignition regimes of forest fuels as a result of the action of focused solar radiation.

The purpose of the work is an experimental study of the ignition regimes of forest fuel by the flux of focused solar radiation in the regime of smoldering and flame ignition.

FIRE SOURCES

Previously, it was proposed to classify all sources of anthropogenic load on linear and point ones (Baranovskiy, 2018; 2020). Linear sources include objects such as railways, highways, river beds. Different technological objects and settlements located on a forested area can be classified as point ones. It should be noted that in the present work, it was proposed to study only linear sources of anthropogenic load, since settlements and technological objects are original points of various linear sources. Studies were conducted in various places of the Timiryazevskiy forestry of Tomsk region (The project…, 1990).

Source 1. Railway. A section of the forested area on the distance from Tomsk to Predtechensk was investigated. Field observations were conducted between two settlements: settlement of geologists (data for georeferencing N56^{0}25'802 and E85^{0}01'329) and the village Prostorniy (N56^{0}25'263 and E85^{0}01'277). It should be noted that the data of GPS-navigators were used to estimate the coordinates of control points for field observations (Weilin et al., 2000). It is necessary to briefly describe the forest growing conditions in this area — the forest is mainly birch, age 50–60 years, deciduous trees litter predominate in the ground cover (Panevin, 2006).

Source 2. Riverbed. The left bank of the Tom River was explored in the vicinity of the village of Takhtamyshevo (georeferencing data N56^{0}22'119 and E84^{0}52'929). Field studies were carried out both on the shore and in the coastal strip of the forest. The flora of the coast is represented by various types of grassy vegetation (dry and yellowed in the autumn and spring). In the coastal forest, among the tree plantations, the vast majority are represented by pines aged 60–80 years, and pine coniferous litter predominates in the ground cover (Panevin, 2006).

Source 3. Highway. A section of the forest along the highway was investigated from the Tomsk to Novosibirsk to the village Kislovka (georeferencing data for control point N56^{0}24'173 and E84^{0}52'792). The forest stand is represented by pine plantations at the age of 50 - 80 years old; in the ground cover, pine coniferous needles are prevailing (Panevin, 2006). In the vicinity of localization of anthropogenic load, burns were recorded after passing the front of a surface forest fire (almost all the trees in this forest area have charred bark at a height of one and a half meters, Fig. 1).

The following technique is proposed. In the vicinity of the reference points of each typical linear object, a forest-covered area is examined for the presence of solar radiation concentrators (glass containers and their fragments). It should be noted right away that during the study of the territory, a negligible number of solar energy concentrators, represented by fragments of glass containers (less than 1%), was revealed. Therefore, these data are excluded from the analysis. Survey routes were laid perpendicular to a linear source of anthropogenic load. Data for all routes (the number of glass objects) were summarized and the arithmetic average for all routes was determined. The location on the route and the color of the glass solar energy concentrator were recorded, since fire danger is most likely represented by transparent, colorless glass containers (Landsberg, 1976). Then, the share of each type of containers in their total number was calculated. This value can be used to assess the probability of a solar energy concentrator in a forested area:

$$P_{con} = \frac{N_{tr}}{N_{tot}}, \tag{1}$$

where P_{con} is the probability of having a solar energy concentrator, N_{tr} is the number of colorless glass containers, N_{tot} is the total number of glass containers of all colors. Moreover, the sum of glass containers of all colors is equal to N_{tot}, since the color of the bottle is an incompatible event from the standpoint of probability theory.

In the case of a superposition of two sources of anthropogenic load in a controlled area, the probability addition theorem should be used, moreover, for joint events (Pakshirajan, 2013). According to probability theory, the formula for assessing the total effect of two sources of anthropogenic load will take the form:

$$P = P(S_1) + P(S_2) - P(S_1S_2), \tag{2}$$

where S_1 is the first linear source, S_2 is the second linear source, and S_1S_2 is the product of events.

In principle, the case of the impact of three linear sources of anthropogenic load on the forested area is also possible. Then, the calculation of the probability of the presence of a solar energy concentrator will also be determined by the theorem on the addition of probabilities, but for three joint events (Pakshirajan, 2013).

As mentioned above, in the forest along the highway to the village. Kislovka recorded burns formed after surface forest fires in the vicinity of the accumulation of localization of solar energy concentrators. At present, forest fire registration books do not include such a factor of occurrence of forest fuel ignition as the effect of focused solar radiation. The analysis of statistical data allows us to conclude that such fires are most likely classified in the statistics as the "unknown cause". In this regard, it is necessary to

Figure 1. Photo of burning after passing the front of the surface forest fire

develop new systems for classifying and assessing the causes of forest fires in controlled forested areas. And also to develop new methods of fire-technical examination of the centers of forest fires.

The results of a quantitative analysis of typical forested areas for the presence of solar energy concentrators are presented in Fig. 2. As a result of field observations, glass containers of three colors were recorded: transparent (colorless), green, and brown (Fig. 3). It must be assumed that fire incidents with colorless containers should lead to fires. Nevertheless, the results of assessing the probability of presence and colored glass containers are presented.

In the scientific literature to date, there have been virtually no studies on the occurrence of forest fires as a result of exposure to focused solar radiation. However, the data on the amplification factor of natural solar radiation for various glass objects are presented in (Babrauskas, 2003). So for glass bottles this value lies in the range from 12 to 22 times (in moderate latitudes, taking into account the attenuation of radiation by the Earth's atmosphere, the amount of natural solar radiation flux is about 1 kW/m^2 [6]). At the same time, it was previously theoretically shown that when the heat flux of focused solar radiation is 15 kW/m^2 or more, ignition of a forest fuel layer is possible. As a result, it can be stated that the presence of solar energy concentrators represents a real fire danger for forests.

Separately, it should be noted that large drops of resin were observed on conifers over the age of 100 years. However, in this article, this question is excluded from consideration, since additional studies are required on the effect of such objects on the concentration of solar energy.

Figure 2. The results of a quantitative assessment of solar energy concentrators at various control points: a) along the railway near the village Prostorniy, b) along the railway near the village settlement of geologist, c) along the Tom River, d) along the highway to the village Kislovka

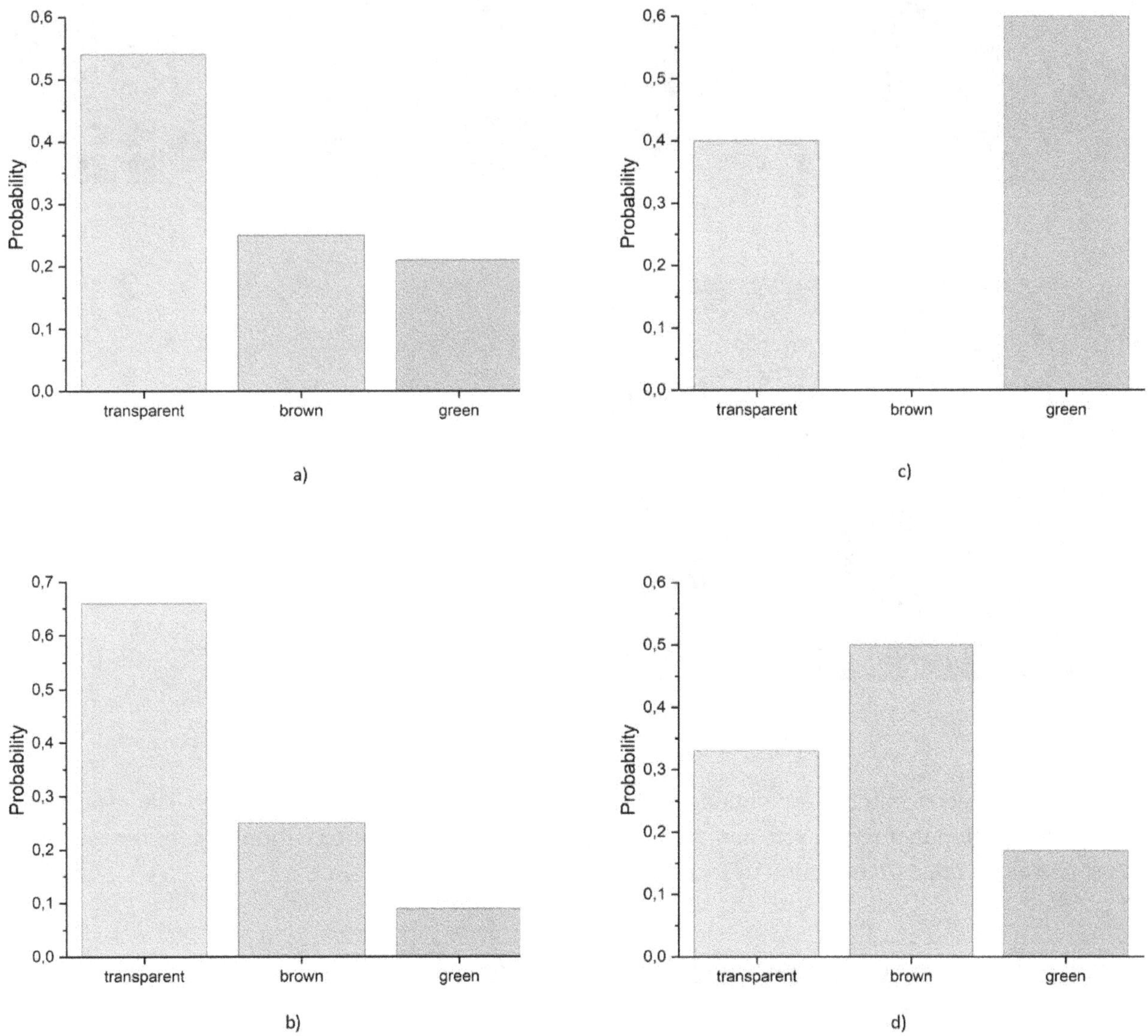

Thus, this article presents the generalized results of field studies of typical territories of the Timiryazevskiy forestry for the presence of solar energy concentrators. Three types of linear sources of anthropogenic load are considered: a railway, a river, and a highway. The probability values of the presence of

Figure 3. Photographs of typical solar energy concentrators: a) in deciduous litter; b) in a grass; c) in coniferous litter

a solar radiation concentrator are obtained using the provisions of probability theory. The results will be used to develop thematic layers and algorithms in the geographic information system for assessing forest fire danger due to the action of focused solar radiation.

SMOLDERING CAUSED BY FOCUSED SUNLIGHT

For the experiments, an experimental setup was made, the circuit diagram of which is shown in Fig. 4. The installation elements are mounted on a bed 1. A tripod 2 provides mounting of the manipulator 3 with an optical lens 4. Using the manipulator 3, the lens 4 focuses the solar radiation on the sample 7, fixed by a tripod 6 and holder 5. The manipulator 3 allows you to receive spots of focused solar radiation of various sizes and, accordingly different intensities (lenses, as a rule, form a radiant heat flux of about 10^4 W/m^2). A system consisting of a holder 5, a tube 8 and a suction device 9 mounted on a tripod

10 is designed to purge air through a porous heated layer of forest combustible material. To record the dynamics of forest fuel ignition during the experiment, a video recording of the processes in the surface layer of a sample of forest fuel was carried out .

At various times of the solar period of the day, a series of 5–7 experiments was carried out to determine the t_{ind} of the process of forest fuel ignition. Under the induction period t_{ind}, in the analysis of the ignition process in the smoldering mode, the time was taken sufficient for the further propagation of combustion in the smoldering mode along a sample of forest fuel without further supplying the energy of focused solar radiation. The averaged t_{ind} values and standard deviations were calculated. Then confidence intervals were determined with a confidence probability of P=0.95 (Pakshirajan, 2013).

Figure 4. Schematic diagram of the experimental setup: 1 is the bed, 2 is the tripod No. 1, 3 is the manipulator, 4 is the optical lens, 5 is the sample holder of forest fuel, 6 is the tripod No. 2, 7 is the sample of forest fuel, 8 is the air purge tube, 9 is the suction device, 10 is the suction tripod

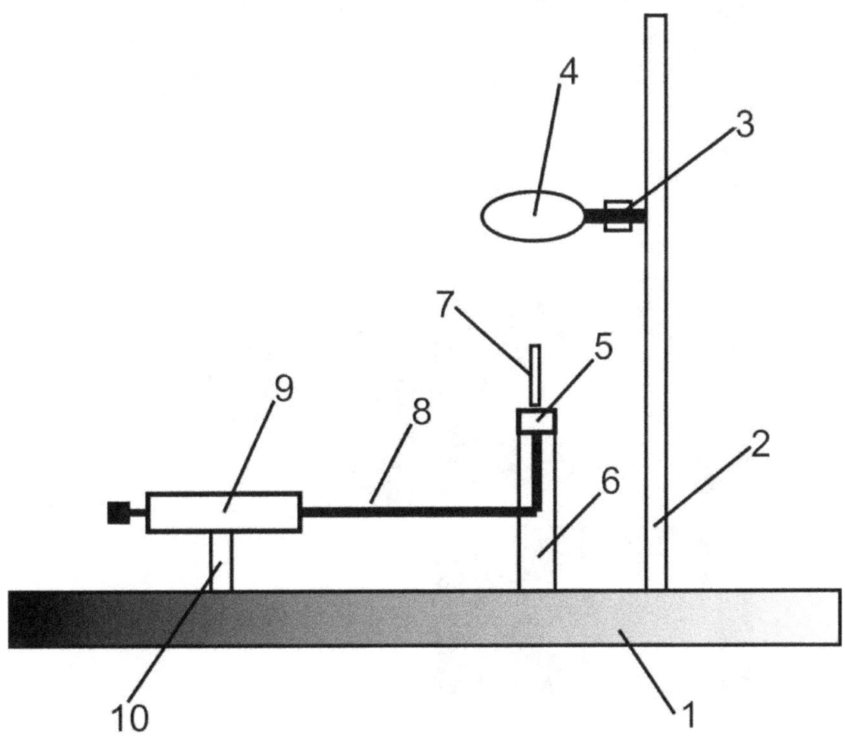

The object of the study was a sample of vegetable combustible material in the form of a cylinder. As such samples, crushed forest fuel was used. This allowed us to simulate forest fuel samples of various diameters. Cylinder briquettes of the following sizes were used:

1. thickness 10 cm, diameter 5 mm;
2. thickness 8.5 cm, diameter 7 mm;
3. thickness 6 cm, diameter 8 mm;
4. thickness 9.5 cm, diameter 12 mm.

Figure 5. Typical video footage of the ignition process in smoldering mode: a) t=0 s is the initial view of the sample; b) t=2 s is the inert heating by focused solar radiation; c) t=10 s is the formation of gaseous pyrolysis products (thermal decomposition of the material); d) t=12 s is the transition of the process into smoldering mode; e) t=15 s is the formation of smoldering focus; f) t=17 s is the formation of a focus of decay along the entire surface of the sample; g) t=19 s is the formation of a smoldering wave during air filtration through a sample (smoldering combustion)

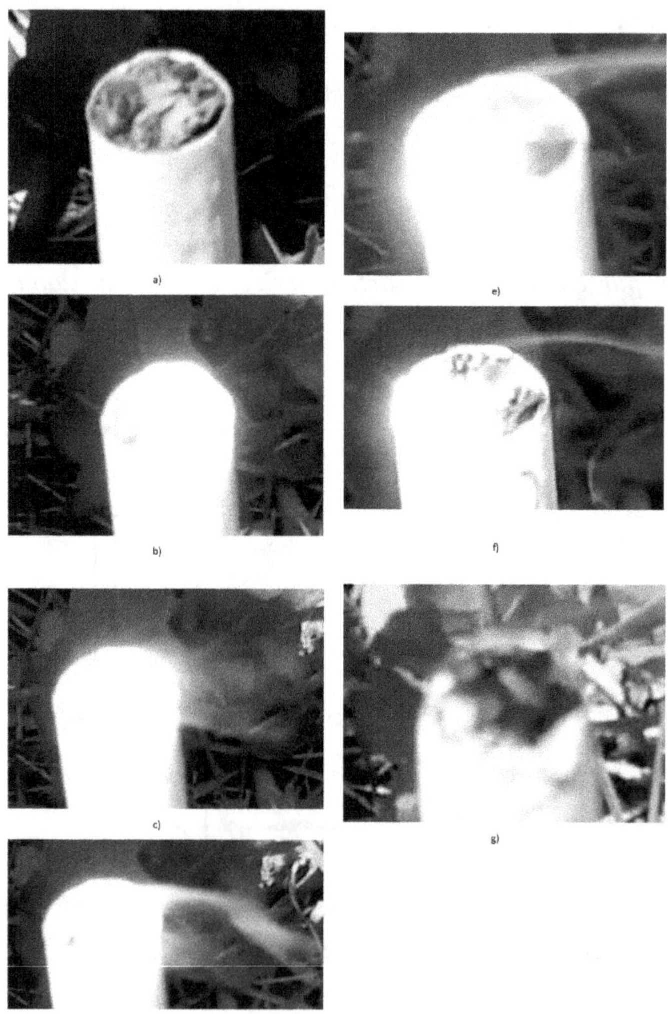

Samples of forest fuel were preliminarily aged in an oven to exclude the possibility of their moisture saturation during storage in unforeseen conditions. A graduation was applied on the outside of the sample with a division value of 1 mm to track the depth of thermal decomposition of the forest fuel during air filtration through the sample.

The physical mechanism of the decay of forest fuel is established. The initial short stage is the inert heating of a sample of forest fuel under the action of focused solar radiation, concentrated by an optical biconvex lens with a diameter of 10 cm (the size of the focused spot of exposure is 2 mm). Then, in the central region of the forest fuel sample, thermal decomposition of the material began. The center of pyrolysis increases in size under the influence of air currents. Over time, the center of thermal decomposition goes into smoldering mode. After this moment, the smoldering process proceeds in the porous layer in the absence of exposure to a focused flux of solar radiation. After the formation of a focus of smoldering, comparable in size to the diameter of the sample of forest fuel, air was pumped through the sample at the facility (Fig. 1). Intensive blowing of an oxidizing agent into a heated region in a layer

Figure 6. t_{ind} of smoldering on the time of day. Samples: a) 10 cm thick and 5 mm in diameter; b) 8.5 cm thick and 7 mm in diameter; c) 6 cm thick and 8 mm in diameter; d) 9.5 cm thick and 12 mm in diameter. In the figure, the numbers indicate: 1 is the average values of the induction period with confidence intervals, 2 is the approximating straight line

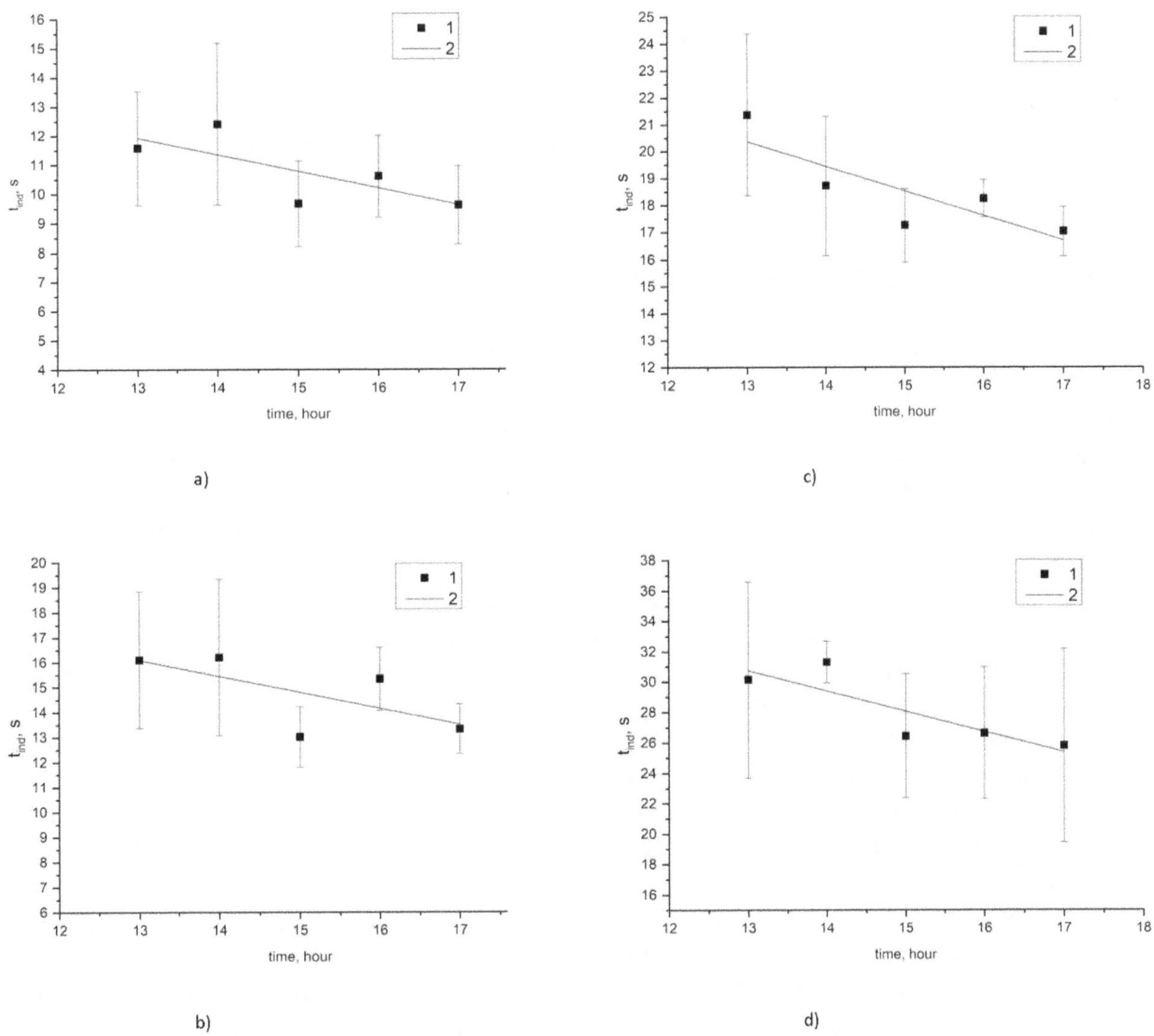

of forest fuel was simulated. The result of air filtration was the burning of a sample of forest fuel to a certain depth (smoldering combustion). Typical video frames of the process are presented in Fig. 5.

Figures 6.a - 6.d show the dependences of the induction period of the smoldering process on various weights of forest fuel on the time of day. The studies were carried out in the period from 13-00 to 17-00 hours of daytime. Common to all samples is the fact that the induction period decreased over time. According to meteorological information, during this period the ambient temperature grew from 25^0C to 31^0C. It has been established that the ambient temperature affects the smoldering processes of forest fuel.

An analysis of the results shows that the dependence of t_{ind} on the time of day can be described by a linear function. It should be noted that there is an analogy between the smoldering process of forest fuel and the ignition of forest fuel by the particle heated up to high temperatures (Baranovskiy and Zakharevich, 2016; 2018). As a result of the action of focused solar radiation by a biconvex lens, a burning out occurs in the layer of forest fuel, which can go into flame combustion or initiate ignition in the gas phase (Alpert et al., 2010; Bukhtoyarov et al., 2008). It can be concluded that the fire danger of focused solar radiation on a layer of forest fuel is a real fire danger.

An experimental study of the decay regime of forest fuel as a result of the action of focused solar radiation is presented. It is established that under the influence of a radiant heat flux focused by a lens with a twofold magnification, under certain conditions a foci of formation are formed. The heating region with a diameter of the order of 2.5 mm is characterized by a smoldering regime. When air is pumped through a sample of forest fuel, a stable combustion wave in a flameless mode (smoldering combustion) is formed.

IGNITION CAUSED BY FOCUSED SUNLIGHT

It is known (Magomedov, 1996), that the value of the solar constant used in solar technology q=1353 W/m^2 $\pm$ 1.5%. Recently, publications have appeared that indicate a more accurate value q = 1373 W/m^2 (Magomedov, 1996). Since the distance between the Earth and the Sun undergoes seasonal changes, the intensity of solar radiation incident on a single area also does not remain constant. It is known that almost all the energy emitted by the Sun falls on a narrow wavelength range in the visible and near infrared (in the range of 0.24–4 µm, 98% of the radiation energy is contained) (Magomedov, 1996). Part of the solar radiation is scattered in the atmosphere (Kabanov, 1997). In temperate latitudes, the heat flux of this radiation up to 1 kW/m^2 affects the Earth's surface (Babrauskas, 2003).

In the experiments, a concave-convex glass lens [25] with a diameter of 11 cm with a focused spot size of 3-3.5 mm in diameter was used. Experimental measurements of the heat flux of focused solar radiation concentrated by the specified lens were carried out. The measurement procedure is based on the optical method of determining the surface temperature T_p of a reference metal cylinder isolated on the lateral surface with characteristic dimensions comparable to the size of the lens exposure spot. Using an optical pyrometer, the surface temperature of this cylinder was measured in the absence of exposure to focused solar radiation, the value of which was subsequently used to set the initial conditions in the calculation process. At subsequent times, the flux of solar radiation focused by the lens acted on the surface of the metal cylinder. At the same time, the optical pyrometer recorded a change in time of its surface temperature. A typical dependence of the surface temperature on the heating time is shown in Fig. 7. After a series of measurements and averaging, the problem of thermal conductivity in a steel cylinder with boundary conditions of the second kind on a horizontal heated surface was solved.

Thermal Conductivity Equation:

$$\rho c \frac{\partial T}{\partial t} = \frac{\lambda}{r} \frac{\partial}{\partial r}\left(r \frac{\partial T}{\partial r} \right) + \lambda \frac{\partial^2 T}{\partial z^2}.$$

Thermal insulation conditions are set on the lateral and end boundaries of the cylinder. On the heating surface by solar radiation, a condition of the second kind:

$$\lambda \frac{\partial T}{\partial z} = q_{sun}.$$

As a result of the calculation cycle, in order to compare the conditions under which the best values of the calculated and measured T_P are achieved, it was found that the heat flux (q_{sun}) of the solar radiation focused by the lens, which affects the heating surface of the reference cylinder, is 17780±1293.5 W/m^2. The confidence interval for determining q_{sun} was calculated with a confidence probability of P=0.95 (Pakshirajan, 2013).

Figure 7. Time dependence of the surface temperature of a reference steel cylinder: 1 - experimentally determined values; 2 - calculated curve at q_{sun}=17780 W/m^2

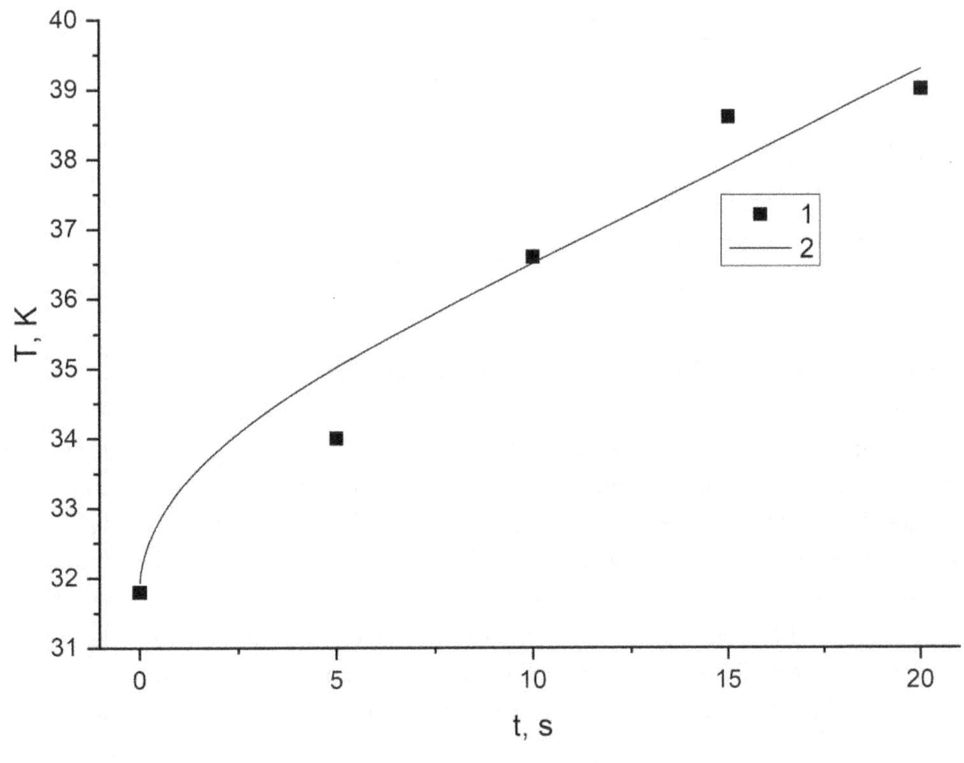

The object of the study was a sample of forest fuel formed from pine needles. The experiments were carried out with brown needles, partially decomposed, previously well dried. The dimensions of individual needles were 7–8 cm in the longitudinal and 0.7–1.2 mm in the transverse direction. Stacking density corresponded to stacking density in a real forest (Grishin et al., 2002). The ignition was initiated by solar radiation focused by a concave-convex lens with a diameter of 11 cm (glass, transparent). The size of the exposure spot is 3-3.5 mm in diameter.

Meteorological conditions on the days of the experiments:

10 June (time from 12-00 to 13-00), clear weather without clouds, air temperature 32°C, wind speed v=4 m/s, east wind, pressure 745 mm.

11 June (time from 12-00 to 14-00), clear weather without clouds, air temperature 29 – 32°C, wind speed v=5 m/s, east wind, pressure 743 mm.

Table 1. Typical results of experiments on ignition of pine needles

Time	T_{air}, °C	Wind speed, m/s	Inert heating, s	Ignition	Spread	Ignition probability	
						Modes 1,2,3	Mode 3
12-00	32	1	5	No	No	0,066	1,0
12-25	32	4	5	Yes	Yes		
12-45	32	2	6	No	No		
12-00	29	2	6,5	No	No	0,667	1,0
12-15	29	1	6	No	No		
12-30	30	0	6	No	No		
12-40	30	5	5,5	Yes	Yes		
12-56	31	4	5,5	Yes	Yes		
13-10	31	0	6	No	No	0,700	1,0
13-15	31	3,5	5,5	Yes	Yes		
13-25	32	4	5	Yes	Yes		
13-35	32	1	6	No	No		
13-50	32	5	5	Yes	Yes		

Table 1 presents the characteristic (typical) results of experiments on the ignition of pine needles.

It was established from the process video that the first short-lived stage is the inert heating of a layer of forest fuels. Then follows the pyrolysis stage of the needles. A necessary condition for ignition is the formation of a thermal decomposition zone on the heating surface of at least 1 cm in diameter. At a wind speed of less than 1 m/s of forest fuels in the impact area and in a small neighborhood around the spot of focused solar radiation, it thermally decomposes in flameless mode (heterogeneous oxidation reactions of coke occur, gaseous pyrolysis products diffuse into the gas phase region and are carried away by convective flows). In the event of a gust of wind (and, as a result, of intense blowing of the oxidizing agent into the region heated to high temperatures), ignition in the gas phase is possible with the formation of a flame microflare. Over time, the flame torch grows and the flame spreads steadily over a layer of forest fuel. The ignition process is random in nature at low wind speeds and strongly depends on the influx of oxidizing agent as a result of the influence of wind on the pyrolysis focus of forest fuel (forced convection of air masses). Typical frames of the ignition process are presented in Fig. 8.

All experimental results can be divided into three groups: a) in the wind at a speed of up to 3 m/s; b) meteorological conditions, which are characterized by the presence of wind at a speed of 3 m/s to 4 m/s; c) wind speed of 4 m/s or more. In the first case, there is the formation of a smoldering center, which over time ceases to increase in size and fades. The second and third regimes are also characterized by an increase in the focus of decay over time. The ignition process at a wind speed of more than 4 m/s is stable. In the range of variation of v from 3 to 4 m/s, ignition is possible, but the process is random. This mode is transient. Under conditions of intense forced convection, an oxidant inflows to the pyrolysis surface and transitions to the gas-phase ignition mode. To achieve ignition conditions, a smoldering lesion of at least 1 cm in diameter must be formed. With smaller sizes of the center of smoldering fires are not obtained.

Figure 8. Typical video footage of the forest fuel ignition (pine needles layer) with focused solar radiation: a) t=0 s is the initial form of the layer, b) t=7 s is the thermal decomposition, c) t=30 s is the formation of the smoldering focus, d) t=40 s is the increase in the size of the focus of decay, e) t=1 min 15 s is the ignition in the gas phase, f) t=1 min 23 s is the formation of a stable flame, g) t=1 min 30 s is the spread of combustion front on layer

Based on the obtained experimental data, the following conclusion can be made. In blown forests or in clearings free of trees, the danger of fires due to the action of focused solar radiation in conditions of even moderate wind has a probability of ignition close to 1. The profile of wind speed in the forest is logarithmic (Grishin, 1997) and, accordingly, at an elevation it will be significant compared to the underlying surface. As a result, the action of sources focusing solar radiation on hills is a more dangerous factor. The difference in the probability of ignition of pine needles located at the level of the earth's surface and at a height of 25 cm can reach 10 times the magnitude. The microrelief features of the area are also a factor in forest fire danger under the action of focused solar radiation.

A comparative analysis of the results of the forest fuel ignition by focused solar radiation and radiant heat flux from a radiation panel (Grishin et al., 2002) shows that in the latter case, ignition occurs at significantly higher heat fluxes. Therefore, experiments with a radiation panel do not correspond to the real mechanism of ignition of forest combustible materials by focused solar radiation. Probably, the emission spectrum and wavelength are important in these processes. In the future, it is advisable to build generalized mathematical models that take into account the spectral composition of radiation.

An analysis of the results of experimental studies also shows that the mechanism of forest fuel ignition by solar radiation is significantly different from the mechanism of forest fuel ignition and other condensed (liquid and solid) substances ignition by a local heating source (e.g., single particle). When a solar energy source is exposed to forest fuel, the supply of an oxidizing agent (due to convection) to the zone heated to significantly lower temperatures plays a much larger role compared to the ignition conditions of condensed substances by single "hot" particles. Under these conditions, it is possible to analyze the ignition processes using a natural and mixed convection models (Kuznetsov and Sheremet, 2006a; 2006b; 2008), developed for a mode of sufficiently moderate gas velocities at elevated temperatures.

CONCLUSION

The generalized results of field studies of typical territories of the Timiryazevskiy forestry for the presence of solar energy concentrators are presented. Three types of linear sources of anthropogenic load are considered: a railway, a river, and a highway. The probability values of the presence of a solar radiation concentrator are obtained using the provisions of probability theory. The results will be used to develop thematic layers and algorithms in geographic information systems for assessing forest fire danger due to the action of focused solar radiation for Tomsk region and Republic of Buryatia (Baranovskiy, 2019; Yankovich et al., 2019). An experimental study of the decay regime of forest fuel as a result of the action of focused solar radiation is presented. It is established that under the influence of a radiant heat flux focused by a lens with a twofold magnification, under certain conditions a foci of formation are formed. The heating region with a diameter of the order of 2.5 mm is characterized by a smoldering regime. When air is pumped through a sample of forest fuel, a stable combustion wave in a flameless mode (smoldering combustion) is formed. An analysis of the results of experimental studies also shows that the mechanism of forest fuel ignition by the solar radiation is significantly different from the mechanism of forest fuel ignition and other condensed (liquid and solid) substances ignition by a local heating source like single hot particle.

ACKNOWLEDGMENT

This work was implemented with the financial support of the Russian Foundation for Basic Research. [Project N 17-29-05093].

REFERENCES

Alpert, H. R., O'Connor, R. J., Spalletta, R., & Connolly, G. N. (2010). Recent Advances in Cigarette Ignition Propensity Research and Development. *Fire Technology*, *46*(2), 275–289. doi:10.100710694-008-0070-8 PMID:20495669

Babrauskas, V. (2002). Ignition of wood: A review of the state of the art. *Journal of Fire Protection Engineering*, *12*(3), 163–189. doi:10.1177/10423910260620482

Babrauskas, V. (2003). *Ignition handbook: principles and applications to fire safety engineering, fire investigation, risk management, and forensic science*. Fire Science Publishers.

Babrauskas, V. (2007). Ignition: A century of research and an assessment of our current status. *Journal of Fire Protection Engineering*, *17*(3), 165–183. doi:10.1177/1042391507059434

Baranovskii, N. V., & Zakharevich, A. V. (2016). Conditions and Characteristics of the Ignition of a Typical Vegetable Combustible Material by a Local Energy Source. *Journal of Engineering Physics and Thermophysics.*, *89*(6), 1538–1544. doi:10.100710891-016-1524-3

Baranovskiy, N. V. (2017). Forest fire danger assessment using SPMD-model of computation for massive parallel system. *International Review on Modeling and Simulations*, *10*(3), 193–201. doi:10.15866/iremos.v10i3.10570

Baranovskiy, N. V. (2018). Mathematical simulation of anthropogenic load from a linear source using partial differential equations in the context of the forest fires occurrence. *Advances in Differential Equations and Control Processes.*, *19*(3), 237–262. doi:10.17654/DE019030237

Baranovskiy, N. V. (2020). Mathematical Simulation of Anthropogenic Load on Forested Territories for Point Source. In N. Baranovskiy (Ed.), *Predicting, Monitoring, and Assessing Forest Fire Dangers and Risks* (pp. 64–88). IGI Global. doi:10.4018/978-1-7998-1867-0.ch003

Baranovskiy, N. V., & Kuznetsov, G. V. (2017). *Forest fire occurrences and ecological impact prediction*. Publishing House of the Siberian Branch of the Russian Academy of Science. doi:10.15372/FOREST2017BNV

Baranovskiy, N. V., & Zakharevich, A. V. (2018). Mechanical, thermophysical and aerothermochemical processes at forest fuel ignition caused by the heated to high temperatures single metal particle. *International Review of Mechanical Engineering*, *12*(6), 522–532. doi:10.15866/ireme.v12i6.15065

Baranovsky, N. V. (2019). The development of application to software origin pro for informational analysis and forecast of forest fire danger caused by thunderstorm activity. *Journal of Automation and Information Sciences*, *51*(4), 12–23. doi:10.1615/JAutomatInfScien.v51.i4.20

Bukhtoyarov, D.V., Kopylov, S.N., Kazakov, A.V., & Popov, A.V. (2008). Smoldering burning: Features. Model Hearth. *Fire Safety*, (3), 63 - 66. (in Russian)

California Department of Forestry and Fire Protection. (n.d.). Available http://www.fire.ca.gov

Country Fire Authority (Victoria state of Australia). (n.d.). Available: https://www.cfa.vic.gov.au/index.htm

Devisilov, V.A. (2010). Russian forest asks for protection! *Safety in the Technosphere*, (6), 3 - 7. (in Russian)

Grishin, A. M. (1997). *Mathematical modeling of forest fire and new methods of fighting them*. Publishing House of the Tomsk State University.

Grishin, A. M., Zima, V. P., Kuznetsov, V. T., & Skorik, A. I. (2002). Ignition of combustible forest materials by a radiant energy flux. *Combustion, Explosion, and Shock Waves*, *38*(1), 24–29. doi:10.1023/A:1014097631884

Hadden, R. M., Scott, S., Lautenberger, Ch., & Fernandez-Pello, C. (2011). Ignition of Combustible Fuel Beds by Hot Particles: An Experimental and Theoretical Study. *Fire Technology*, *47*(2), 341–355. doi:10.100710694-010-0181-x

Kabanov, M. V. (1997) Regional monitoring of the atmosphere. Part I. Scientific and methodological foundations: Monograph. Tomsk: Publishing house "Spectrum" IOA SB RAS. (in Russian)

Kuznetsov, G. V., Mamontov, G. Ya., & Taratushkina, G. V. (2004a). Ignition of condensed substance with 'hot' particle. *Khimicheskaya Fizika*, *23*(3), 67–73.

Kuznetsov, G. V., Mamontov, G. Ya., & Taratushkina, G. V. (2004b). Numerical Simulation of Ignition of a Condensed Substance by a Particle Heated to High Temperatures. *Combustion, Explosion, and Shock Waves*, *40*(1), 70–76. doi:10.1023/B:CESW.0000013669.57057.42

Kuznetsov, G. V., & Rudzinskii, V. P. (2000). High-temperature heat and mass transfer in a layer of coke of heat-shielding materials. *High Temperature*, *38*(4), 629–635. doi:10.1007/BF02755811

Kuznetsov, G.V., & Sheremet, M.A. (2006a). The two-dimensional problem of natural convection in a rectangular region with local heating and heat-conducting boundaries of finite thickness. *Bulletin of the Russian Academy of Sciences. Mechanics of Fluid and Gas*, (6), 29 - 39. (in Russian)

Kuznetsov, G. V., & Sheremet, M. A. (2006b). Conjugate heat transfer in a closed domain with a locally lumped heat-release source. *Journal of Engineering Physics and Thermophysics*, *79*(1), 57–64. doi:10.100710891-006-0066-5

Kuznetsov, G. V., & Sheremet, M. A. (2008). About one approach to mathematical modeling of thermal conditions of electronic equipment and electronic equipment. [in Russian]. *Microelectronics*, *37*, 150–158.

Kuznetsov, G. V., & Strizhak, P. A. (2009a). Numerical Solution of the Problem of Ignition of a Combustible Liquid by a Single Hot Particle. *Combustion, Explosion, and Shock Waves*, *45*(5), 543–550. doi:10.100710573-009-0066-9

Kuznetsov, G. V., & Strizhak, P. A. (2009b). Simulation of the ignition of a liquid fuel with a hot particle. *Russian Journal of Physical Chemistry B*, *3*(3), 441–447. doi:10.1134/S199079310903018X

Landsberg, G. S. (1976). *Optics*. Nauka. (in Russian)

Magomedov, A. M. (1996). Unconventional renewable energy sources. Makhachkala: AOZT Jupiter. (in Russian)

Moskalev, P. V., & Shitov, V. V. (2007). *Mathematical modeling of porous structures*. FIZMATLIT. (in Russian)

On the approval of the Fire Safety Rules in forests: Decree of the Government of the Russian Federation of June 30, 2007 No. 417. Fire Safety. 2007. No. 4.

Pakshirajan, R. P. (2013). Probability Theory. Springer.

Panevin, V. S. (2006). *Forests and forestry of the Tomsk region*. Publishing House of Tomsk University. (in Russian)

Samuilov, E. V., Faminskaya, M. V., & Golovina, E. S. (2004). Model and calculation of the gasification process of single carbon particle. *Fizika Gorenia i Vzryva*, *40*(1), 86–94.

Strakhov, V. L., Garashchenko, A. N., Kuznetsov, G. V., & Rudzinskii, V. P. (2001). Mathematical simulation of thermophysical and thermo chemical processes during combustion of intumescent fire-protective coatings. *Combustion, Explosion, and Shock Waves*, *37*(2), 178–186. doi:10.1023/A:1017557726294

Vilyunov, V. N. (1984). *Theory of ignition of condensed matter*. Science, Sib. Department.

Weilin, L., Buo, X., & Yu, L. (2000). Application of RS, GPS and GIS to forest management in China. *Journal of Forestry Research*, *11*(1), 69–71. doi:10.1007/BF02855502

Yankovich, K.S., Yankovich, E.P., & Baranovskiy, N.V. (2019). Classification of Vegetation to Estimate Forest Fire Danger Using Landsat 8 Images: Case Study. *Mathematical Problems in Engineering*. doi:10.1155/2019/6296417

ADDITIONAL READING

Antonov, D. V., Volkov, R. S., Voitkov, I. S., Zhdanova, A. O., & Kuznetsov, G. V. (2018). Influence of Special Additives in a Water Aerosol on the Suppression of a Forest Fire with it. *Journal of Engineering Physics and Thermophysics.*, *91*(5), 1250–1259. doi:10.100710891-018-1855-3

Baranovskiy, N. V., Sherstnyov, V. S., Yankovich, E. P., Engel, M. V., & Belov, V. V. (2016) Focused sunlight factor of forest fire danger assessment using Web-GIS and RS technologies. Proceedings of SPIE - The International Society for Optical Engineering, 9688, article 968823, 6 P. 10.1117/12.2240378

Baranovskiy, N. V., & Zakharevich, A. V. (2015). *Experimental research of grassy rags ignition by heated up to high temperatures carbon particle*. Multifunctional Materials and Modeling., doi:10.1201/b18552

Chen, T., Zhang, J., & Wu, J. (2016). Kinetic and energy production analysis of pyrolysis of lignocellulosic biomass using a three-parallel Gaussian reaction model. *Bioresource Technology*, *211*, 502–508. doi:10.1016/j.biortech.2016.03.091 PMID:27035484

Goman, P. N. (2020). The Dynamics of Surface Forest Fire and Forest Fuel Ignition Under the Heat Radiation From the Fire Line. In N. Baranovskiy (Ed.), *Predicting, Monitoring, and Assessing Forest Fire Dangers and Risks* (pp. 1–47). IGI Global. doi:10.4018/978-1-7998-1867-0.ch001

Grishin, A. M., & Shipulina, O. V. (2002). Mathematical model for spread of crown fires in homogeneous forests and along openings. *Combustion, Explosion, and Shock Waves*, *38*(6), 622–632. doi:10.1023/A:1021127924913

Hameed, S., Sharma, A., Pareek, V., Wu, H., & Yu, Y. (2019). A review on biomass pyrolysis models: Kinetic, network and mechanistic models. *Biomass and Bioenergy*, *123*, 104–122. doi:10.1016/j.biombioe.2019.02.008

Kansa, E., Perlee, H., & Chaiken, R. (1977). Mathematical model of wood pyrolysis. *Combustion and Flame*, *29*, 311–324. doi:10.1016/0010-2180(77)90121-3

Liu, H., Wang, C., Zhao, W., Yang, S., & Hou, X. (2018). Pyrolysis characteristics and kinetic modeling of Artemisia apiacea by thermogravimetric analysis. *Journal of Thermal Analysis and Calorimetry*, *131*(2), 1783–1792. doi:10.100710973-017-6599-3

Thomas, J. C., Hadden, R. M., & Simeoni, A. (2017). Experimental investigation of the impact of oxygen flux on the burning dynamics of forest fuel bed. *Fire Safety Journal*, *91*, 855–863. doi:10.1016/j.firesaf.2017.03.086

KEY TERMS AND DEFINITIONS

Anthropogenic Load: Different human activities on forested territories lead to forest fire occurrence and characterized by presence of fire sources.

Focused Sunlight: Sunlight concentrated by natural or artificial lens.

Forest Fire: Uncontrolled aerothermochemical phenomenon characterized by step-by-step mechanism which includes following stages: inert heating, moisture evaporation, high temperature terpens evaporation, dry organic matter pyrolysis, flammable combustion, and smoldering.

Forest Fuel: It can be considered like dead and live forest fuel. Main types of forest fuel which can be involved in combustion during forest fire: ground forest fuel (needles, leaves and dry grass, small branches) and crown forest fuel (needles, small branches).

Ignition: Inflammation of forest fuel caused by definite source of high temperature or energy.

Ignition Delay: Time before flame flash after forest fuel heating.

Prediction: Under the prediction of forest fires is the calculation of the parameters of forest fire danger with a certain projection in advance in order to have enough time to anticipate an emergency. The calculation in this case is carried out in a mode ahead of the real time of the development of the catastrophe - the occurrence of a forest fire.

Smoldering: Smoldering is ignition without flame characterized by heterogeneous chemical reactions.

Chapter 12
Mathematical Modeling of Forest Fuel Ignition by the Heated Up to High Temperatures Particle

ABSTRACT

This chapter provides information on the physical and mathematical models of forest fuel layer ignition by the metal or non-metal particle heated up to high temperatures. Mathematically, the forest fuel layer ignition is described by a system of equations of heat conduction and diffusion with the corresponding initial and boundary conditions. Scenario modeling of forest fuel layer ignition was carried out in the temperature range characteristic to the occurrence of forest fires. A scenario of a catastrophic forest fire danger is considered when there is no moisture in the forest fuel layer.

INTRODUCTION

Forest fires in the last decade are increasingly occurring (Brushlinsky et al., 2019). The burnt area is also increasing. All the causes of forest fires can be divided into natural and man-made (Baranovskiy, 2020a). A natural cause is usually understood as thunderstorm activity. The generalized mechanism of forest fire from a thunderstorm suggests the occurrence of secondary ignition sources. These are particles formed as a result of thermal destruction and mechanical damage of a tree trunk after an impact of a cloud-to-ground lightning discharge. Such particles can fall onto a layer of forest fuel near a tree and cause a fire. Analysis of the generalized mechanism of forest fire from anthropogenic causes also shows that particles heated to high temperatures are one of the common sources of ignition. For example, carbon particles are formed by cracking firewood in unburnt fires. Metal particles can form during cutting and welding of metals in forested areas.

To date, various approaches to the assessment of forest fire danger have been developed, both on the basis of soft calculations, and using probabilistic and deterministic mathematical models (Al Janabi

DOI: 10.4018/978-1-7998-7250-4.ch012

et al., 2017; Grishin and Filkov, 2011). It must be assumed that the most promising approach is the deterministic-probabilistic approach.

It should be noted that the existing forest fire danger prediction systems (CWFIS, 2020; EFFIS, 2020; WFAS, 2020; Podolskaya et al., 2011) do not take into account the sedimentation of particles heated up to high temperatures onto a layer of forest fuel. First, it is necessary to determine which scenario of forest fire danger is most likely (Baranovskiy, 2007). The second step is to carry out a numerical simulation of the process of forest fuel layer ignition by the particle heated up to high temperatures under certain meteorological conditions.

The purpose of the work is a mathematical simulation of forest fuel layer ignition by the heated up to high temperatures metal and non-metal particle.

PHYSICAL MODEL

The following physical model is considered. On the underlying surface is a forest fuel layer, on which a particle falls, heated to high temperatures. The forest fuel layer is heated and thermally decomposed with the release of gaseous pyrolysis products (for example, CO), which diffuse into the air containing the oxidizing agent (oxygen) and inert components. In the gas phase, at certain temperatures and concentrations of the reacting gases, the oxidation reaction of gaseous combustible pyrolysis products proceeds and gas-phase ignition of the oxidizer-fuel mixture occurs. In a one-dimensional formulation, it is assumed that the gaseous pyrolysis products instantly appear above the particle, where they mix and ignite at a certain temperature and concentration of reagents. When considering the problem in two-dimensional and three-dimensional formulations, the pyrolysis products diffuse into the region of the gas phase on the sides of the particle, where they are mixed with an oxidizing agent. Ignition criteria: 1) the heat input from the chemical reaction exceeds the heat input from the heated particle; 2) the temperature in the gas mixture reaches a certain critical value.

The main assumptions adopted in the formulation of the problem: 1) A particle heated to high temperatures is modeled by a plate of finite thickness (one-dimensional formulation), a square (two-dimensional formulation), and a cube (three-dimensional formulation). 2) Carbon particles are considered (this option corresponds to the scenario of the occurrence of massive forest fires as a result of the transfer and release of charred hot twigs and forest fuel residues onto the forest fuel layer untouched by the fire) and steel particles are considered (this variant corresponds to the scenario of an anthropogenic fire source, for example, electric welding). 3) The gas mixture is taken as a three-component, containing an oxidizing agent (oxygen), combustible components (gaseous products of pyrolysis - carbon monoxide) and inert components (nitrogen, carbon dioxide). 4) It is believed that there is no moisture in the forest fuel (the assumption corresponds to a scenario of catastrophic fire danger quite typical for many territories).

Designations: where T_i, ρ_i, c_i, λ_i are the temperature, density, heat capacity, thermal conductivity (1 is the air, 2 is the particle, 3 is the forest fuel layer); C_i,M_i are the concentration and molar mass (4 is the oxidizing agent, 5 is the combustible gas, 6 is the inert components of air); q_p is the thermal effect of the pyrolysis of forest fuel; k_1 is the pre-exponent of the forest fuel pyrolysis reaction; E_1 is the activation energy of the forest fuel pyrolysis reaction; R is the universal gas constant; φ is the volume fraction of dry organic matter of forest fuel; q_5 is the thermal effect of the oxidation reaction of carbon monoxide; ν_5 is the fraction of heat absorbed by the forest fuel layer; R_5 is the mass rate of the oxidation reaction of carbon monoxide; α_1 is the heat transfer coefficient; α_2 is the heat transfer coefficient; k_5 is the pre-exponent of the oxidation reaction of carbon monoxide; E_5 is the activation energy of the oxidation reaction of carbon monoxide; D is the diffusion coefficient, Y_5 is the mass flow of combustible pyrolysis products, x_i are the auxiliary variables; x,y,z are the spatial variables. t is the time. The indices es, ea, n correspond to environmental parameters in the soil, air, and at the initial time.

For a numerical study of the problem of igniting an forest fuel layer with a single particle heated up to high temperatures, the following initial data were used: ρ_1=0.1 kg/m^3; ρ_2=1500 kg/m^3 (carbon); ρ_2=7900 kg/m^3(steel); ρ_3=500 kg/m^3; c_1=1200 J/(kg×K); c_2=1150 J/(kg×K) (carbon); c_2=460 J/(kg×K) (steel); c_3=1400 J/(kg×K); λ_1=0.1 W/(m×K); λ_2=1.5 W/(m×K) (carbon); λ_2=71 W/(m×K) (steel); λ_3=0.102 W/(m×K); q_1=1000 J/kg; k_1=3.63×10^4; E_1/R=9400 K; $\varphi_{1н}$=1; q_5=10^7 J/kg; k_5=3×10^{13} s^{-1}; E_5/R=11500 K; ν_5=0.3; α_1=80 W/(m^2×K); α_2=20 W/(m^2×K);D=10^{-6}; M_4=0.032; M_5=0.028; M_6=0.028.

1D MATHEMATICAL MODEL

Geometrically, the problem is posed as follows - a three-layer plate is considered: 1 is the gas mixture, 2 is the particle heated up to high temperatures, 3 is the forest fuel layer. The scheme of the solution domain is presented in Fig. 1. The boundary conditions of the fourth kind are set at the interfaces of the layers B1, B2.

The mathematical formulation of the problem of gas-phase forest fuel layer ignition is developed. The system of non-stationary differential equations of heat conduction and diffusion for the "gas mixture-particle-forest fuel" system (Fig. 1), corresponding to the formulated physical model, has the following form:

The energy equation for the forest fuel layer:

$$\rho_3 c_3 \frac{\partial T_3}{\partial t} = \lambda_3 \frac{\partial^2 T_3}{\partial x^2} + q_p \cdot k_1 \cdot \rho_3 \cdot \varphi_1 \cdot \exp\left(-\frac{E_1}{RT_3}\right), \quad (1)$$

The energy equation for a carbon (or steel) particle:

$$\rho_2 c_2 \frac{\partial T_2}{\partial t} = \lambda_2 \frac{\partial^2 T_2}{\partial x^2}, \quad (2)$$

The energy equation for a gas mixture:

$$\rho_1 c_1 \frac{\partial T_1}{\partial t} = \lambda_1 \frac{\partial^2 T_1}{\partial x^2} + q_5 (1-\nu_5) \cdot R_5 . \quad (3)$$

The boundary conditions for equations (1) - (3):

$$x=0,\ \alpha_1 (T - T_e) = \lambda_1 \frac{\partial T_1}{\partial x}, \quad (4.1)$$

Figure 1. Initial temperature distribution in the "Gas mixture-particle-forest fuel" system

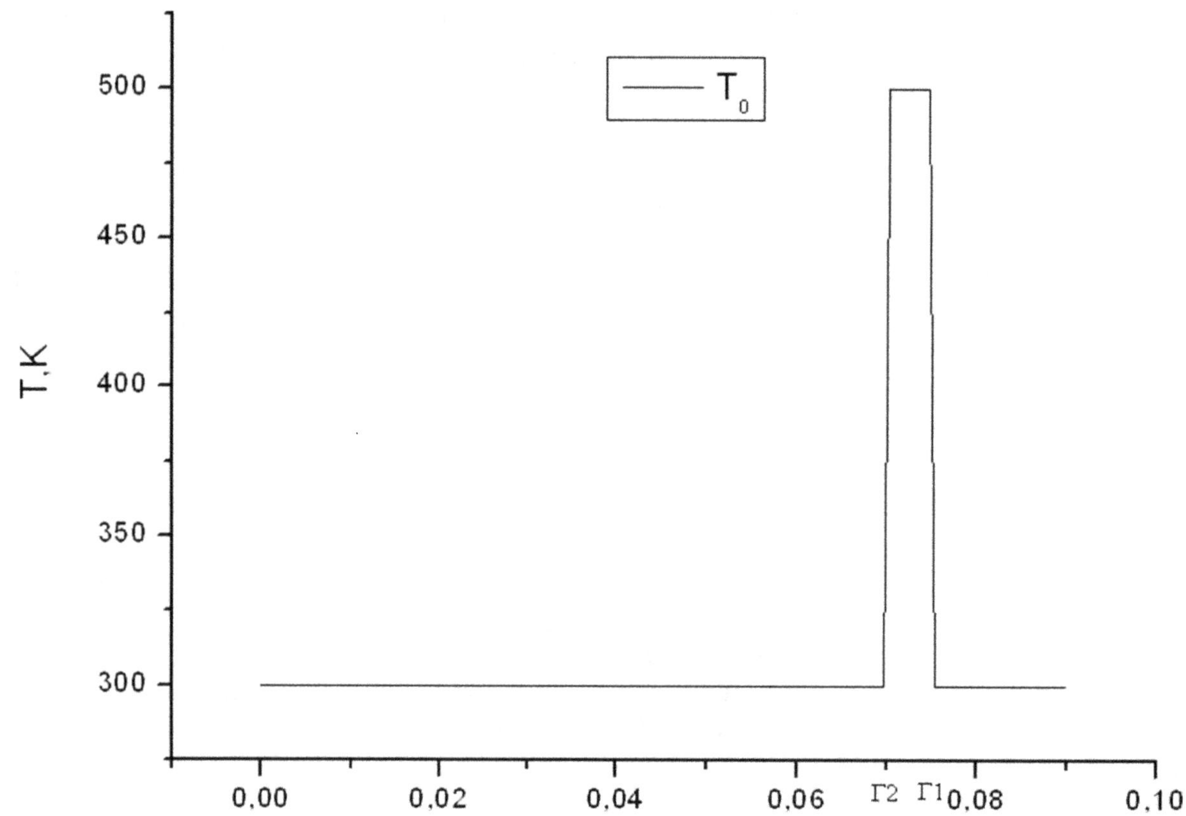

$$x=\Gamma_2,\ \lambda_1 \frac{\partial T_1}{\partial x} = \lambda_2 \frac{\partial T_2}{\partial x},\ T_1=T_2, \tag{4.2}$$

$$x=\Gamma_1,\ \lambda_2 \frac{\partial T_2}{\partial x} = \lambda_3 \frac{\partial T_3}{\partial x},\ T_2=T_3 \tag{4.3}$$

$$x=L, \alpha_2 (T - T_e) = \lambda_3 \frac{\partial T_3}{\partial x} \tag{4.4}$$

The initial conditions for equations (1) - (3):

$$t=0,\ T_3=300K,\ T_1=300K,\ T_2=900K \tag{5}$$

The kinetic equation and initial condition:

$$\rho_3 \frac{\partial \varphi}{\partial t} = -k_1 \rho_3 \varphi \exp\left(-\frac{E_1}{RT}\right), \ t=0,\ \varphi=\varphi 0 \qquad (6)$$

The diffusion equation for an oxidizing agent:

$$\frac{\partial C_4}{\partial t} = D\frac{\partial^2 C_4}{\partial x^2} - \frac{M_4}{M_5} R_5 \qquad (7)$$

The boundary and initial conditions for equation (7):

$$x=0,\ \rho D\frac{\partial C_4}{\partial x} = 0, \qquad (8.1)$$

$$x=\Gamma_2,\ \rho D\frac{\partial C_4}{\partial x} = 0, \qquad (8.2)$$

$$t=0,\ C_4=0.3 \qquad (8.3)$$

The diffusion equation for gaseous combustible pyrolysis products:

$$\frac{\partial C_5}{\partial t} = D\frac{\partial^2 C_5}{\partial x^2} - R_5 \qquad (9)$$

The boundary and initial conditions for equation (9):

$$x=0,\ \rho D\frac{\partial C_5}{\partial x} = 0, \qquad (10.1)$$

$$x=\Gamma_2,\ \rho D\frac{\partial C_5}{\partial x} = Y_5, \qquad (10.2)$$

$$t=0,\ C_5=0.0 \qquad (10.3)$$

The mass balance equation:

$$C_{N_2} = 1 - C_{O_2} - C_{CO}, \qquad (11)$$

The expression for the mass reaction rate R_5 (Grishin and Shipulina, 2002):

$$R_5 = k_5 M_5 T^{-2,25} \exp\left(-\frac{E_5}{RT}\right) \cdot \begin{cases} x_1^{0,25} x_2, x_1 > 0,05 \\ x_1 x_2, x_1 < 0,05 \end{cases} \tag{12}$$

$$x_i = \frac{C_i}{\sum_{k=4}^{6} \frac{C_k}{M_k} M_i}, \tag{13}$$

This section presents the results of mathematical modeling of forest fuel layer ignition. In the general case, the temperature of hot carbon or steel particles can vary in the range from 330 K to 1800 K. Relatively low values of the initial particle temperature were considered, which are of the greatest interest for practice. It is assumed that at the initial time, the particle has a temperature from 900 K to 1200 K. Figure 2 shows the temperature distribution of the "Gas mixture-particle-forest fuel" system at various times before ignition. Ignition occurs when the heat transfer from the oxidation reaction of gaseous pyrolysis products exceeds the heat transfer from the hot particle. Tab. 1 shows ignition delay for various temperatures of a carbon particle.

The lower limit of ignition is a particle temperature of 900 K. At temperatures below 900 K, ignition does not occur. The selected temperature range in which the ignition of dry forest fuel layer is possible corresponds to the actual temperatures achieved in practically significant fire danger situations. Such, for example, include the blowing of extinguished fires by the wind and the expansion of individual particles of unburned wood (coal) at a distance of 1-3 m from an open fire source. The presence of the temperature limit value at which forest fuel layer is still ignited is in this case not only due to the kinetics of the processes of thermal decomposition and gas-phase ignition of the mixture of fuel and oxidizer. An important factor is also that the particle, unlike a massive heated body (Vilyunov, 1984), has a finite heat reserve.

Figure 2 shows the ignition delay for various temperatures of a steel particle.

The lower ignition limit is a particle temperature of 900 K.

A parametric analysis of the influence of the main parameters on the ignition delay of the forest fuel layer is carried out. The density of the forest fuel layer varied between 400 - 700 kg/m3. An analysis of the results showed that an increase in such a parameter as the density of the forest fuel leads to an increase in the ignition delay. The thermal conductivity of the forest fuel layer varied within 1.4–1.8 W/(m·K). An increase in the thermal conductivity leads to a decrease in the ignition delay. The heat capacity of the forest fuel layer varied in the range of 1300 - 1600 J/(kg·K). It was found that a change in this parameter does not significantly affect the ignition delay of the forest fuel layer. In general, parametric analysis allows us to conclude that the mathematical model responds adequately to changes in input parameters.

Analysis of the temperature distribution in the "Gas mixture-particle-forest fuel" system allows us to conclude that ignition occurs in the gas phase in the immediate vicinity of the particle surface. In this zone, where gaseous pyrolysis products are injected, the concentration of carbon monoxide is maximal and close to 1 (Fig. 3). The concentration of the oxidizing agent (oxygen) as it approaches the surface of the particle decreases due to the injection of gaseous pyrolysis products, as well as the consumption

Figure 2. Temperature of the "Gas mixture-particle-forest fuel" system until the forest fuel layer is ignited (carbon particle), the initial particle temperature is 900 K, t=2.63 s

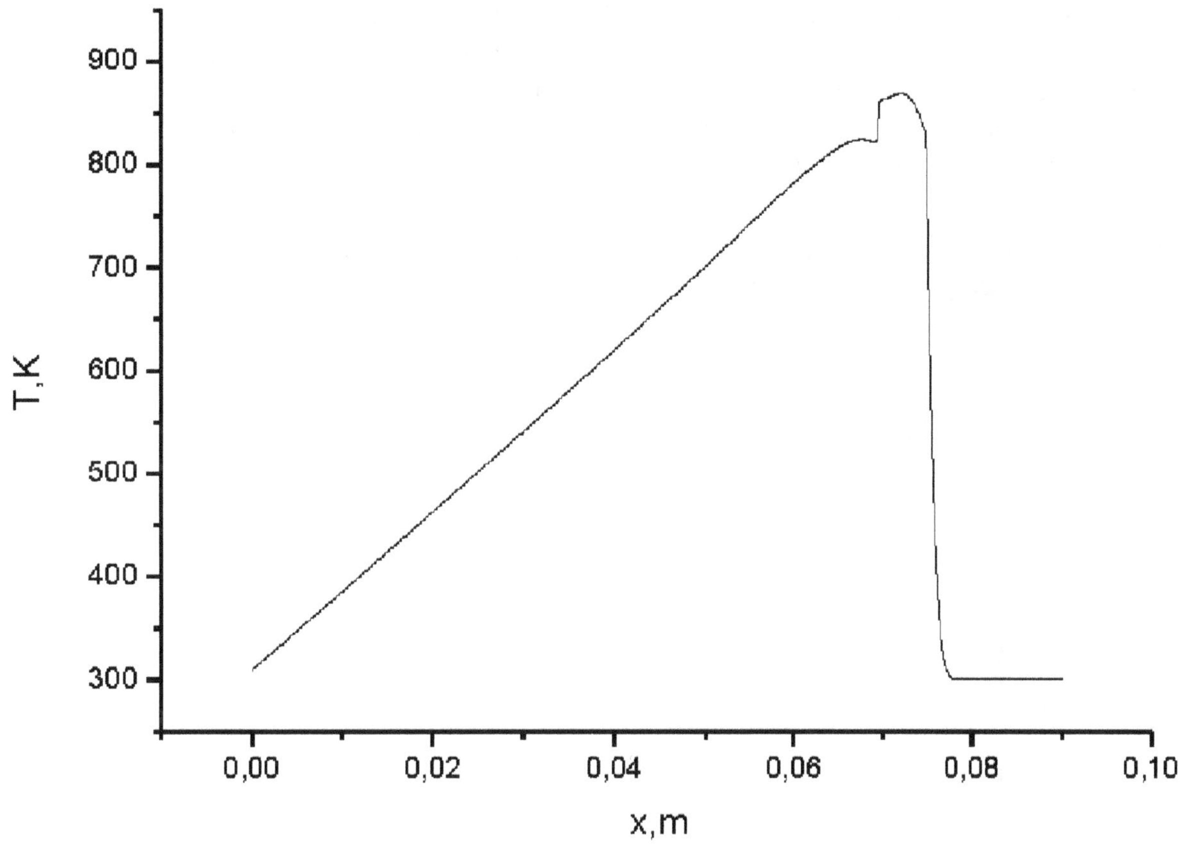

Table 1. Ignition delay for a carbon particle

Particle Temperature, K	Ignition Delay, s
900	2.63
1000	0.46
1100	0.19
1200	0.099

Table 2. Ignition delay for a steel particle

Particle Temperature, K	Ignition Delay, s
900	0.3
1000	0.07
1100	0.05
1200	0.02

of the oxidizing agent for the oxidation reaction of carbon monoxide (Fig. 4). The concentration of inert components adequately changes with changing concentrations of carbon monoxide and oxidizing agent. The certain maximum concentration of the inert component can be distinguished in the ignition region in the gas phase due to the influx of carbon dioxide from the oxidation reaction of gaseous pyrolysis products (Fig. 5). The distinguished regularities are similar for both carbon and steel particles. The differences are that heating and all other processes in the case of a steel particle occur faster due to the greater thermal conductivity of the steel.

Figure 3. Distribution of the oxidizing agent in the gas mixture behind the particle, the initial particle temperature of 900 K

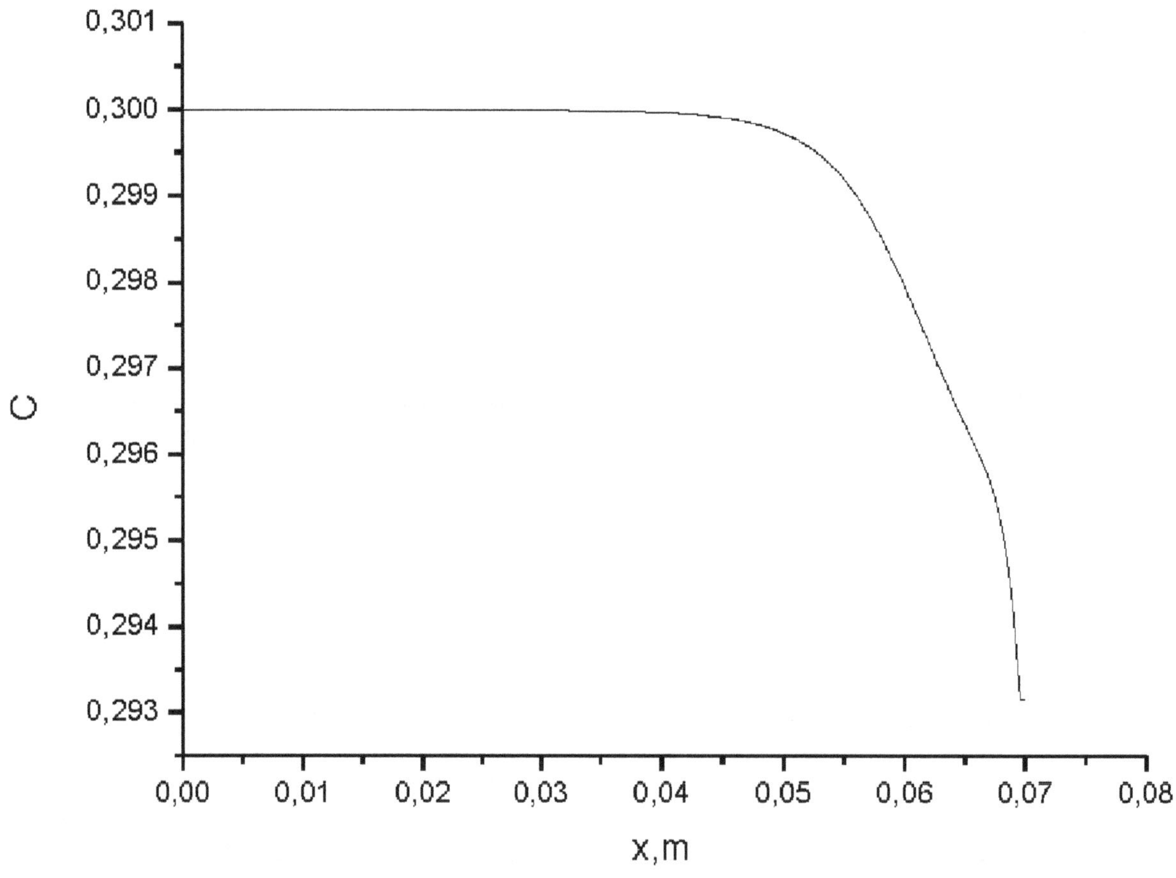

The effect of the particle thickness on the ignition delay of the forest fuel layer is considered. As an analysis of the results showed, a change in this parameter does not significantly affect the ignition delay. This difference is about several 0.1 percents.

Figure 4. Distribution of combustible gas in the gas mixture behind the particle, the initial particle temperature of 900 K

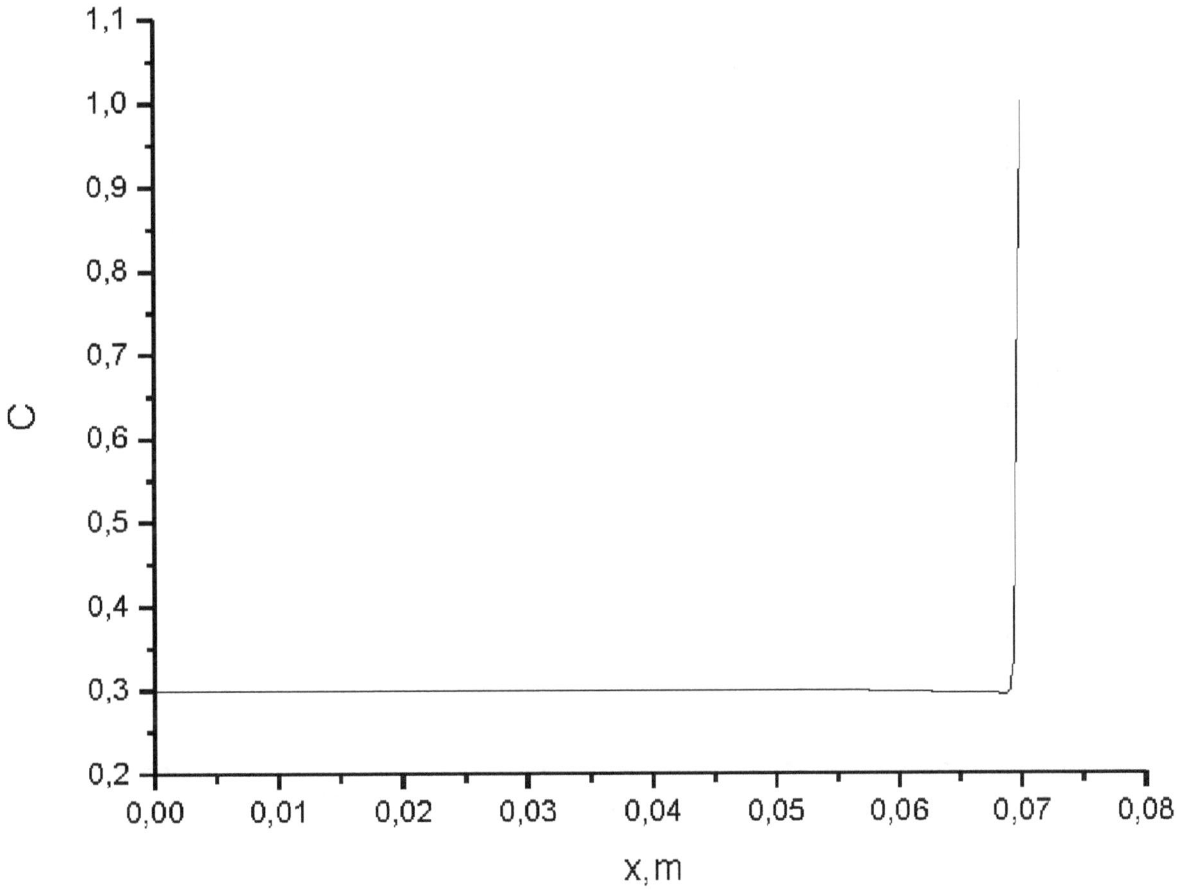

2D MATHEMATICAL MODEL

Fig. 6 shows the geometry of the solution domain. Symbols G indicate the boundaries of the solution region and various layers.

The process of forest fuel layer ignition by the particle heated up to high temperatures is described by a system of two-dimensional non-stationary nonlinear equations of heat conduction and diffusion with the corresponding initial and boundary conditions. For numerical implementation, the finite-difference method is used. Two-dimensional and three-dimensional difference equations are solved by the locally one-dimensional method (Samarskiy, 1983). Difference analogues of the one-dimensional heat conduction and diffusion equations are solved by the marching method in combination with the simple iteration method (Samarskiy and Vabishchevich, 2001). The algorithm of the program was tested on various less complicated problems compared with the solved heat and mass transfer.

The energy equation for a gas mixture:

Figure 5. Distribution of inert components in the gas mixture behind the particle, the initial particle temperature of 900 K

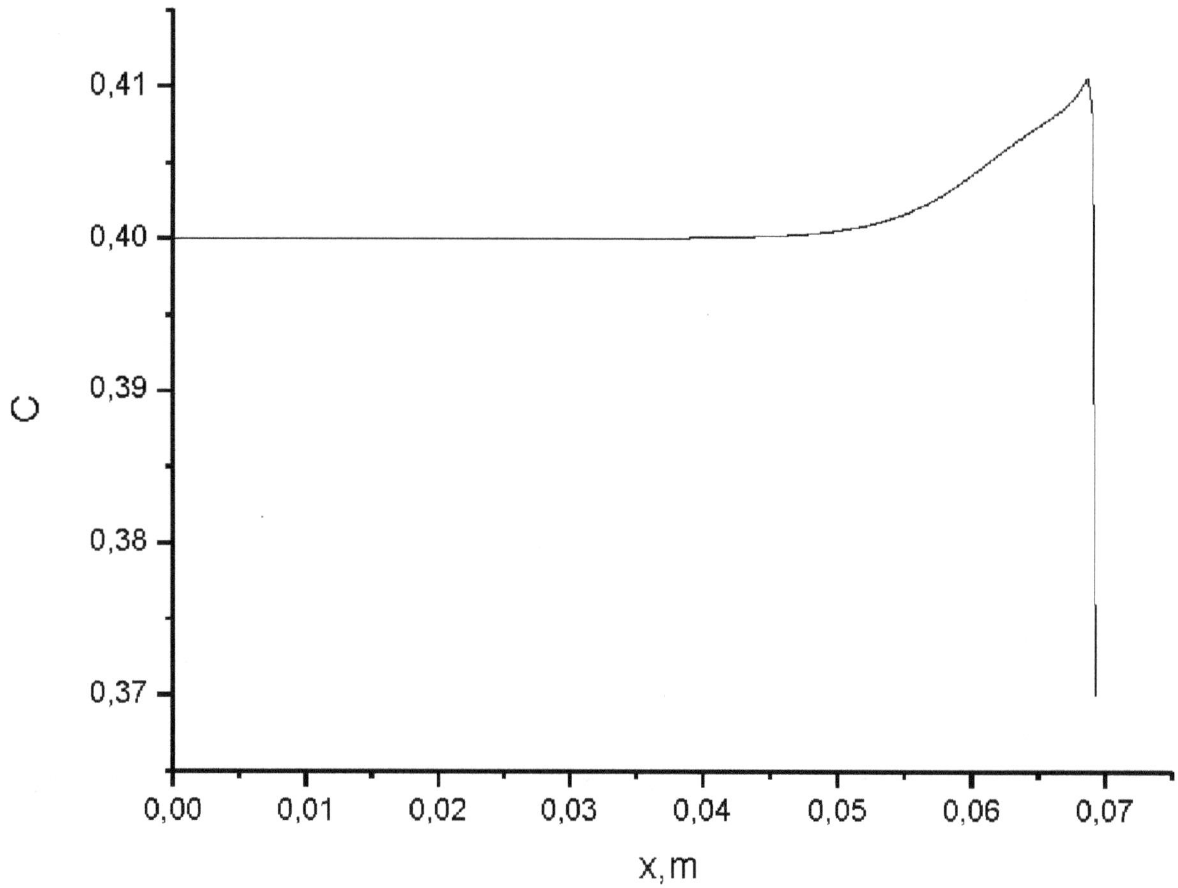

$$\rho_1 c_1 \frac{\partial T_1}{\partial t} = \lambda_1 \left(\frac{\partial^2 T_1}{\partial x^2} + \frac{\partial^2 T_1}{\partial z^2} \right) + q_5 (1 - \nu_5) R_5 , \tag{14}$$

The energy equation for a particle:

$$\rho_2 c_2 \frac{\partial T_2}{\partial t} = \lambda_2 \left(\frac{\partial^2 T_2}{\partial x^2} + \frac{\partial^2 T_2}{\partial z^2} \right), \tag{15}$$

The energy equation for the forest fuel layer:

$$\rho_3 c_3 \frac{\partial T_3}{\partial t} = \lambda_3 \left(\frac{\partial^2 T_3}{\partial x^2} + \frac{\partial^2 T_3}{\partial z^2} \right) + q_1 k_1 \rho_3 \phi \exp\left(-\frac{E_1}{R T_3} \right), \tag{16}$$

Figure 6. Solution Area

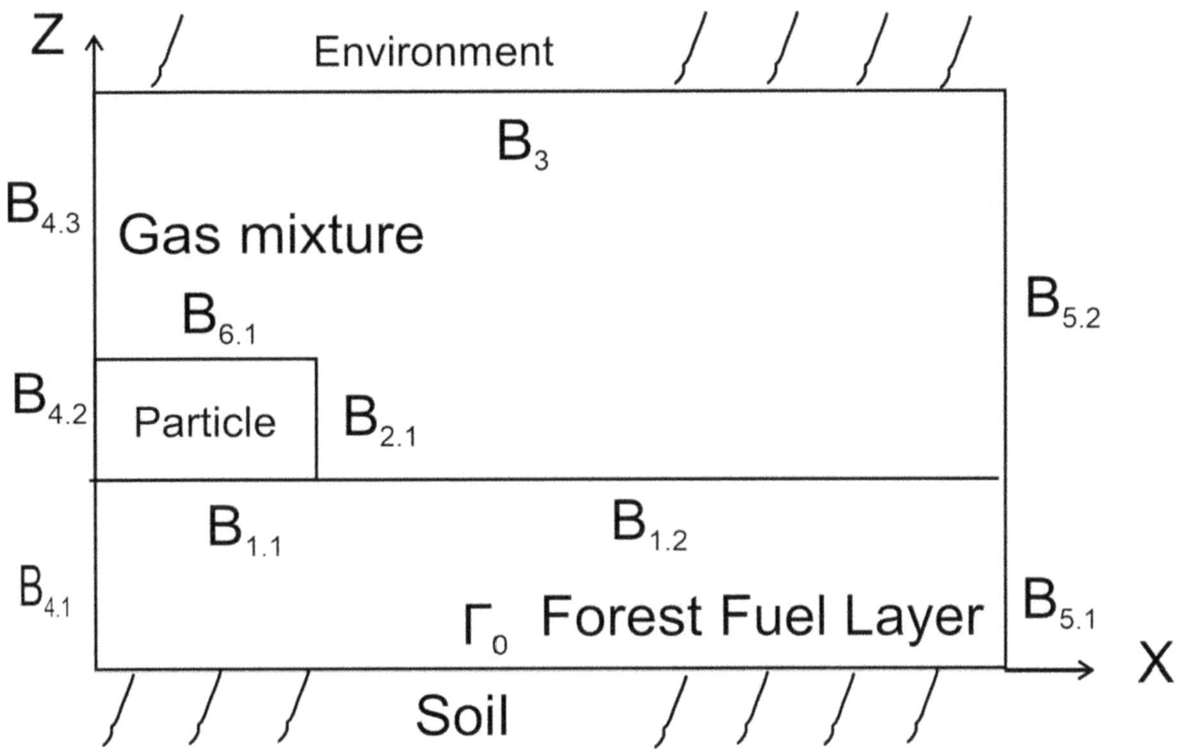

The boundary conditions for equations (14) - (16):

$$B_0\, \alpha_1 (T - T_{es}) = \lambda_3 \frac{\partial T_3}{\partial z}, \tag{17.1}$$

$$B_{1.1}\, \lambda_3 \frac{\partial T_3}{\partial z} = \lambda_2 \frac{\partial T_2}{\partial z},\ T_1 = T_2, \tag{17.2}$$

$$B_{1.2}\, \lambda_1 \frac{\partial T_1}{\partial z} = \lambda_3 \frac{\partial T_3}{\partial z},\ T_1 = T_3, \tag{17.3}$$

$$B_{2.1}\, \lambda_2 \frac{\partial T_2}{\partial x} = \lambda_1 \frac{\partial T_1}{\partial x},\ T_2 = T_1, \tag{17.4}$$

$B_3 \alpha_2 (T - T_{ea}) = \lambda_1 \frac{\partial T_1}{\partial z}$, (17..5)

$$B_{4.1} \lambda_3 \frac{\partial T_3}{\partial x} = 0, \quad (17.6)$$

$$B_{4.2} \lambda_2 \frac{\partial T_2}{\partial x} = 0, \quad (17.7)$$

$$B_{4.3} \lambda_1 \frac{\partial T_1}{\partial x} = 0, \quad (17.8)$$

$$B_{5.1} T_3 = T_{3e}, \quad (17.9)$$

$$B_{5.2} T_1 = T_{1e}, \quad (17.10)$$

$$B_{6.1} \lambda_2 \frac{\partial T_2}{\partial z} = \lambda_1 \frac{\partial T_1}{\partial z}. \quad (17.11)$$

The initial conditions for equations (14) - (16):

$$T_i \big|_{t=0} = T_{i0}, \ i=1,2,3 \quad (18)$$

The kinetic equation and initial condition:

$$\rho_3 \frac{\partial \phi}{\partial t} = -k_1 \rho_3 \phi \exp\left(-\frac{E_1}{RT_3}\right), \ \phi\big|_{t=0} = \phi_0, \quad (19)$$

The diffusion equation for an oxidizing agent:

$$\frac{\partial C_4}{\partial t} = D\left(\frac{\partial^2 C_4}{\partial x^2} + \frac{\partial^2 C_4}{\partial z^2}\right) - \frac{M_4}{M_5} R_5, \quad (20)$$

The boundary conditions for equation (20):

$$\mathrm{B}_{1.2}\rho D\frac{\partial C_4}{\partial z} = 0, \tag{21.1}$$

$$\mathrm{B}_{2.1}\rho D\frac{\partial C_4}{\partial x} = 0, \tag{21.2}$$

$$\mathrm{B}_{3}\rho D\frac{\partial C_4}{\partial z} = 0, \tag{21.3}$$

$$\mathrm{B}_{4.3}\rho D\frac{\partial C_4}{\partial x} = 0, \tag{21.4}$$

$$\mathrm{B}_{5.2}\rho D\frac{\partial C_4}{\partial x} = 0 \tag{21.5}$$

$$\mathrm{B}_{6.1}\rho D\frac{\partial C_4}{\partial z} = 0. \tag{21.6}$$

The initial conditions for equation (20):

$$C_4\big|_{t=0} = C_{4.0}, \tag{22}$$

The diffusion equation for the combustible components of pyrolysis:

$$\frac{\partial C_5}{\partial t} = D\left(\frac{\partial^2 C_5}{\partial x^2} + \frac{\partial^2 C_5}{\partial z^2}\right) - R_5, \tag{23}$$

The boundary conditions for equation (23):

$$\mathrm{B}_{1.2}\rho D\frac{\partial C_5}{\partial z} = Y_5, \tag{24.1}$$

$$B_{2.1}\rho D\frac{\partial C_5}{\partial x}=0, \tag{24.2}$$

$$B_{3}\rho D\frac{\partial C_5}{\partial z}=0, \tag{24.3}$$

$$B_{4.3}\rho D\frac{\partial C_5}{\partial x}=0, \tag{24.4}$$

$$B_{5.2}\rho D\frac{\partial C_5}{\partial x}=0, \tag{24.5}$$

$$B_{6.1}\rho D\frac{\partial C_5}{\partial y}=0. \tag{24.6}$$

The initial conditions for equation (23):

$$C_5\big|_{t=0}=C_{5.0} \tag{25}$$

The mass balance equation:

$$\sum_{i=4}^{6}C_i=1. \tag{26}$$

The initial conditions for equation (26):

$$C_6\big|_{t=0}=C_{6.0}. \tag{27}$$

The expression for the mass reaction rate R_5 (Grishin and Shipulina, 2002):

$$R_5=k_5M_5T^{-2.25}\exp\left(-\frac{E_5}{RT_1}\right)\cdot\begin{cases}x_1^{0.25}x_2, x_1>0.05\\ x_1x_2, x_1\le 0.05\end{cases}, \tag{28}$$

$$x_i = \frac{C_i}{\sum_{k=4}^{6} \frac{C_k}{M_k} M_i}, \tag{29}$$

Parametric analysis showed that the mathematical model adequately describes the processes of heat and mass transfer during ignition of the forest fuel layer by a single particle heated up to high temperatures. As input parameters, the thermophysical characteristics of the forest fuel layer were varied. This will make it possible in the future to consider the physicochemical ignition processes of not only pine forest fuel, but also other conifers and larch.

The events of the summer of 2010 showed that the catastrophic effects of forest fires can cause enormous economic and environmental damage. It becomes clear that the existing methodology for predicting forest fires used in the Russian Federation is not perfect (Nesterov, 1949). A modern forest fire danger assessment system must satisfy a number of requirements and specifications. The main condition for creating an adequate methodology for predicting the occurrence of forest fires is the presence of forest fuel ignition models in the system. Such systems can be based on the results of processing experimental data or physical and mathematical models for the numerical study of the processes of forest fuel ignition by various sources. Generalized mechanisms have already been developed for the occurrence of forest fires from thunderstorm activity and anthropogenic load, which include the scenario of ignition of an forest fuel layer by the particle. The physical model should be clarified. It is assumed that the gas mixture is represented by three components: the oxidizing agent is oxygen, the combustible gas is carbon monoxide, the inert components are nitrogen and carbon dioxide. We believe that the pyrolysis products instantly appear in the region of the gas phase. The processes of their filtration in the forest fuel layer to the surface of the layer from under the particles are not taken into account. This approach has proven itself, for example, in the study of physicochemical processes of thermal degradation of polymeric materials and in solving thermal protection issues. The boundary conditions of a nonzero mass flow are set in a narrow region around the particle. The diffusion of pyrolysis products is carried out in a narrow wall region of the particle. The output of the pyrolysis products from the forest fuel layer is carried out at the location of elevated temperature.

The mass flow in the region below the particle and from the side was summed and distributed in proportion to the factor T[i,iz]/T[xz,iz], where xz is the lateral boundary of the particle and iz is the lower boundary of the particle. This approximation provides a decrease in the release of gaseous pyrolysis products with a decrease in the temperature of the forest fuel layer. Note that this paper does not address the thermal decomposition of cellulose and other components of forest fuel. Effective thermokinetic constants of the pyrolysis process are used, which are obtained for each type of forest fuel. In this work, forest fuel layer is considered consisting of litter of pine needles. Moisture in forest fuel is absent in accordance with the scenario of catastrophic forest fire danger. Exposure to an elevated particle temperature provides a significant decomposition of the forest fuel layer beneath it. A chemical reaction on the side of the particle forms an additional heat flux to the forest fuel layer. On the side under the particle, the layer decomposes more strongly.

Consider the processes of mass transfer in the vicinity of a particle. The peak concentration of gaseous combustible pyrolysis components is observed in the near-wall region of the particle. In addition, horizontal diffusion of carbon monoxide was observed. Over time, the peak concentration of carbon monoxide shifts in the opposite direction from the particle. The concentration distribution of gaseous

Figure 7. Distribution of the concentration of carbon monoxide in the vertical coordinate in various sections at the time of ignition

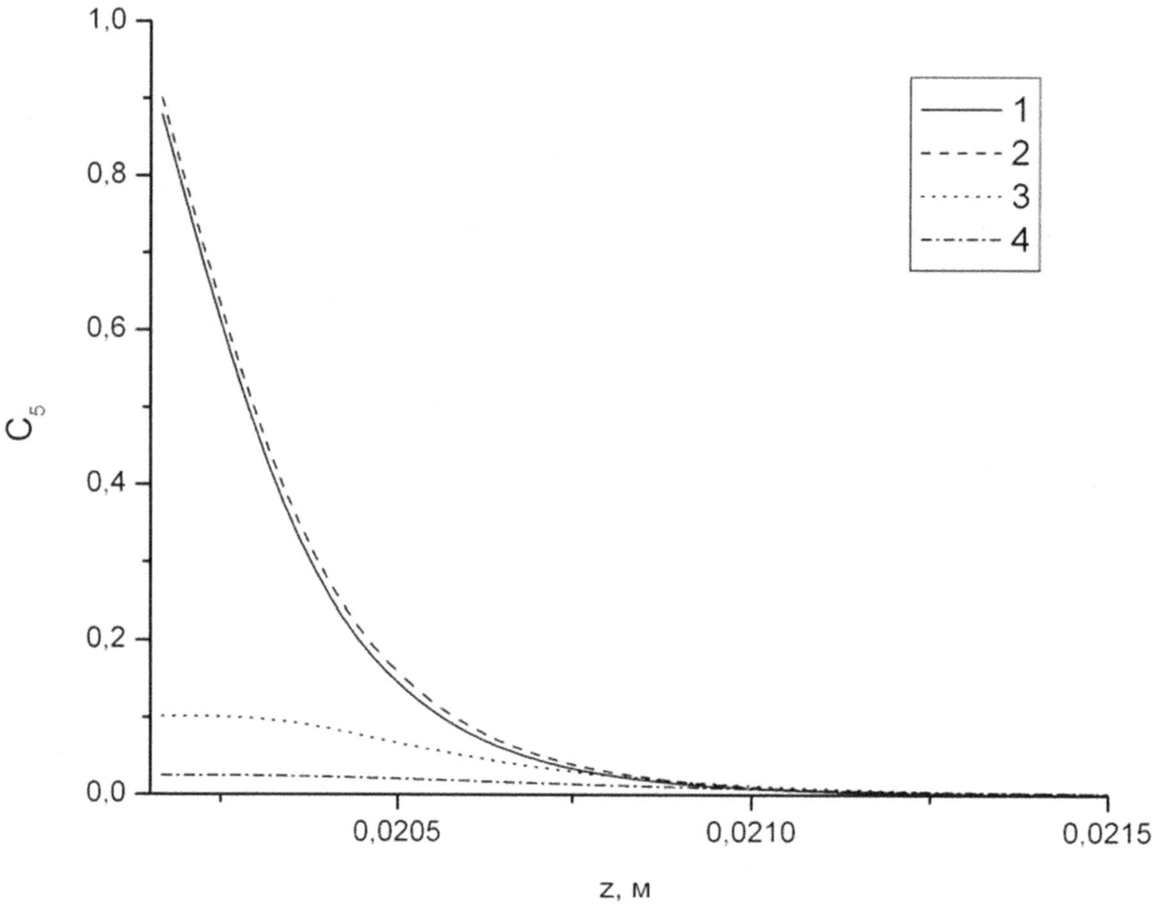

products of pyrolysis, oxygen, and inert components in the gas phase at the time of ignition is shown in Fig. 7-9. In the figures, curves 1,2,3,4 correspond to the coordinates x = 0.0026 m; 0.0031 m; 0.0036 m; 0.0041 m.

In the near-wall region, the pyrolysis products are injected and carbon monoxide is oxidized to carbon dioxide. As a result of these physicochemical processes, the oxygen concentration in the near-wall region tends to zero over time. The first of the described processes helps to reduce the concentration of inert components, which are also displaced from the near-wall region. At a certain distance from the particle, a peak is observed on the concentration curve of inert components. This is the location of the chemical reaction of the oxidation of carbon monoxide to carbon dioxide.

Fig. 10 shows the temperature distribution in the "Gas mixture-particle-forest fuel" system at the time of ignition. A peak can be highlighted on the temperature curve that corresponds to the location of the chemical reaction. At this point, critical ignition conditions are achieved, i.e. the ignition criteria are met.

Tables 3.4 show the ignition delay times for various particle temperatures. The minimum temperature of a carbon or steel particle at which ignition is still observed corresponds to 900 K. At temperatures below this critical value, the ignition criteria are not fulfilled. The ignition delay times differing from the one-

Figure 8. Distribution of oxidizer concentration along the vertical coordinate in various sections at the time of ignition

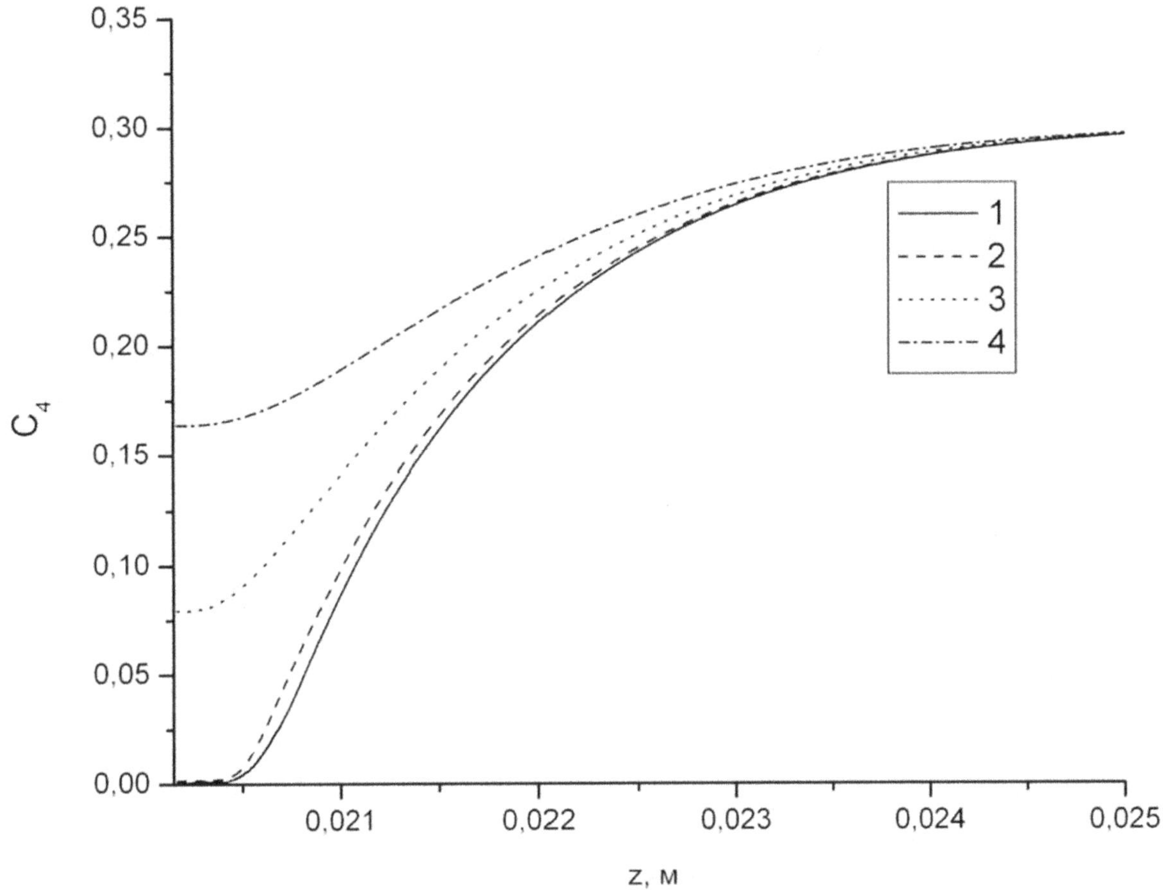

dimensional formulation are obtained. This is due to the features of mass transfer in a two-dimensional formulation and the kinetics of the chemical reaction. In a two-dimensional formulation, it is possible to obtain more physically meaningful results compared to a one-dimensional one.

3D MATHEMATICAL MODEL

As a result of cracking of wood in the burning and not extinguished bonfires, particles similar to cubes can form. Based on a one-dimensional physical and mathematical model, the very possibility of igniting a layer of forest fuel by the particle heated up to high temperatures is shown. This formulation made it possible to qualitatively study the process of ignition of the forest fuel layer by a heated particle. The next step was the development of a flat physical and mathematical formulation of the indicated problem. It was found that the ignition of the forest fuel layer occurs in the gas phase in the immediate vicinity to the side of the particle.

Figure 9. Distribution of the concentration of inert components along the vertical coordinate in various sections at the time of ignition

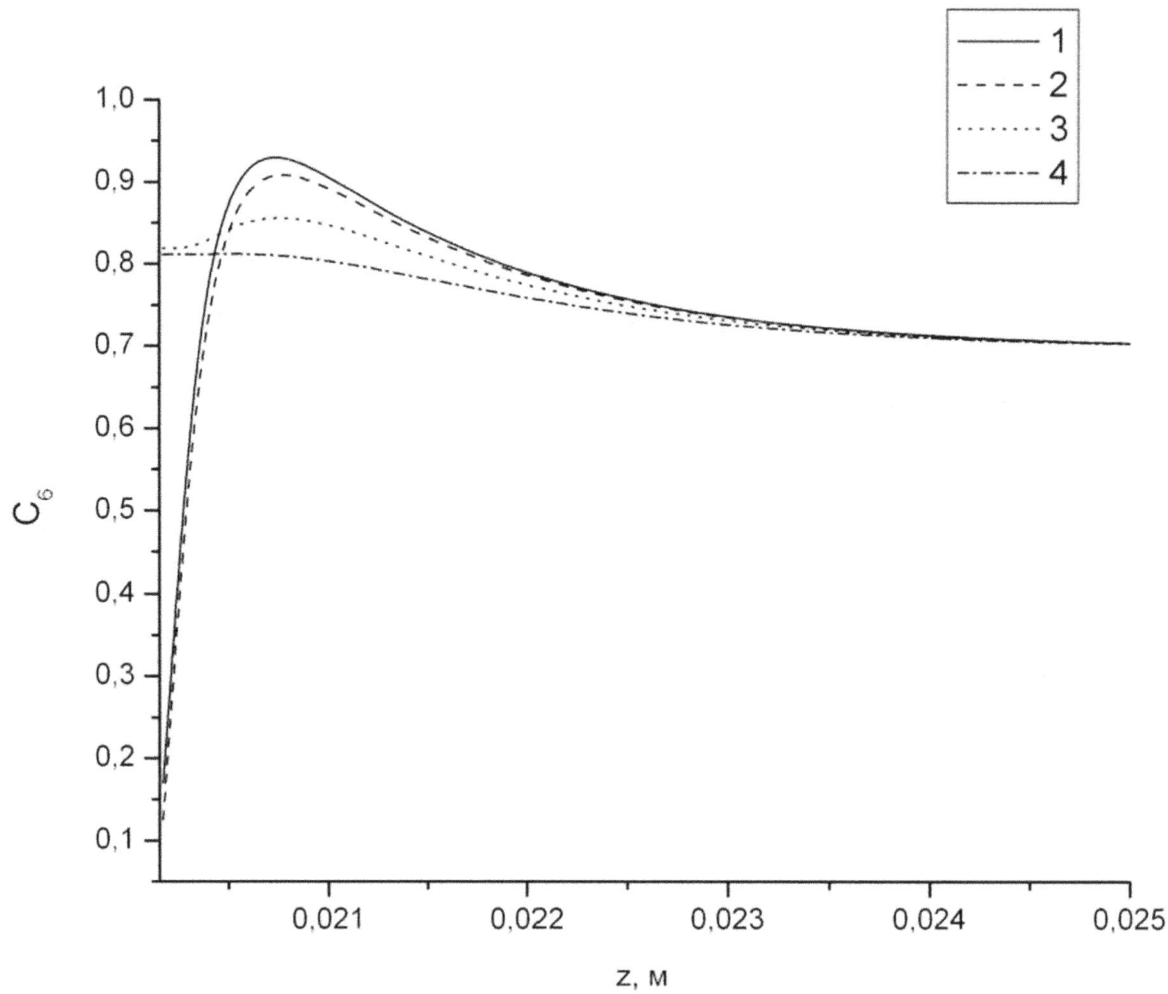

It is of interest to create a three-dimensional formulation of the problem of forest fuel ignition by the particle heated up to high temperatures, since the specific features of the particle geometry do not make it possible to reasonably use both the one-dimensional and two-dimensional approximations of the problem. The geometry of the solution domain is shown in Fig. 11.

The process of forest fuel layer ignition by the particle heated up to high temperatures is described by a system of three-dimensional non-stationary non-linear equations of heat conduction and diffusion. For numerical implementation, the finite-difference method is used. Two-dimensional and three-dimensional difference equations are solved by the locally one-dimensional method (Samarskiy, 1983). Difference analogues of the one-dimensional heat conduction and diffusion equations are solved by the marching method in combination with the simple iteration method (Samarskiy and Vabishchevich, 2001). The algorithm of the program was tested on various less complicated problems compared with the solved heat and mass transfer.

Figure 10. Temperature distribution in the Oxz plane at the time of ignition

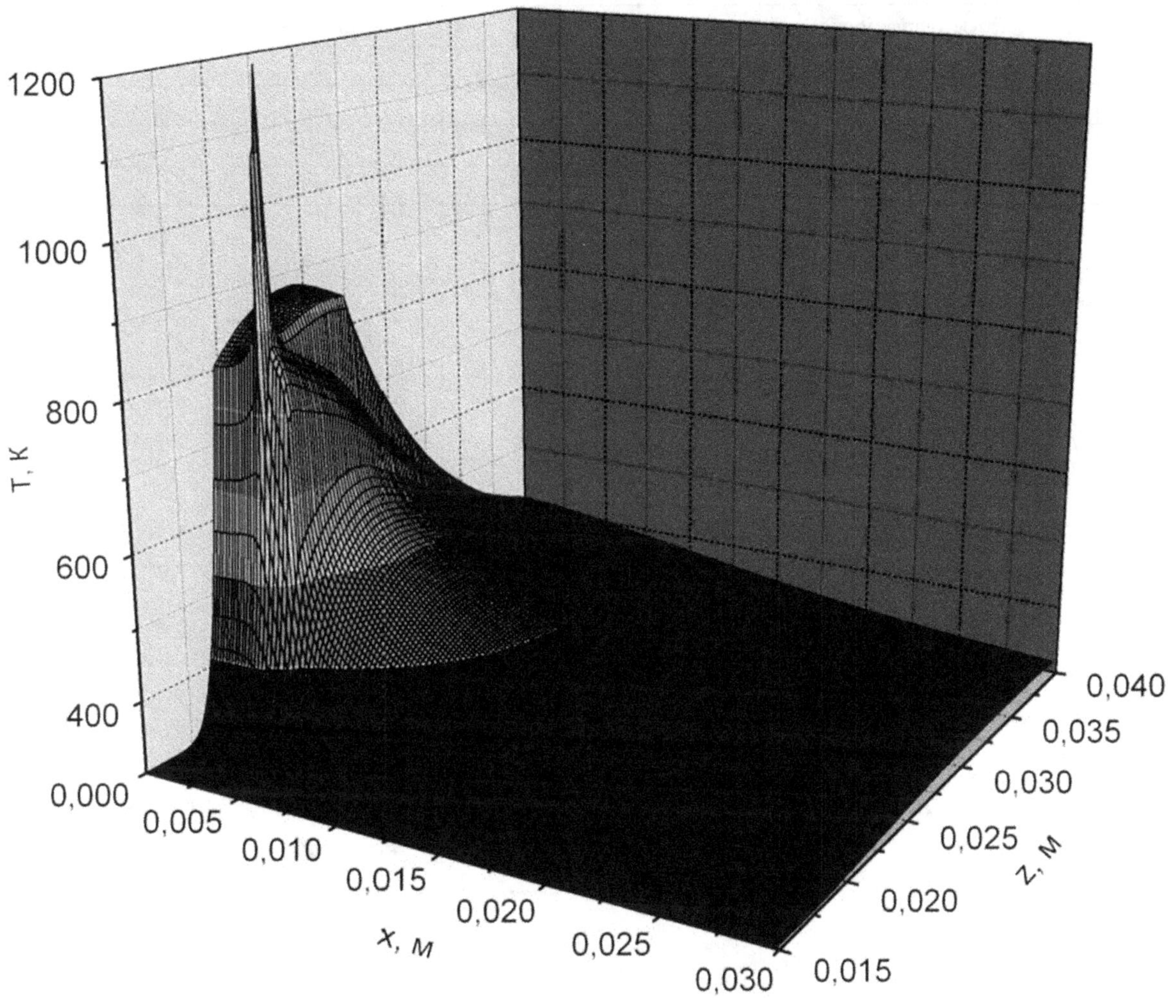

The energy equation for a gas mixture:

$$\rho_1 c_1 \frac{\partial T_1}{\partial t} = \lambda_1 \left(\frac{\partial^2 T_1}{\partial x^2} + \frac{\partial^2 T_1}{\partial y^2} + \frac{\partial^2 T_1}{\partial z^2} \right) + q_5 (1 - \nu_5) R_5, \tag{30}$$

Table 3. Ignition delay for a carbon particle

Particle Temperature, K	Ignition Delay, s
900	2.69
1000	1.38
1100	0.64
1200	0.24

Table 4. Ignition delay for a steel particle

Particle Temperature, K	Ignition Delay, s
900	1.35
1000	0.76
1100	0.38
1200	0.18

The energy equation for a particle:

Figure 11. Solution Area

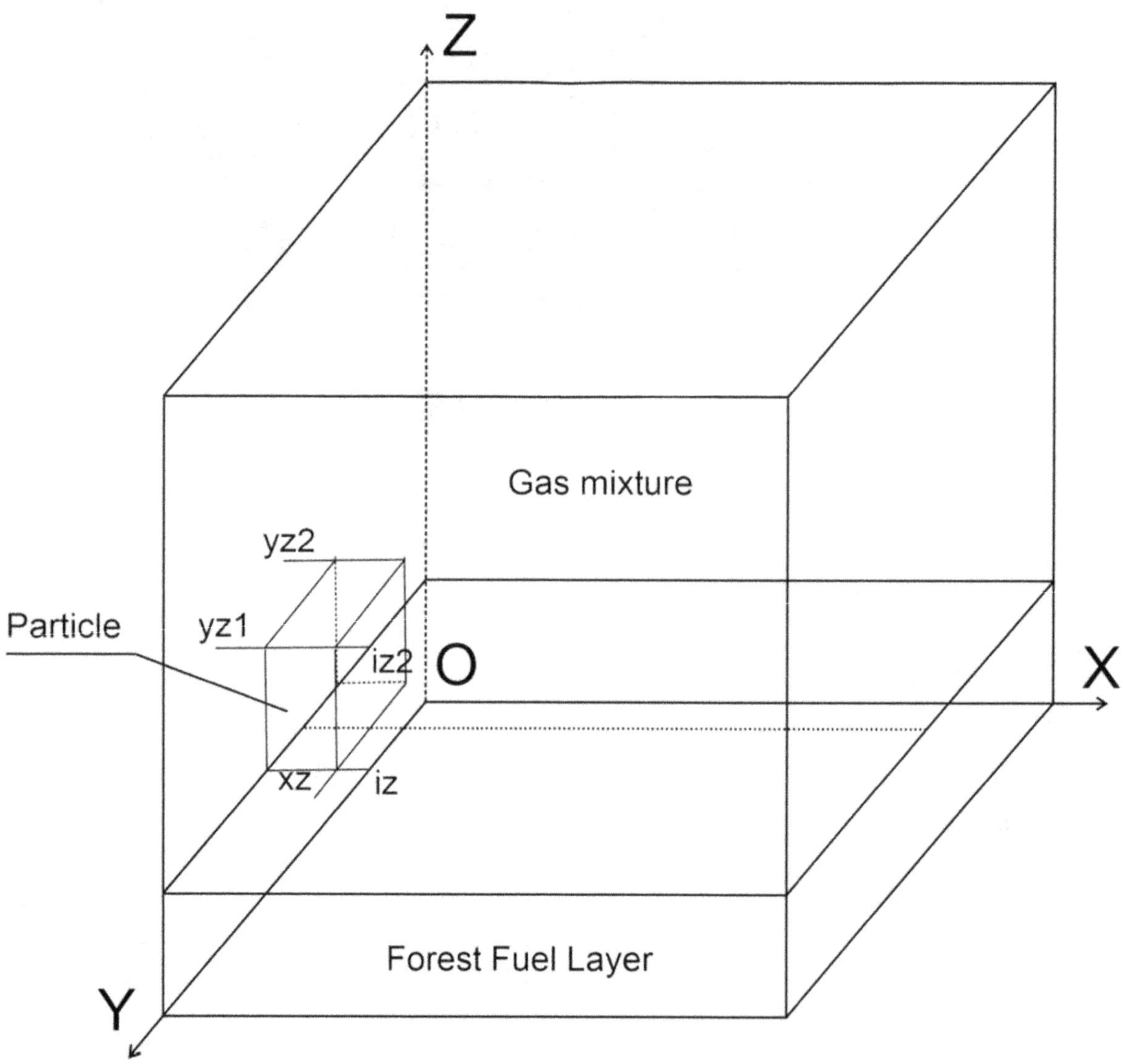

$$\rho_2 c_2 \frac{\partial T_2}{\partial t} = \lambda_2 \left(\frac{\partial^2 T_2}{\partial x^2} + \frac{\partial^2 T_2}{\partial y^2} + \frac{\partial^2 T_2}{\partial z^2} \right), \tag{31}$$

The energy equation for the forest fuel layer:

$$\rho_3 c_3 \frac{\partial T_3}{\partial t} = \lambda_3 \left(\frac{\partial^2 T_3}{\partial x^2} + \frac{\partial^2 T_3}{\partial y^2} + \frac{\partial^2 T_3}{\partial z^2} \right) + q_1 k_1 \rho_3 \phi \exp\left(-\frac{E_1}{RT_3} \right). \tag{32}$$

The boundary conditions for equations (30) - (32):

$$\mathrm{B}_0 \ \alpha_1 (T - T_{es}) = \lambda_3 \frac{\partial T_3}{\partial z}, \tag{33.1}$$

$$\mathrm{B}_{1.1} \ \lambda_3 \frac{\partial T_3}{\partial z} = \lambda_2 \frac{\partial T_2}{\partial z}, \ T_1 = T_2, \tag{33.2}$$

$$\mathrm{B}_{1.2} \ \lambda_1 \frac{\partial T_1}{\partial z} = \lambda_3 \frac{\partial T_3}{\partial z}, \ T_1 = T_3, \tag{33.3}$$

$$\mathrm{B}_{2.1} \ \lambda_2 \frac{\partial T_2}{\partial x} = \lambda_1 \frac{\partial T_1}{\partial x}, \ T_2 = T_1, \tag{33.4}$$

$$\mathrm{B}_3 \ \alpha_2 (T - T_{ea}) = \lambda_1 \frac{\partial T_1}{\partial z}, \tag{33.5}$$

$$\mathrm{B}_{4.1} \ \lambda_3 \frac{\partial T_3}{\partial x} = 0, \tag{33.6}$$

$$\mathrm{B}_{4.2} \ \lambda_2 \frac{\partial T_2}{\partial x} = 0, \tag{33.7}$$

$$\mathrm{B}_{4.3}\ \lambda_1 \frac{\partial T_1}{\partial x} = 0\,, \tag{33.8}$$

$$\mathrm{B}_{5.1}\ T_3 = T_{3e}\,, \tag{33.9}$$

$$\mathrm{B}_{5.2}\ T_1 = T_{1e}\,, \tag{33.10}$$

$$\mathrm{B}_{6.1}\ \lambda_2 \frac{\partial T_2}{\partial z} = \lambda_1 \frac{\partial T_1}{\partial z}\,. \tag{33.11}$$

The initial conditions for equations (30) - (32):

$$T_i\big|_{t=0} = T_{i0}\,,\ \text{i=1,2,3} \tag{34}$$

The kinetic equation and initial condition:

$$\rho_3 \frac{\partial \phi}{\partial t} = -k_1 \rho_3 \phi \exp\left(-\frac{E_1}{RT_3}\right), \quad \phi\big|_{t=0} = \phi_0\,, \tag{35}$$

The diffusion equation for an oxidizing agent:

$$\frac{\partial C_4}{\partial t} = D\left(\frac{\partial^2 C_4}{\partial x^2} + \frac{\partial^2 C_4}{\partial y^2} + \frac{\partial^2 C_4}{\partial z^2}\right) - \frac{M_4}{M_5} R_5\,. \tag{36}$$

The boundary conditions for equation (36):

$$\mathrm{B}_{1.2}\ \rho D \frac{\partial C_4}{\partial z} = 0\,, \tag{37.1}$$

$$\mathrm{B}_{2.1}\ \rho D \frac{\partial C_4}{\partial x} = 0\,, \tag{37.2}$$

$$B_3 \rho D \frac{\partial C_4}{\partial z} = 0, \tag{37.3}$$

$$B_{4.3} \rho D \frac{\partial C_4}{\partial x} = 0, \tag{37.4}$$

$$B_{5.2} \rho D \frac{\partial C_4}{\partial x} = 0 \tag{37.5}$$

$$B_{6.1} \rho D \frac{\partial C_4}{\partial z} = 0. \tag{37.6}$$

The initial conditions for equation (36):

$$C_4\big|_{t=0} = C_{4.0}. \tag{38}$$

The diffusion equation for the combustible components of pyrolysis:

$$\frac{\partial C_5}{\partial t} = D\left(\frac{\partial^2 C_5}{\partial x^2} + \frac{\partial^2 C_5}{\partial y^2} + \frac{\partial^2 C_5}{\partial z^2}\right) - R_5. \tag{39}$$

The boundary conditions for equation (39):

$$B_{1.2} \rho D \frac{\partial C_5}{\partial z} = Y_5, \tag{40.1}$$

$$B_{2.1} \rho D \frac{\partial C_5}{\partial x} = 0, \tag{40.2}$$

$$B_3 \rho D \frac{\partial C_5}{\partial z} = 0, \tag{40.3}$$

$$\mathrm{B}_{4.3}\rho D\frac{\partial C_5}{\partial x}=0, \tag{40.4}$$

$$\mathrm{B}_{5.2}\rho D\frac{\partial C_5}{\partial x}=0, \tag{40.5}$$

$$\mathrm{B}_{6.1}\rho D\frac{\partial C_5}{\partial y}=0. \tag{40.6}$$

The initial conditions for equation (39):

$$C_5\Big|_{t=0}=C_{5.0}. \tag{41}$$

The mass balance equation:

$$\sum_{i=4}^{6}C_i=1. \tag{42}$$

The initial conditions for equation (42):

$$C_6\Big|_{t=0}=C_{6.0}. \tag{43}$$

The expression for the mass reaction rate R_5 (Grishin and Shipulina, 2002):

$$R_5=k_5M_5T^{-2.25}\exp\left(-\frac{E_5}{RT_1}\right)\cdot\begin{cases}x_1^{0.25}x_2, x_1>0.05\\ x_1x_2, x_1\le 0.05\end{cases}, \tag{44}$$

$$x_i=\frac{C_i}{\sum_{k=4}^{6}\frac{C_k}{M_k}M_i}, \tag{45}$$

It is believed that the gaseous pyrolysis products instantly come from the forest fuel layer into the region of the gas phase. Moreover, combustible gases flow in a narrow zone to the side of the particle and diffuse in the near-wall region of the particle. The release area of these substances is the localization of elevated temperature in the forest fuel layer and its thermal decomposition. The mass flow in the region below the particle and from the side was summed and distributed in proportion to the factor

T[i,iz]/T[xz,iz], where xz is the lateral boundary of the particle and iz is the lower boundary of the particle. Thus, a decrease in the release of gaseous pyrolysis products with a decrease in the temperature of the forest fuel layer was ensured. Fig. 12 shows the distribution of the volume fraction of dry organic matter at the time of ignition. It can be seen that forest fuel decomposes almost in half in the region of localization of the chemical reaction. Under the particle, its almost complete decomposition occurs. However, the depth of decomposition is not great. In the lateral region below the particle, more significant decomposition occurs. Here, the forest fuel layer borders on the location of the chemical reaction, from which additional heat is supplied.

Figure 12. Distribution of the volume fraction of dry organic matter in the forest layer at the time of ignition in the OYZ section

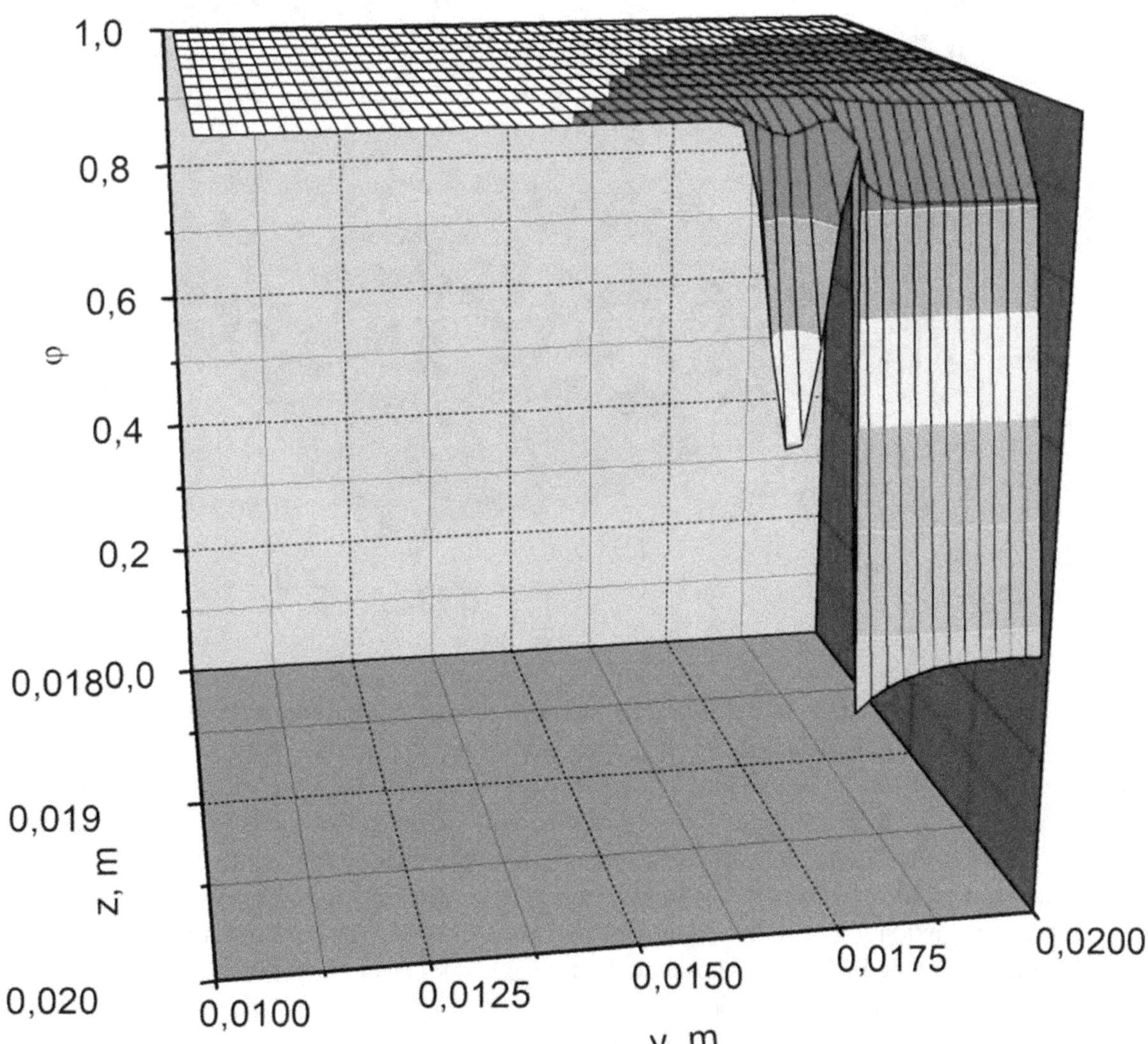

Figure 13. Distribution of the concentration of gaseous pyrolysis products in the gas phase at the time of ignition in the OXZ section

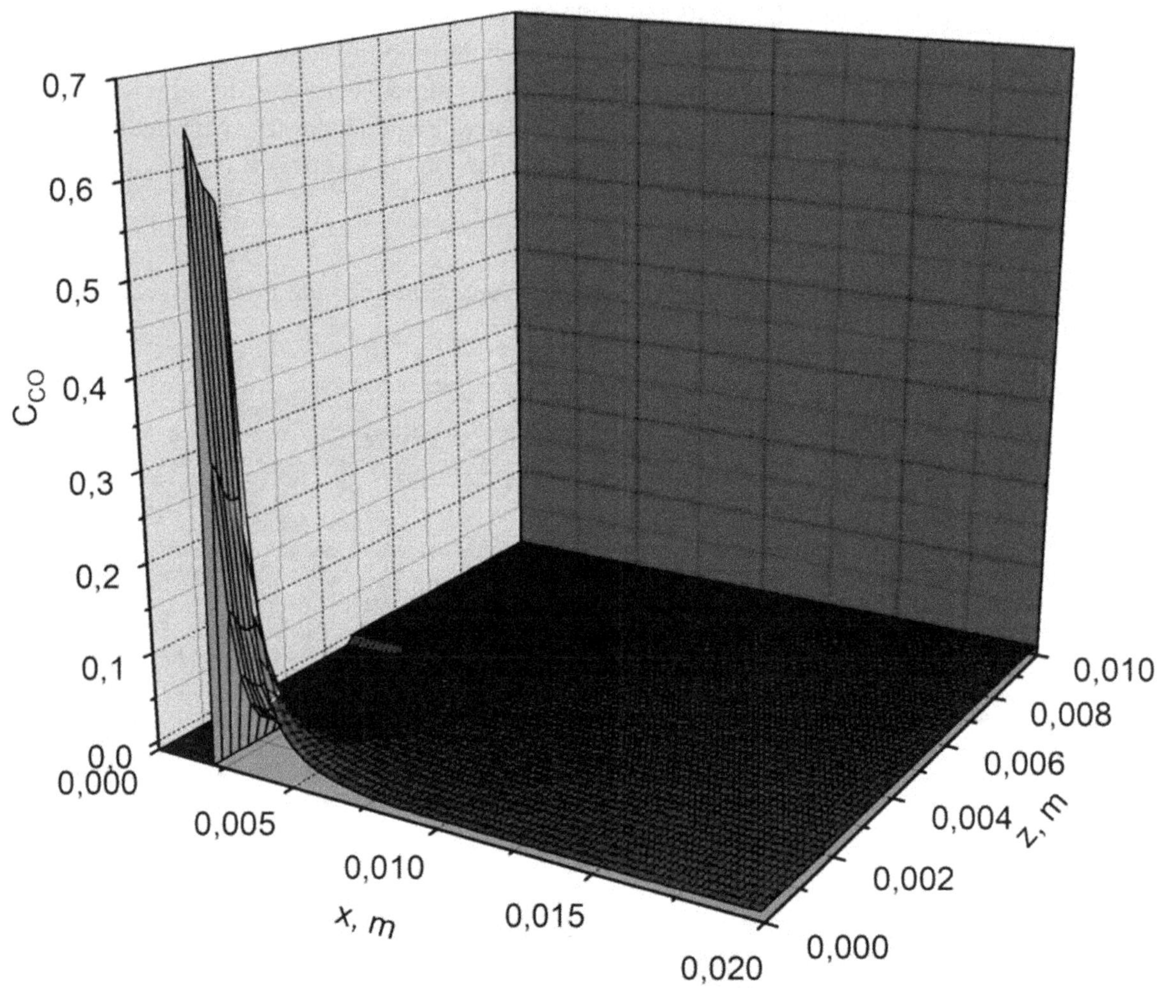

The maximum concentration of pyrolysis components in the wall region of the particle. Moreover, there is a diffusion of combustible gases in the horizontal direction and a shift in the concentration maximum to the side from the immediate boundary of the particle. The concentration distribution of gaseous products of pyrolysis, oxygen, and inert components in the gas phase at the time of ignition is shown in Fig. 13-15. The figures show the distribution of the gas phase components in the OXZ section.

The concentration of oxygen in the near-wall region decreases due to the injection of combustible components and the oxidation of carbon monoxide to carbon dioxide. The concentration of inert components in the immediate vicinity decreases due to their displacement by combustible components. A peak is observed in the distribution of the concentration of inert components, which is explained by the intake of carbon dioxide as a result of the oxidation of carbon monoxide. Compared with the one-dimensional formulation, the concentration of carbon monoxide in the near-wall region is somewhat lower. This is explained by the diffusion of this component in the horizontal direction and the redistribution of the mass flow in the region of injection of the pyrolysis products. Fig. 16 shows the temperature distribution

Figure 14. Distribution of oxygen concentration in the gas phase at the time of ignition in the OXZ section

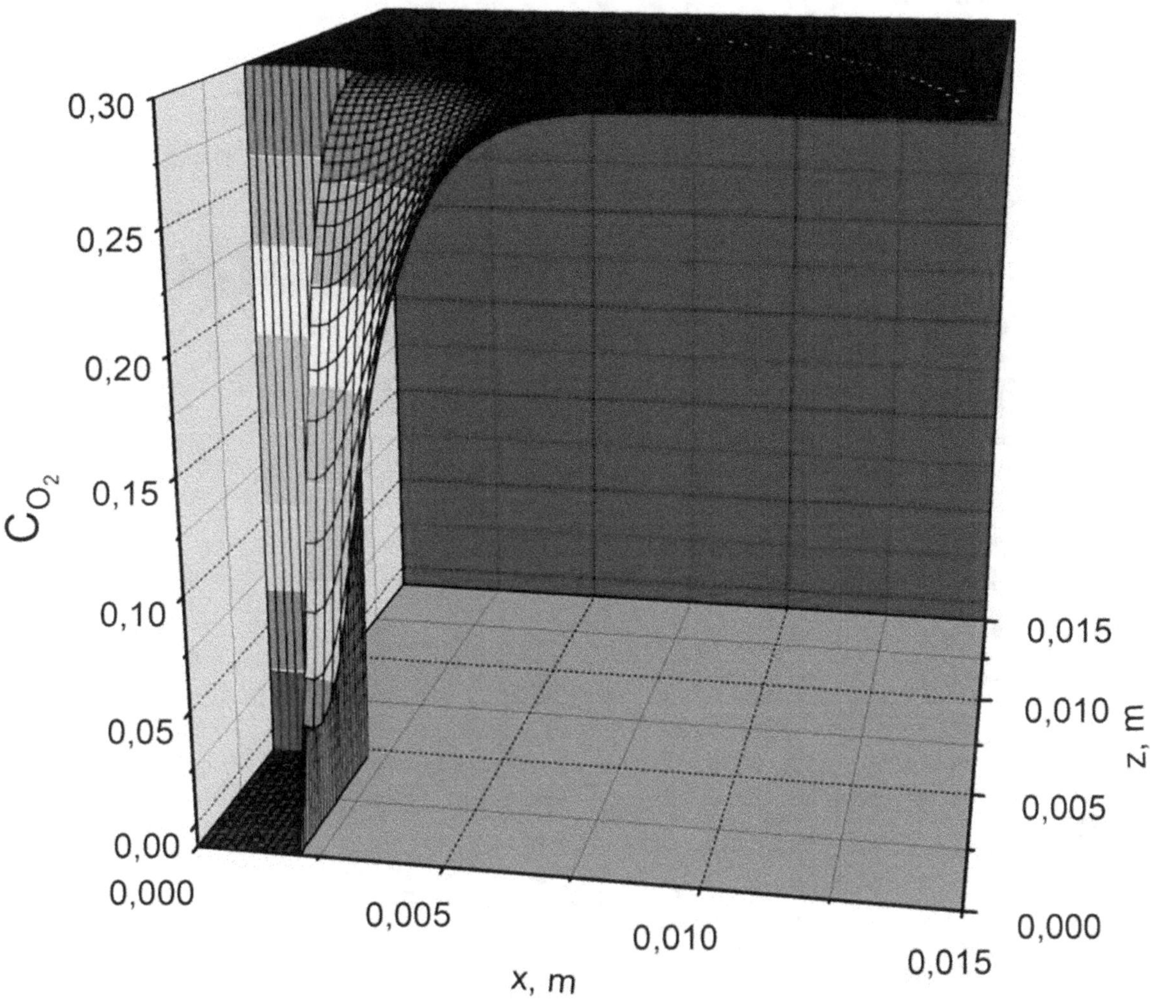

in the "Gas mixture-particle-forest fuel" system at the moment of ignition in the sections OYZ (a) and OXZ (b). In the near-wall region of the particle at a certain distance from the boundary of the "forest fuel-gas mixture layer", a peak is observed in the temperature distribution. It is in this place that the critical ignition conditions are achieved. As the results shown in fig. 16a, ignition can occur on either side of the particle.

Table 5 presents the ignition delay for various particle temperatures. The minimum particle temperature at which ignition is still observed is 900 K. At lower temperatures, ignition does not occur. The presence of a minimum temperature at which ignition of the forest fuel layer still occurs is caused not only by the kinetics of thermal decomposition processes and reactions in the gas phase. In addition, a particle, in contrast to a massive heated body, has a finite heat reserve (Vilyunov, 1984). The ignition delay times differ from those obtained in one-dimensional and two-dimensional formulations. This is primarily due to the greater degree of cooling of the particle in real 3D geometry, in contrast to the "plate" and "rod" approximations. However, this cooling is not so significant as to ignore the possibility

Figure 15. Distribution of the concentration of inert components in the gas phase at the time of ignition in the OXZ section

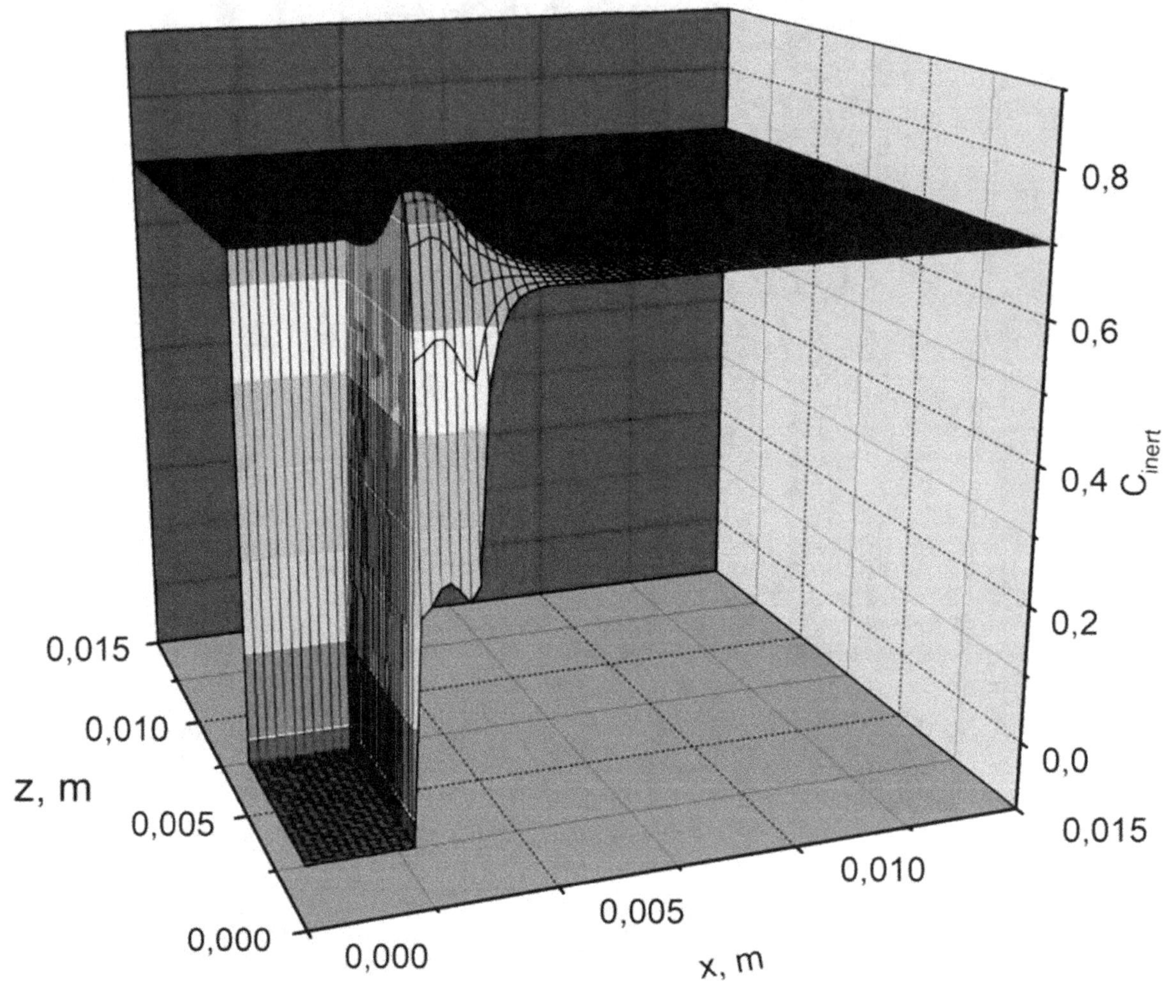

of using a one-dimensional formulation for the operational forecast of forest fire danger. The legitimacy of using simplified settings is confirmed by the obtained data on the determination of the ignition delay time, which are presented in table 6 for various temperatures of the steel particle. The differences in the calculated values of the ignition delay times are determined by the intensification of heat transfer processes in the case of the action of a steel particle on the forest fuel layer.

CONCLUSION

Thus, the current work presents a group of mathematical models of forest fuel layer ignition by the particle heated up to high temperatures. The software implementation of these mathematical models can be performed using up-to-date computing tools (Matlab, 2020; MS Visual Studio, 2020; RAD Studio, 2020). These mathematical models can be used both for the development of new forest fire danger

Figure 16. Temperature distribution in the "Gas mixture-particle-forest fuel" system at the moment of ignition: section OXY (a) and OXZ (b)

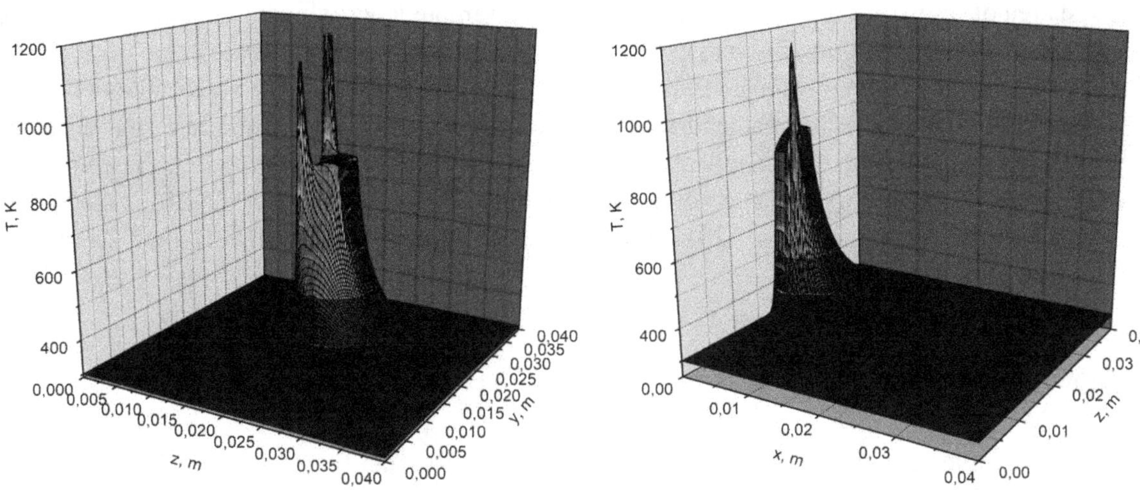

(a) (b)

predicting systems and for the modernization of existing systems (CWFIS, 2020; EFFIS, 2020; WFAS, 2020; Podolskaya et al., 2011).

The Tomsk Polytechnic University is intensively developing a deterministic-probabilistic approach to predicting forest fire danger (Baranovskiy, 2007), which takes into account both thunderstorm activity (Belikova and Glebova, 2020) and anthropogenic load (Baranovskiy, 2020b), as well as meteorological conditions and properties of the forested area (Yankovich and Yankovich, 2020). For scenario modeling, data are used for two regions of Siberia (Tomsk Region and Republic of Buryatia), which are typical representatives of the boreal forest zone.

Table 5. Ignition delay for a carbon particle

Particle Temperature, K	Ignition Delay, s
900	2.73
1000	1.40
1100	0.65
1200	0.24

Table 6. Ignition delay for a steel particle

Particle Temperature, K	Ignition Delay, s
900	1.37
1000	0.77
1100	0.38
1200	0.18

The developed mathematical models of forest fuel layer ignition can be used for scenario modeling in these territories, both for a thunderstorm activity factor and for anthropogenic reasons. It is known that a large area of forests in the Tomsk Region belongs to remote areas where the main cause of forest fires is cloud-to-ground lightning discharges (Belikova and Glebova, 2020). At the same time, most of the Baikal forests are affected by humans, namely, recreational loads (Yankovich and Yankovich, 2020), both for individual tourists and organized recreation at various campsites. Therefore, the developed mathematical model will find direct application in these two regions. Also, this model can be integrated with solutions of other authors (Adab et al., 2013; Amatulli et al., 2007; Bui et al., 2017; Chen et al., 2017; Duan et al., 2001; Ghimire et al., 2012; Grala et al., 2017; Hong et al., 2018; Hu and Zhou, 2014; Jimenez-Ruano et al., 2017).

ACKNOWLEDGMENT

This work was implemented with the financial support of the Russian Foundation for Basic Research. [Project N 16-41-700831 and Project N 17-29-05093]. Scenario calculations on forest fuel layer ignition by the carbonaceous particle were conducted within project N 16-41-700831. Scenario calculations on forest fuel layer ignition by the metal particle were conducted within project N 17-29-05093.

REFERENCES

Adab, H., Kanniah, K. D., & Solaimani, K. (2013). Modeling forest fire in the northeast of Iran using remote sensing and GIS techniques. *Natural Hazards*, *65*(3), 1723–1743. doi:10.100711069-012-0450-8

Al Janabi, S., Al Shourbaji, I., & Salman, M. A. (2017). *Assessing the suitability of soft computing approaches for forest fires prediction*. Applied Computing and Informatics. doi:10.1016/j.aci.2017.09.006

Amatulli, G., Perez-Cabello, F., & Du la Riva, J. (2007). Mapping lightning/human-caused wildfires occurrence under ignition point location uncertainty. *Ecological Modelling*, *200*(3-4), 321–333. doi:10.1016/j.ecolmodel.2006.08.001

Baranovskiy, N. V. (2007). *Mathematical modeling of the most probable scenarios and conditions of forest fires* (PhD Thesis). Tomsk: TSU.

Baranovskiy, N. V. (2020). *Predicting, Monitoring, and Assessing Forest Fire Dangers and Risks*. IGI Global. doi:10.4018/978-1-7998-1867-0

Baranovskiy, N. V. (2020). Mathematical Simulation of Anthropogenic Load on Forested Territories for Point Source. In N. Baranovskiy (Ed.), *Predicting, Monitoring, and Assessing Forest Fire Dangers and Risks* (pp. 64–88). IGI Global. doi:10.4018/978-1-7998-1867-0.ch003

Belikova, M. Y., & Glebova, A. V. (2020). Cluster Analysis Algorithms for RS and WWLLN Data Processing. In N. Baranovskiy (Ed.), *Predicting, Monitoring, and Assessing Forest Fire Dangers and Risks* (pp. 151–181). IGI Global. doi:10.4018/978-1-7998-1867-0.ch007

Brushlinsky, N. N., Ahrens, M., Sokolov, S. V., & Wagner, P. (2019). World fire statistics. International Association of Fire and Rescue Services.

Bui, D. T., Bui, Q.-T., Nguyen, Q.-P., Pradhan, B., Nampak, H., & Trinh, P. T. (2017). A hybrid artificial intelligence using GIS-based neural-fuzzy inference system and particle swarm optimization for forest fire susceptibility modeling at a tropical area. *Agricultural and Forest Meteorology*, *233*, 32–44. doi:10.1016/j.agrformet.2016.11.002

Canadian Wildland Fire Information System. (n.d.). Access: https://cwfis.cfs.nrcan.gc.ca/home

Chen, W., Xie, X., Wang, J., Pradhan, B., Hong, H., Tien Bui, D., Duan, Z., & Ma, J. (2017). A comparative study of logistic model tree, random forest and classification and regression tree models for spatial prediction of landslide susceptibility. *Catena*, *151*, 147–160. doi:10.1016/j.catena.2016.11.032

Duan, K., Keerthi, S., & Poo, A. (2001). *Evaluation of simple performance measures for tuning SVM hyperparameters (Technical report)*. National University of Singapore. Department of Mechanical Engineering.

European Forest Fire Information System. (n.d.). Access: http://effis.jrc.ec.europa.eu

Ghimire, B., Rogan, J., Galiano, V. R., Panday, P., & Neeti, N. (2012). An evaluation of bagging, boosting and random forests for land-cover classification in CapeCod, Massachusetts, USA. *GIScience & Remote Sensing*, *49*(5), 623–643. doi:10.2747/1548-1603.49.5.623

Grala, K., Grala, R. K., Hussain, A., Cooke, W. H. III, & Varner, J. M. III. (2017). Impact of human factors on wildfire occurrence in Mississippi, USA. *Forest Policy and Economics*, *81*, 38–47. doi:10.1016/j.forpol.2017.04.011

Grishin, A. M., & Filkov, A. I. (2011). A deterministic-probabilistic system for predicting forest fire hazard. *Fire Safety Journal*, *46*(1-2), 56–62. doi:10.1016/j.firesaf.2010.09.002

Grishin, A. M., & Shipulina, O. V. (2002). Mathematical model for spread of crown fires in homogeneous forests and along openings. *Combustion, Explosion, and Shock Waves*, *38*(6), 622–632. doi:10.1023/A:1021127924913

Hong, H., Tsangaratos, P., Ilia, I., Liu, J., Zhu, A.-X., & Xu, C. (2018). Applying genetic algorithms to set the optimal combination of forest fire related variables and model forest fire susceptibility based on data mining models. The case of Dayu Country, China. *The Science of the Total Environment*, *630*, 1044–1056. doi:10.1016/j.scitotenv.2018.02.278 PMID:29554726

Hu, T., & Zhou, G. (2014). Drivers of lightning- and human-caused fire regimes in the Great Xing'an Mountains. *Forest Ecology and Management*, *329*, 49–58. doi:10.1016/j.foreco.2014.05.047

Jimenez-Ruano, A., Rodrigues Mimbrero, M., & de la Riva Fernandez, J. (2017). Understanding wildfires in mainland Spain. A comprehensive analysis of fire regime features in a climate-human context. *Applied Geography (Sevenoaks, England)*, *89*, 100–111. doi:10.1016/j.apgeog.2017.10.007

Matlab. (2020). Available https://www.mathworks.com/products/matlab.html

Microsoft Visual Studio. (2020). Available https://visualstudio.microsoft.com

Nesterov, V. G. (1949). *Forest combustibility and methods for its determination.* Goslesbumizdat.

Podolskaya, A.S., Ershov, D.V., & Shulyak, P.P. (2011). Application of the method for assessing the probability of occurrence of forest fires in ISDM-Rosleskhoz. *Modern Problems of Remote Sensing of the Earth From Space, 8*, 118 - 126. (in Russian)

Samarskiy, A. A. (1983). *Theory of difference schemes.* Nauka. (in Russian)

Samarskiy, A. A., & Vabishchevich, P. N. (2001). *Additive schemes for problems of mathematical physics.* Nauka.

StudioR. A. D. (2020). Available https://www.embarcadero.com/ru/products/rad-studio

Vilyunov, V. N. (1984). *Theory of ignition of condensed matter.* Science, Sib. Department.

WFAS Wildland Fire Assessment System. (n.d.). Access: http://www.wfas.net

Yankovich, E. P., & Yankovich, K. S. (2020). Classification of Territory on Forest Fire Danger Level Using GIS and Remote Sensing. In N. Baranovskiy (Ed.), *Predicting, Monitoring, and Assessing Forest Fire Dangers and Risks* (pp. 258–267). IGI Global. doi:10.4018/978-1-7998-1867-0.ch011

ADDITIONAL READING

Antonov, D. V., Volkov, R. S., Voitkov, I. S., Zhdanova, A. O., & Kuznetsov, G. V. (2018). Influence of Special Additives in a Water Aerosol on the Suppression of a Forest Fire with it. *Journal of Engineering Physics and Thermophysics., 91*(5), 1250–1259. doi:10.100710891-018-1855-3

Baranovskii, N. V., & Zakharevich, A. V. (2016). Conditions and Characteristics of the Ignition of a Typical Vegetable Combustible Material by a Local Energy Source. *Journal of Engineering Physics and Thermophysics., 89*(6), 1538–1544. doi:10.100710891-016-1524-3

Baranovskiy, N. V., Sherstnyov, V. S., Yankovich, E. P., Engel, M. V., & Belov, V. V. (2016) Focused sunlight factor of forest fire danger assessment using Web-GIS and RS technologies. Proceedings of SPIE - The International Society for Optical Engineering, 9688, article 968823, 6 P. 10.1117/12.2240378

Baranovskiy, N. V., & Zakharevich, A. V. (2015). *Experimental research of grassy rags ignition by heated up to high temperatures carbon particle.* Multifunctional Materials and Modeling., doi:10.1201/b18552

Baranovskiy, N. V., & Zakharevich, A. V. (2018). Mechanical, thermophysical and aerothermochemical processes at forest fuel ignition caused by the heated to high temperatures single metal particle. *International Review of Mechanical Engineering, 12*(6), 522–532. doi:10.15866/ireme.v12i6.15065

Chen, T., Zhang, J., & Wu, J. (2016). Kinetic and energy production analysis of pyrolysis of lignocellulosic biomass using a three-parallel Gaussian reaction model. *Bioresource Technology, 211*, 502–508. doi:10.1016/j.biortech.2016.03.091 PMID:27035484

Hameed, S., Sharma, A., Pareek, V., Wu, H., & Yu, Y. (2019). A review on biomass pyrolysis models: Kinetic, network and mechanistic models. *Biomass and Bioenergy*, *123*, 104–122. doi:10.1016/j.biombioe.2019.02.008

Kansa, E., Perlee, H., & Chaiken, R. (1977). Mathematical model of wood pyrolysis. *Combustion and Flame*, *29*, 311–324. doi:10.1016/0010-2180(77)90121-3

Liu, H., Wang, C., Zhao, W., Yang, S., & Hou, X. (2018). Pyrolysis characteristics and kinetic modeling of Artemisia apiacea by thermogravimetric analysis. *Journal of Thermal Analysis and Calorimetry*, *131*(2), 1783–1792. doi:10.100710973-017-6599-3

Thomas, J. C., Hadden, R. M., & Simeoni, A. (2017). Experimental investigation of the impact of oxygen flux on the burning dynamics of forest fuel bed. *Fire Safety Journal*, *91*, 855–863. doi:10.1016/j.firesaf.2017.03.086

KEY TERMS AND DEFINITIONS

Anthropogenic Load: Different human activities on forested territories lead to forest fire occurrence and characterized by presence of fire sources.

Forest Fire: Uncontrolled aerothermochemical phenomenon characterized by step-by-step mechanism which includes following stages: inert heating, moisture evaporation, high temperature terpens evaporation, dry organic matter pyrolysis, flammable combustion and smoldering.

Forest Fuel: It can be considered like dead and live forest fuel. Main types of forest fuel which can be involved in combustion during forest fire: ground forest fuel (needles, leaves and dry grass, small branches) and crown forest fuel (needles, small branches).

Ignition: Inflammation of forest fuel caused by definite source of high temperature or energy.

Ignition Delay: Time before flame flash after forest fuel heating.

Lightning Activity: The production of a computer model of forest fire conditions and prerequisites, especially for the purpose of study.

Monitoring: Monitoring refers to the periodic calculation of the parameters of forest fire danger with a portion of information available in real time.

Prediction: Under the prediction of forest fires is the calculation of the parameters of forest fire danger with a certain projection in advance in order to have enough time to anticipate an emergency. The calculation in this case is carried out in a mode ahead of the real time of the development of the catastrophe - the occurrence of a forest fire.

Chapter 13
Mathematical Modeling of Forest Fuel Ignition by the Focused Sunlight

ABSTRACT

This chapter is devoted to numerical simulation of forest fuel ignition by focused solar radiation. A physical model of the forest fuel ignition by focused solar radiation is presented. Three mathematical models of the studied process are presented (one-dimensional, two-dimensional, and three-dimensional). Mathematically, the ignition of the forest fuel layer by focused solar radiation is described by a system of equations of heat conduction and diffusion with the corresponding initial and boundary conditions. The results of scenario modeling are presented.

INTRODUCTION

The occurrence of forest fires is due to various reasons (Baranovskiy and Kuznetsov, 2017). For example, thunderstorm activity and anthropogenic load are the causes of forest fires in many countries of the world community (Read et al., 2018; Ye et al., 2017). Thunderstorm activity is a typical natural cause of forest fires. The anthropogenic load is due to human activity in forested areas. However, forest fires are also possible for the so-called mixed causes. The ignition source may be focused solar radiation (Kuznetsov and Baranovskiy, 2013). Concentrators of solar radiation can be glass containers and their fragments. Such objects remain in the forest after recreational loads.

Radiant heat flux depends on a number of parameters, for example, time of year, latitude, time of day. In temperate regions, the heat flux is about 1 kW/m^2 (Babrauskas, 2003). Existing forest fire danger prediction systems do not take this factor into account for forest fires. In addition, such systems do not take into account the physicochemical processes occurring during the ignition of forest fuel (Babrauskas, 2002). It is necessary to develop mathematical models of the forest fuel layer ignition by focused solar radiation.

DOI: 10.4018/978-1-7998-7250-4.ch013

The purpose of the work is a mathematical simulation of forest fuel layer ignition by focused solar radiation.

PHYSICAL MODEL

The following scheme of the process under study was adopted. On the underlying surface there is a forest fuel layer, on a small area of which the flux of solar radiation is focused. The processes occurring in the focusing element are not modeled due to the lack of both experimental data and the results of theoretical studies on this problem. A layer of forest fuel is heated and thermally decomposed with the formation of gaseous pyrolysis products. The composition of the gas mixture is taken in three components (fuel - carbon monoxide, oxidizing agent - oxygen, inert components). The pyrolysis products diffuse into the region of the gas mixture. At certain temperatures and concentrations of the reacting gases, the mixture ignites. The following ignition criteria are adopted: 1) the heat input from the chemical reaction exceeds the heat flux from the heated surface to the region of the gas mixture; 2) the temperature in the gas mixture reaches a critical value. The convective transfer of heat and matter is not taken into account, since during the ignition period, the thermal and diffusion relaxation lengths are several orders of magnitude longer than the convective (Vilyunov, 1984).

Designations: where T_i, ρ_i, c_i, λ_i are temperature, density, heat capacity, thermal conductivity (1 is the layer of forest fuel, 2 is the air); C_i, M_i are concentration and molar mass (4 is the oxygen, 5 is the combustible gas, 6 is the inert components of air); q_p is the thermal effect of the pyrolysis reaction of forest fuel; k_1 is the pre-exponent of the pyrolysis reaction of forest fuel; E_1 is the activation energy of the pyrolysis reaction of forest fuel; R is the universal gas constant; φ is the volume fraction of dry organic matter of forest fuel; q_5 is the thermal effect of the oxidation reaction of carbon monoxide; ν_5 is the fraction of heat absorbed by a layer of forest fuel; R_5 is the mass rate of the oxidation reaction of carbon monoxide; α_1 is the heat transfer coefficient; α_2 is the heat transfer coefficient; k_5 is the pre-exponent of the oxidation reaction of carbon monoxide; E_5 is the activation energy of the oxidation reaction of carbon monoxide; D is the diffusion coefficient, Y_5 is the mass flow of combustible pyrolysis products, x_i are auxiliary variable; q_s is the flux of focused solar radiation. x,y,z are spatial coordinate. t is the time coordinate. The indices es, ea, n correspond to environmental parameters in the soil, air, and at the initial time.

Initial data: ρ_1=500 kg/m^3; ρ_2=0.1 kg/m^3; c_1=1400 J/(kg×K); c_2=1200 J/(kg×K); λ_1=0.102 W/(m×K); λ_2=0.1 W/(m×K); q_p=1000 J/kg; k_1=3.63×10^4; E_1/R=9400 K; $\varphi_{1н}$=1; q_5=10^7 J/kg; k_5=3×10^{13} s^{-1}; E_5/R=11500 K; ν_5=0.3; α_1=20 W/(m^2×K); α_2=80 W/(m^2×K); D=10^{-6}; M_4=0.032; M_5=0.028; M_6=0.044.

1D MATHEMATICAL MODEL

In Figure 1 shows the geometry of the solution domain. Symbols B denote the boundaries of the solution region and various layers.

The process of forest fuel layer ignition by a focused solar flux is described by a system of one-dimensional non-stationary non-linear equations of heat conduction and diffusion with the corresponding initial and boundary conditions. The numerical implementation was carried out using the finite-difference method. Difference analogues of the one-dimensional equations of heat conduction and diffusion are

Figure 1. Solution Area

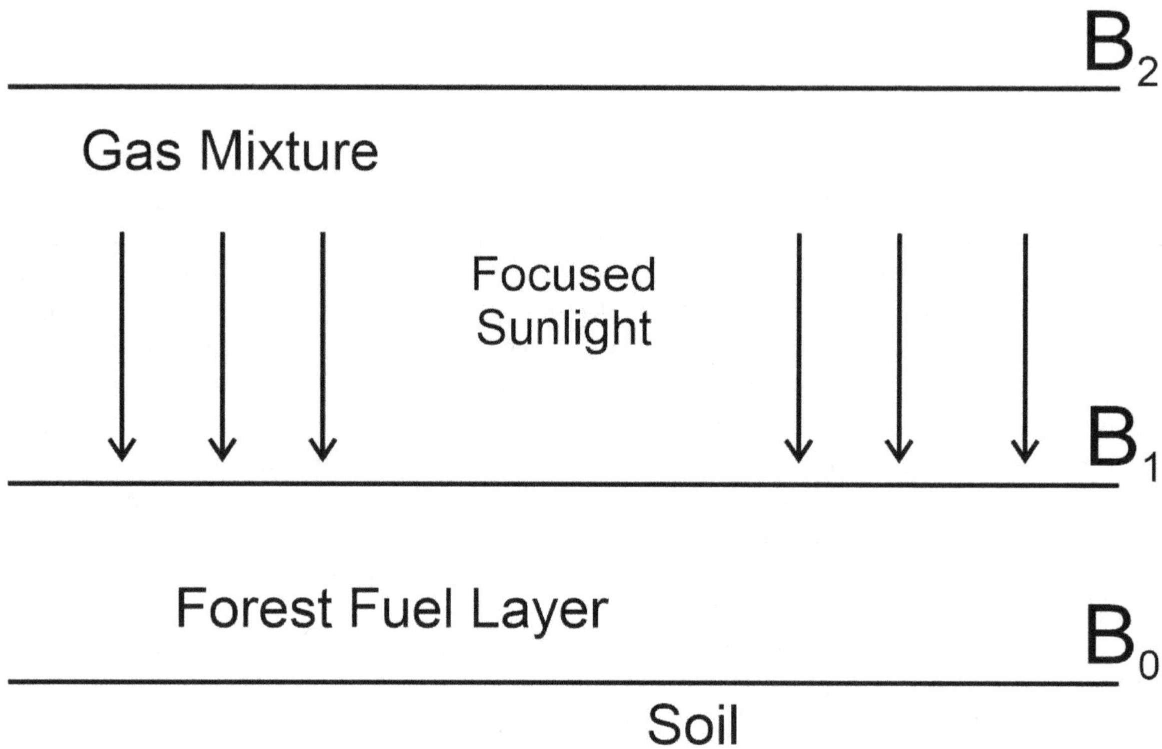

solved by the marching method in combination with the simple iteration method (Samarskiy, 1983). The algorithm of the program has been tested on various problems of thermal conductivity.

The energy equation for the forest fuel layer:

$$\rho_1 c_1 \frac{\partial T_1}{\partial t} = \lambda_1 \frac{\partial^2 T_1}{\partial z^2} + q_p k_1 \rho_3 \phi \exp\left(-\frac{E_1}{RT_1}\right), \tag{1}$$

The energy equation for a gas mixture:

$$\rho_2 c_2 \frac{\partial T_2}{\partial t} = \lambda_2 \frac{\partial^2 T_2}{\partial z^2} + q_5 (1 - \nu_5) R_5 . \tag{2}$$

The boundary conditions for equations (1) - (2):

$$\mathrm{B}_0 \, \alpha_1 (T - T_{es}) = \lambda_1 \frac{\partial T_1}{\partial z}, \tag{3.1}$$

$$\mathrm{B}_1 \lambda_1 \frac{\partial T_1}{\partial z} = \lambda_2 \frac{\partial T_2}{\partial z} + q_s \,, T_1 = T_2, \tag{3.2}$$

$$\mathrm{B}_2 \alpha_2 (T_{ea} - T) = \lambda_2 \frac{\partial T_2}{\partial z} \,. \tag{3.3}$$

The initial conditions for equations (1) - (2):

$$T_i \Big|_{t=0} = T_{i0} \,, i=1,2 \tag{4}$$

Kinetic equation and initial condition:

$$\rho_1 \frac{\partial \phi}{\partial t} = -k_p \rho_1 \phi \exp\left(-\frac{E_1}{RT_1}\right), \quad \phi\Big|_{t=0} = \phi_0 \,, \tag{5}$$

The diffusion equation for an oxidizing agent:

$$\frac{\partial C_4}{\partial t} = D \frac{\partial^2 C_4}{\partial z^2} - \frac{M_4}{M_5} R_5 \,. \tag{6}$$

The boundary conditions for equation (6):

$$\mathrm{B}_1 \rho D \frac{\partial C_4}{\partial z} = 0 \,, \tag{7.1}$$

$$\mathrm{B}_2 \rho D \frac{\partial C_4}{\partial z} = 0 \,. \tag{7.2}$$

The initial conditions for equation (6):

$$C_4 \Big|_{t=0} = C_{4.0} \,. \tag{8}$$

The diffusion equation for the combustible components of pyrolysis:

$$\frac{\partial C_5}{\partial t} = D \frac{\partial^2 C_5}{\partial z^2} - R_5 \,. \tag{9}$$

The boundary conditions for equation (9):

$$B_1 \rho D \frac{\partial C_5}{\partial z} = Y_5, \tag{10.1}$$

$$B_2 \rho D \frac{\partial C_5}{\partial z} = 0. \tag{10.2}$$

The initial conditions for equation (9):

$$C_5\big|_{t=0} = C_{5.0}. \tag{11}$$

Mass balance equation:

$$\sum_{i=4}^{6} C_i = 1. \tag{12}$$

The initial conditions for equation (12):

$$C_6\big|_{t=0} = C_{6.0}. \tag{13}$$

The expression for the mass reaction rate R_5 (Grishin and Shipulina, 2002):

$$R_5 = k_5 M_5 T^{-2.25} \exp\left(-\frac{E_5}{RT_1}\right) \cdot \begin{cases} x_1^{0.25} x_2, x_1 > 0.05 \\ x_1 x_2, x_1 \le 0.05 \end{cases}, \tag{14}$$

$$x_i = \frac{C_i}{\sum_{k=4}^{6} \frac{C_k}{M_k} M_i}, \tag{15}$$

The objective of the study was to determine the lower limit of the flux of focused solar radiation, at which the ignition of a layer of forest fuel is possible. Table 1 presents the results of a numerical calculation of the ignition delay and known experimental data (Kasperov and Goman, 2010). The lower limit q_s at which ignition of the forest fuel layer is possible, according to the results of a numerical study, was 15 kW/m^2. This value is approximately 10 times higher than the value of the unfocused heat flux of solar radiation (Scientific-applied reference…, 1993). The average deviation of the results of numerical simulation from experimental data (Kasperov and Goman, 2010) was about 42%.

Table 1. Ignition delay of forest fuel layer by radiant heat flux

Heat flux, q_s, kW/m²	Ignition delay (calculation), t_{ign}, s	Ignition delay (experiment (Kasperov and Goman, 2010), t_{ign}, s
15	96	No Ignition
20	59	197-207
25	31	93-100
30	24	40-42
35	20	27-30
40	18	18-26

According to (Grishin, 1997), the error in solving the problems of the theory of forest fires lies in the range from 15 to 580%. Thus, the results obtained can be considered satisfactory from the point of view of their correspondence to experimental data (Kasperov and Goman, 2010). It should be noted that the difference between the theoretical and experimental values of the ignition delay decreases with an increase in the flux of focused radiation.

It can also be noted that the experimental values of the ignition delays in the entire range of q_s exceed the theoretical ones. This is obviously due to the presence of moisture in the forest fuel in experiments (Kasperov and Goman, 2010). In addition, all forest fuels differ significantly in composition. Accordingly, the thermochemical characteristics of, for example, pine needles from different regions of Russia and Belarus differ. The various kinetic parameters of the needles pyrolysis processes with which experiments were carried out (Kasperov and Goman, 2010) and needles for which numerical simulation was carried out are a source of certain deviations in the ignition delays of forest fuel.

A typical temperature distribution along the vertical coordinate in the "forest fuel-gas mixture" system at the time of ignition is shown in Figure 2. The peak in the temperature curve corresponds to the location of the chemical reaction of the carbon monoxide oxidation to carbon dioxide. Figure 3 shows the distribution of component concentrations in the gas phase at the time of ignition.

It should be noted that for more accurate forecasts and assessments of the fire danger of forest fuels under the considered conditions, special studies of the scale of solar radiation concentration as it passes through large resin droplets and glass containers of various sizes, partially or completely filled with water, are also required. In this case, it is advisable to analyze the influence of the orientation of the "concentrator" of radiant energy relative to the surface of the forest fuel and the direction of the flow of solar radiation. Based on the results obtained, it can be concluded that there is a high probability of fires of dry forest fuelss when exposed to concentrated fluxes of solar radiation. Moreover, the degree of energy concentration is not very high and seems quite achievable in practice if solar radiation passes not only through empty or partially filled containers (glass jars or bottles) or their fragments, but, possibly, through large drops of coniferous resin.

2D MATHEMATICAL MODEL

When modeling the processes of heating and thermal decomposition of forest combustible material, it was assumed that possible processes of shrinkage or expansion did not occur in it (Strakhov et al., 2001),

Figure 2. Temperature distribution in the system "forest fuel layer-gas mixture" at the moment of ignition at q_s=15000 W/m²

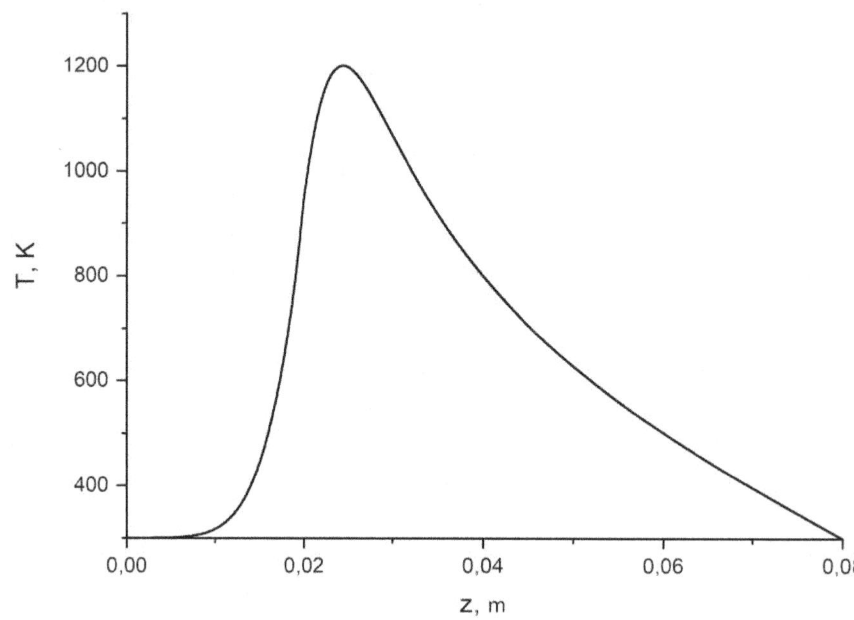

as well as any thermomechanical processes (Kuznetsov, 1998) accompanied by dispersion. Typical forest fuels (for example, needles) are characterized by very high values of initial porosity, which virtually eliminates the possibility of deformation of a layer of heated and decaying material (expansion or

Figure 3. Distribution of gas phase components at the moment of ignition at q_s=15000 W/m²

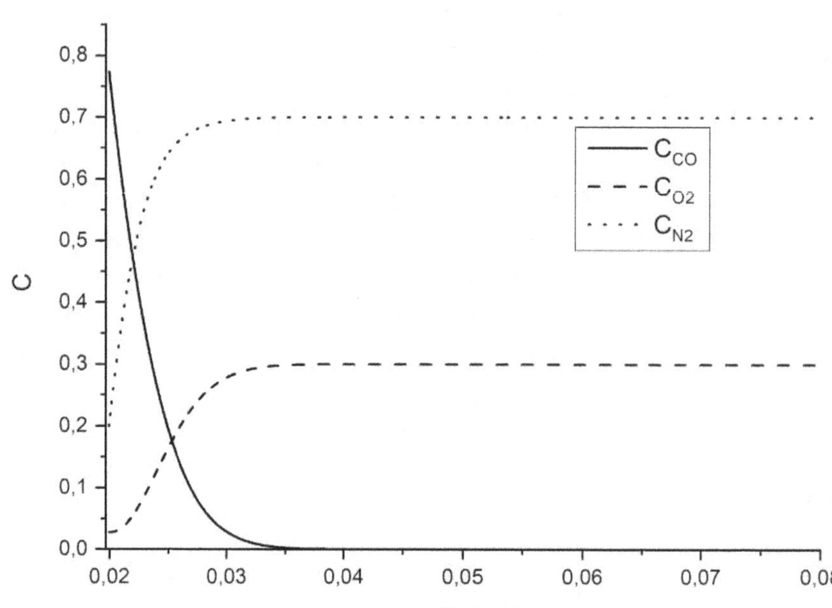

shrinkage). Such processes are characteristic, for example, of heat-shielding materials that decompose upon heating to high temperatures (Strakhov et al., 2001; Kuznetsov, 1998).

When setting the task, it was also assumed that the transfer of fuel (gaseous pyrolysis products) in the air layer above the surface of the forest fuel material is carried out only due to diffusion. In contrast to the conditions for ignition of combustible liquids by local sources (Kuznetsov and Strizhak, 2008a,2008b), when convection is decisive, and for solving the problems of heat and mass transfer in a gas mixture above the heating and evaporation surface of a fuel, it is necessary to take into account the processes of natural or mixed convection (Kuznetsov and Sheremet, 2006; 2009), the intensity of injection of combustible The products of pyrolysis of forest fuel into the reaction zone are 100 or more times lower than the mass flow rates of flammable liquid vapors upon their ignition (Kuznetsov and Strizhak, 2008a,2008b).

Figure 4 shows the geometry of the solution domain. Symbols B denote the boundaries of the solution region and various layers.

Figure 4. Solution Area

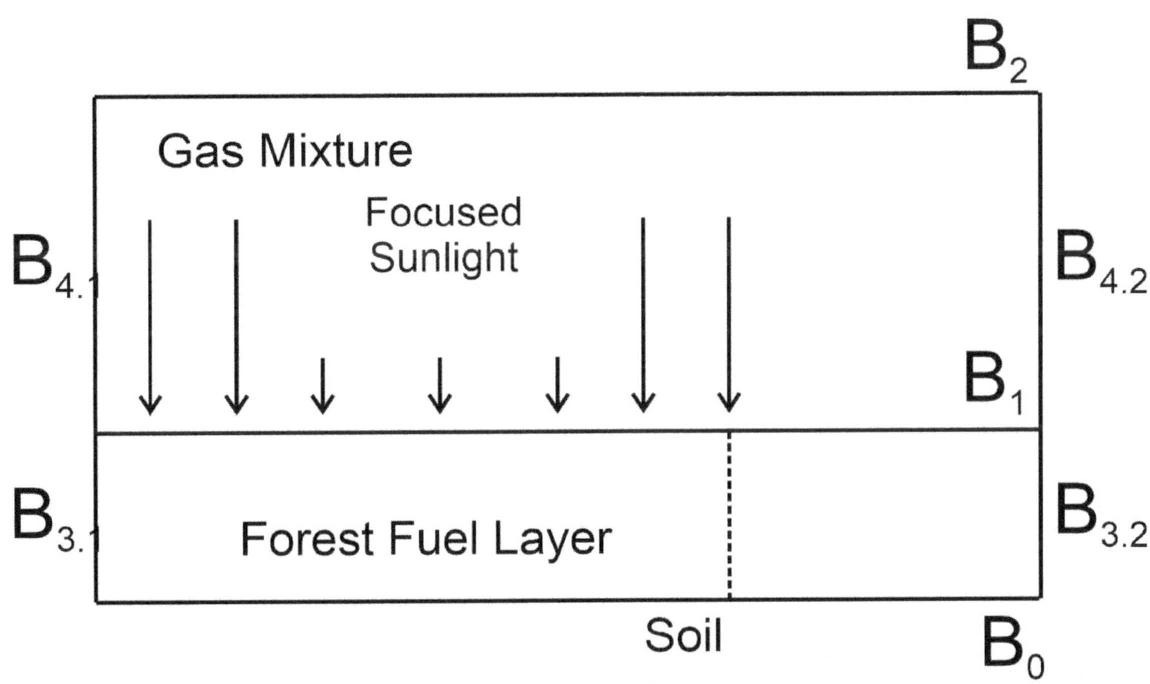

The process of ignition of forest fuel layer by a focused solar flux is described by a system of two-dimensional non-stationary non-linear equations of heat conduction and diffusion with the corresponding initial and boundary conditions. For numerical implementation, the finite-difference method is used. Two-dimensional difference equations are solved by the locally one-dimensional method (Samarskiy and Vabishchevich, 2001). Difference analogues of the one-dimensional equations of heat conduction and diffusion are solved by the marching method in combination with the simple iteration method (Samarskiy, 1983). The algorithm of the program was tested on various less complicated problems compared with the solved heat and mass transfer.

The energy equation for the forest fuel layer:

$$\rho_1 c_1 \frac{\partial T_1}{\partial t} = \lambda_1 \left(\frac{\partial^2 T_1}{\partial x^2} + \frac{\partial^2 T_1}{\partial z^2} \right) - q_p k_1 \rho_3 \phi \exp\left(-\frac{E_1}{RT_1} \right), \tag{16}$$

The energy equation for a gas mixture:

$$\rho_2 c_2 \frac{\partial T_2}{\partial t} = \lambda_2 \left(\frac{\partial^2 T_2}{\partial x^2} + \frac{\partial^2 T_2}{\partial z^2} \right) + q_5 (1 - \nu_5) R_5 . \tag{17}$$

The boundary conditions for equations (16) - (17):

$$\mathrm{B}_0 \alpha_1 (T - T_{es}) = \lambda_1 \frac{\partial T_1}{\partial z}, \tag{18.1}$$

$$\mathrm{B}_1 \lambda_1 \frac{\partial T_1}{\partial z} = \lambda_2 \frac{\partial T_2}{\partial z} + q_s,\ T_1 = T_2, \tag{18.2}$$

$$\mathrm{B}_2 \alpha_2 (T_{ea} - T) = \lambda_2 \frac{\partial T_2}{\partial z}, \tag{18.3}$$

$$\mathrm{B}_{3.1} \lambda_1 \frac{\partial T_1}{\partial x} = 0, \tag{18.4}$$

$$\mathrm{B}_{3.2} T_1 = T_{1S}, \tag{18.5}$$

$$\mathrm{B}_{4.1} \lambda_2 \frac{\partial T_2}{\partial x} = 0, \tag{18.6}$$

$$\mathrm{B}_{4.2} \alpha_2 (T_{ea} - T) = \lambda_2 \frac{\partial T_2}{\partial x}. \tag{18.7}$$

The initial conditions for equations (16) - (17):

$$T_i \big|_{t=0} = T_{i0},\ i=1,2 \tag{19}$$

The kinetic equation and initial condition:

$$\rho_1 \frac{\partial \phi}{\partial t} = -k_p \rho_1 \phi \exp\left(-\frac{E_1}{RT_1}\right), \quad \phi\big|_{t=0} = \phi_0 . \tag{20}$$

The diffusion equation for an oxidizing agent:

$$\frac{\partial C_4}{\partial t} = D\left(\frac{\partial^2 C_4}{\partial x^2} + \frac{\partial^2 C_4}{\partial z^2}\right) - \frac{M_4}{M_5} R_5 . \tag{21}$$

The boundary conditions for equation (21):

$$\mathrm{B}_1 \rho D \frac{\partial C_4}{\partial z} = 0 , \tag{22.1}$$

$$\mathrm{B}_2 \rho D \frac{\partial C_4}{\partial z} = 0 , \tag{22.2}$$

$$\mathrm{B}_{4.1} \rho D \frac{\partial C_4}{\partial x} = 0 , \tag{22.3}$$

$$\mathrm{B}_{4.2} \rho D \frac{\partial C_4}{\partial x} = 0 . \tag{22.4}$$

The initial conditions for equation (21):

$$C_4\big|_{t=0} = C_{4.0} . \tag{23}$$

The diffusion equation for the combustible components of pyrolysis:

$$\frac{\partial C_5}{\partial t} = D\left(\frac{\partial^2 C_5}{\partial x^2} + \frac{\partial^2 C_5}{\partial z^2}\right) - R_5 . \tag{24}$$

The boundary conditions for equation (24):

$$\mathrm{B}_1 \rho D \frac{\partial C_5}{\partial z} = Y_5 , \tag{25.1}$$

$$\mathrm{B}_2 \rho D \frac{\partial C_5}{\partial z} = 0 , \tag{25.2}$$

$$\mathrm{B}_{4.1} \rho D \frac{\partial C_5}{\partial x} = 0 , \tag{25.3}$$

$$\mathrm{B}_{4.2} \rho D \frac{\partial C_5}{\partial x} = 0 . \tag{25.4}$$

The initial conditions for equation (24):

$$C_5 \big|_{t=0} = C_{5.0} . \tag{26}$$

The mass balance equation:

$$\sum_{i=4}^{6} C_i = 1 . \tag{27}$$

The initial conditions for equation (27):

$$C_6 \big|_{t=0} = C_{6.0} . \tag{28}$$

The expression for the mass reaction rate R_5 (Grishin and Shipulina, 2002):

$$R_5 = k_5 M_5 T^{-2.25} \exp\left(-\frac{E_5}{RT_1} \right) \cdot \begin{cases} x_1^{0.25} x_2, x_1 > 0.05 \\ x_1 x_2, x_1 \le 0.05 \end{cases} , \tag{29}$$

$$x_i = \frac{C_i}{\sum_{k=4}^{6} \frac{C_k}{M_k} M_i} , \tag{30}$$

As a result of modeling, the lower limit of the focused solar radiation flux is theoretically established, at which the ignition of a layer of forest combustible material is possible. Table 2 presents the results of numerical determination of ignition delays and known experimental data (Kasperov and Goman, 2010). The minimum flux of focused solar radiation was 15 kW/m^2. That is, a 10-fold excess of the flux of natural natural solar radiation (Scientific-applied reference…, 1993) can lead to a forest fire.

The average deviation of the results of numerical simulation from experimental data was less than 40%. Similar numerical studies were carried out to determine the ignition delay times of forest fuel using a one-dimensional model of heat and mass transfer, ceteris paribus. It has been established that there is an increase in the ignition delay in the range of 15-35 kW/m^2 compared to the one-dimensional setting. This suggests a more adequate simulation of the ignition process in a flat setting. When the flux of focused solar radiation is equal to or greater than 40 kW/m^2, the differences in the ignition delay for one-dimensional and two-dimensional settings are about several 0.01 s. It should also be noted that the differences in the ignition delay, determined theoretically and experimentally (Kasperov and Goman, 2010), decrease with an increase in the flux of focused solar radiation. This fact is probably due to the fact that under the influence of relatively high heat fluxes, differences in the structural and physical characteristics of the forest fuel layer are leveled out by intensification of thermophysical and physico-chemical processes. The obtained t_{ign} values can be considered satisfactory from the point of view of correspondence of theoretical results to experimental data.

Table 2. Ignition delay of forest fuel layer by radiant heat flux

Heat flux, q_s, kW/m^2	Ignition delay (1D calculation), t_{ign}, s	Ignition delay (2D calculation), t_{ign}, s	Ignition delay (experiment (Kasperov and Goman, 2010), t_{ign}, s
15	96	99	No Ignition
20	59	62	197-207
25	31	33	93-100
30	24	26	40-42
35	20	21	27-30
40	18	18	18-26

It should be noted that the experimental values of the ignition delay in the entire range of q_s exceed the theoretical ones. This is probably due to the presence of moisture in the forest combustible material or the difference in the density of the layer in the experiments (Kasperov and Goman, 2010).

A typical temperature distribution in the Oxz plane in the "forest fuel-gas mixture" system at the time of ignition is shown in Figure 5. The peak on the temperature surface corresponds to the location of the chemical reaction of the oxidation of carbon monoxide to carbon dioxide. Ignition criteria are met in the center of the focused solar radiation exposure area. Acceleration of the chemical reaction is also observed at the edges of the strip and in its vicinity. However, heat in these areas is not enough to ignite. Figure 6 shows the distribution of the volume fraction of dry organic matter in the Oxz plane at the time of ignition. Almost complete decomposition of forest fuel is observed in the area affected by the flow of focused solar radiation. However, a partial decomposition of forest fuel is also observed behind the impact site.

Figure 5. Temperature distribution in the system "forest fuel layer-gas mixture" at the moment of ignition at q_s=15000 W/m²

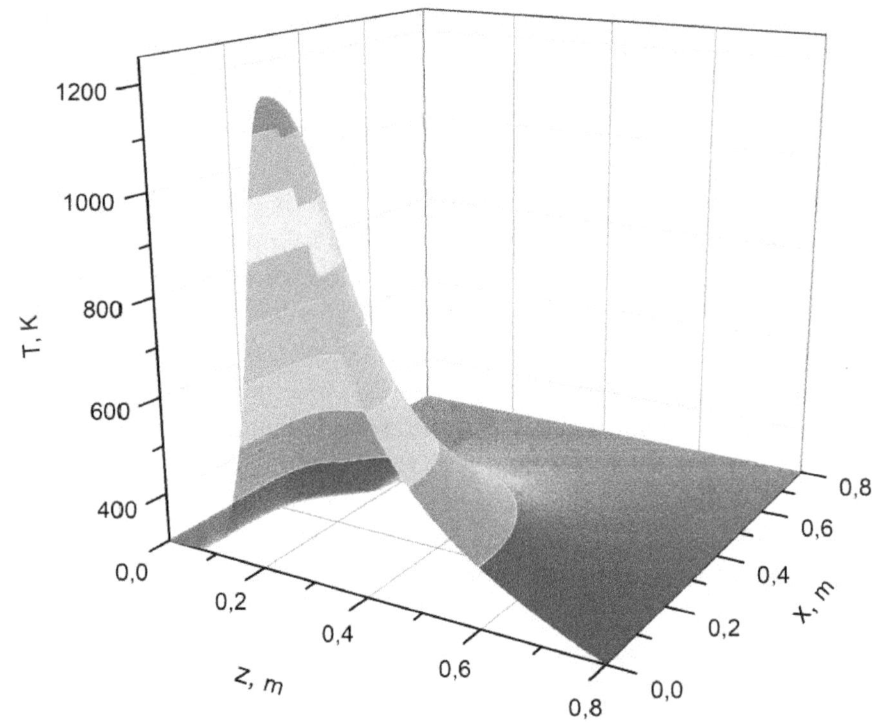

Figures 7-9 show the distribution of concentrations of the components of the gas mixture at the time of ignition. In contrast to the one-dimensional formulation, in the planar case, diffusion of carbon monoxide is observed in the horizontal direction, which leads to an insignificant decrease in the concentration in the center of the site of exposure to radiant heat flux. Therefore, the ignition delay time increases slightly (later, the required fuel concentration for ignition is reached). Carbon monoxide diffuses beyond the boundaries of the exposure zone and, as a result, additional heating of the mixture in this area. However, the ignition criteria are not met here.

3D MATHEMATICAL MODEL

Figure 10 shows the geometry of the solution domain. Symbols B indicate the boundaries of the solution region and various layers.

Ignition of the forest fuel layer by a focused solar flux is described by a system of three-dimensional non-stationary non-linear equations of heat conduction and diffusion with the corresponding initial and boundary conditions. For numerical implementation, the finite-difference method is used. Two-dimensional and three-dimensional difference equations are solved by the locally one-dimensional method (Samarskiy and Vabishchevich, 2001). Difference analogues of the one-dimensional equations of heat conduction and diffusion are solved by the marching method in combination with the simple iteration

Figure 6. Distribution of the volume fraction of dry organic matter at the time of ignition at q_s=15000 W/m^2

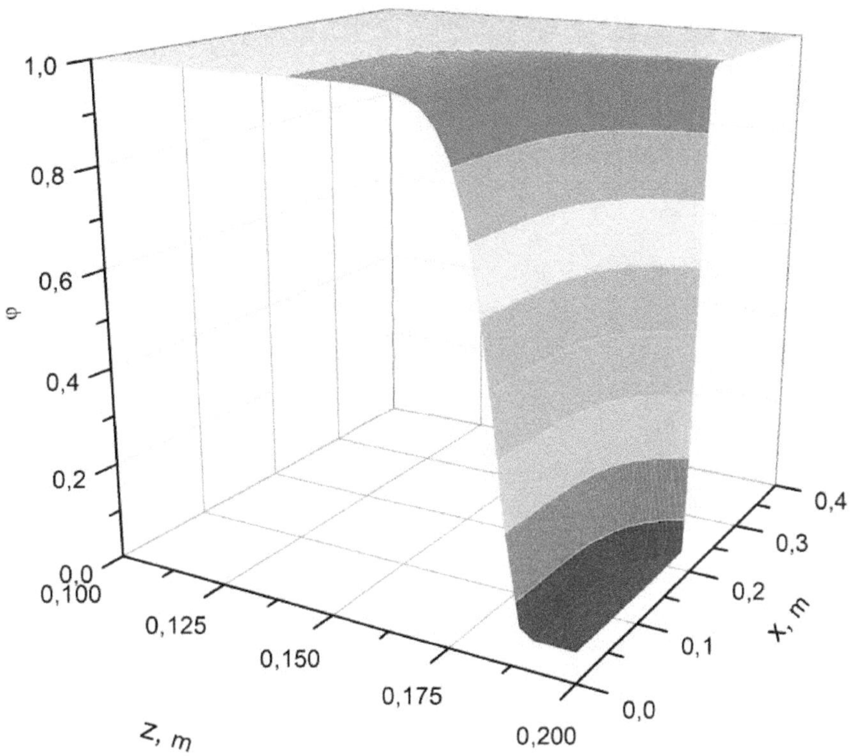

Figure 7. Distribution of combustible components of the gas phase at the time of ignition at q_s=15000 W/m^2

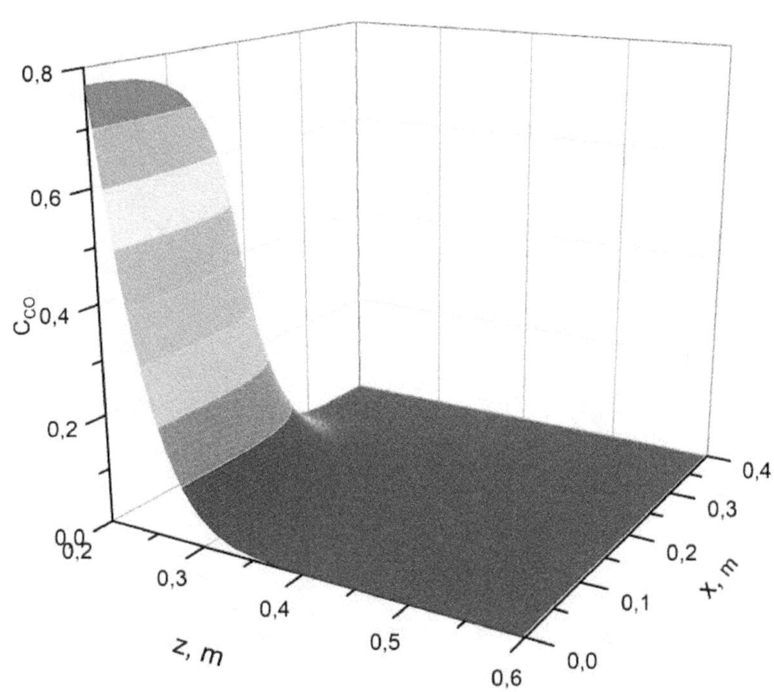

Figure 8. Distribution of the oxidizer of the gas phase at the time of ignition at q_s=15000 W/m²

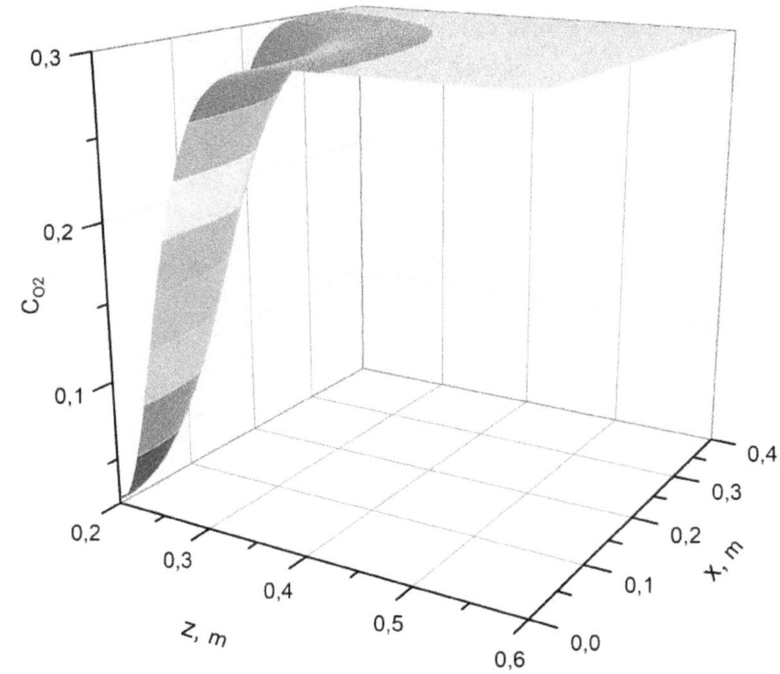

Figure 9. Distribution of inert components of the gas phase at the time of ignition at q_s=15000 W/m²

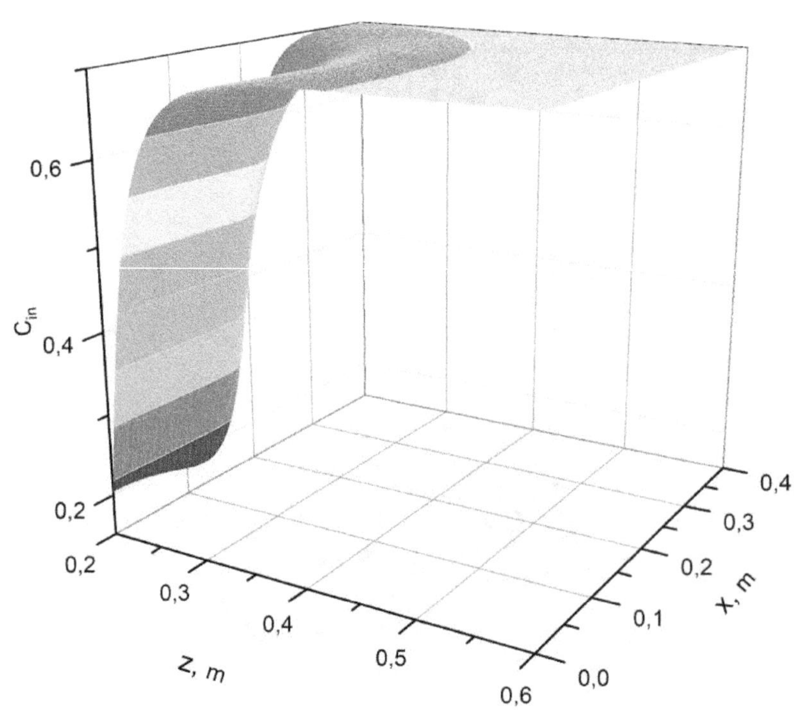

Figure 10. Solution Area

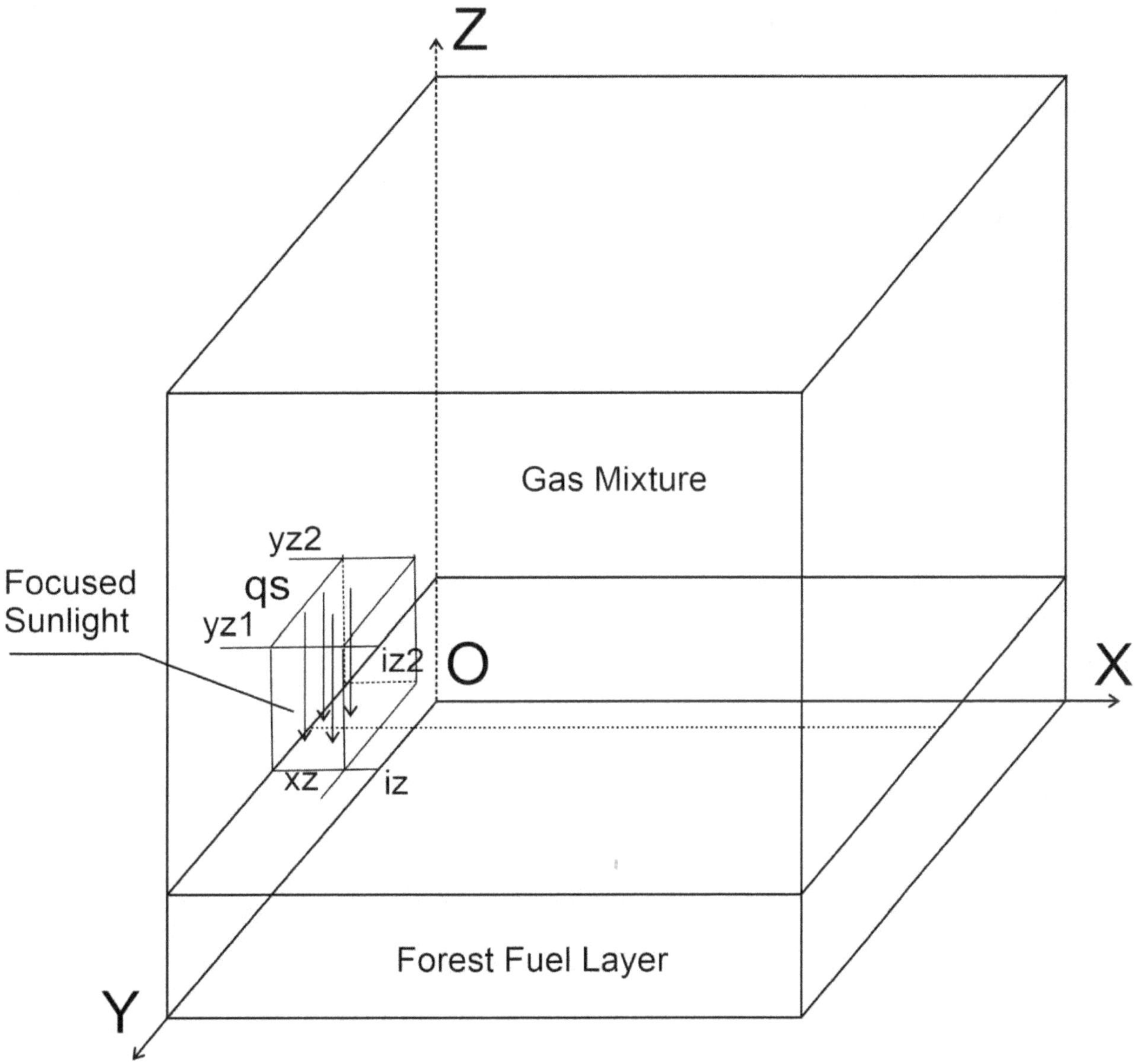

method (Samarskiy, 1983). The algorithm of the program was tested on various less complicated problems compared with the solved heat and mass transfer.

The energy equation for the LGM layer:

$$\rho_1 c_1 \frac{\partial T_1}{\partial t} = \lambda_1 \left(\frac{\partial^2 T_1}{\partial x^2} + \frac{\partial^2 T_1}{\partial y^2} + \frac{\partial^2 T_1}{\partial z^2} \right) - q_p k_1 \rho_3 \phi \exp\left(-\frac{E_1}{RT_1} \right), \tag{31}$$

The energy equation for a gas mixture:

$$\rho_2 c_2 \frac{\partial T_2}{\partial t} = \lambda_2 \left(\frac{\partial^2 T_2}{\partial x^2} + \frac{\partial^2 T_2}{\partial y^2} + \frac{\partial^2 T_2}{\partial z^2} \right) + q_5 (1 - \nu_5) R_5 . \tag{32}$$

The boundary conditions for equations (31) - (32):

$$\mathrm{B}_0 \, \alpha_1 (T - T_{es}) = \lambda_1 \frac{\partial T_1}{\partial z} , \tag{33.1}$$

$$\mathrm{B}_1 \, \lambda_1 \frac{\partial T_1}{\partial z} = \lambda_2 \frac{\partial T_2}{\partial z} + q_s \, , \, T_1 = T_2 , \tag{33.2}$$

$$\lambda_1 \frac{\partial T_1}{\partial z} = \lambda_2 \frac{\partial T_2}{\partial z} \, , \, T_1 = T_2 , \tag{33.3}$$

$$\mathrm{B}_2 \, \alpha_2 (T_{ea} - T) = \lambda_2 \frac{\partial T_2}{\partial z} , \tag{33.4}$$

$$\mathrm{B}_{3.1} \, \lambda_1 \frac{\partial T_1}{\partial x} = 0 , \tag{33.5}$$

$$\mathrm{B}_{3.2} \, T_1 = T_{1S} , \tag{33.6}$$

$$\mathrm{B}_{3.3} \, \lambda_2 \frac{\partial T_2}{\partial x} = 0 , \tag{33.7}$$

$$\mathrm{B}_{3.4} \, \alpha_2 (T_{ea} - T) = \lambda_2 \frac{\partial T_2}{\partial x} , \tag{33.8}$$

$$\mathrm{B}_{4.1} \, \alpha_1 (T - T_{es}) = \lambda_1 \frac{\partial T_1}{\partial y} , \tag{33.9}$$

$$\mathrm{B}_{4.2}\,\alpha_2(T - T_{ea}) = \lambda_2 \frac{\partial T_2}{\partial y}, \tag{33.10}$$

$$\mathrm{B}_{4.3}\,\alpha_2(T_{es} - T) = \lambda_1 \frac{\partial T_1}{\partial y}, \tag{33.11}$$

$$\mathrm{B}_{4.4}\,\alpha_2(T_{ea} - T) = \lambda_2 \frac{\partial T_2}{\partial y},. \tag{33.12}$$

The initial conditions for equations (31) - (32):

$$T_i\big|_{t=0} = T_{i0},\ i=1,2 \tag{34}$$

The kinetic equation and initial condition:

$$\rho_1 \frac{\partial \phi}{\partial t} = -k_p \rho_1 \phi \exp\left(-\frac{E_1}{RT_1}\right),\quad \phi\big|_{t=0} = \phi_0\,. \tag{35}$$

The diffusion equation for an oxidizing agent:

$$\frac{\partial C_4}{\partial t} = D\left(\frac{\partial^2 C_4}{\partial x^2} + \frac{\partial^2 C_4}{\partial y^2} + \frac{\partial^2 C_4}{\partial z^2}\right) - \frac{M_4}{M_5} R_5\,. \tag{36}$$

The boundary conditions for equation (36):

$$\mathrm{B}_1\,\rho D \frac{\partial C_4}{\partial z} = 0, \tag{37.1}$$

$$\mathrm{B}_2\,\rho D \frac{\partial C_4}{\partial z} = 0, \tag{37.2}$$

$$\mathrm{B}_{3.3}\,\rho D \frac{\partial C_4}{\partial x} = 0, \tag{37.3}$$

$$B_{3.4}\rho D\frac{\partial C_4}{\partial x}=0, \quad (37.4)$$

$$B_{4.2}\rho D\frac{\partial C_4}{\partial y}=0, \quad (37.5)$$

$$B_{4.4}\rho D\frac{\partial C_4}{\partial y}=0. \quad (37.6)$$

The initial conditions for equation (36):

$$C_4\Big|_{t=0}=C_{4.0}. \quad (38)$$

The diffusion equation for the combustible components of pyrolysis:

$$\frac{\partial C_5}{\partial t}=D\left(\frac{\partial^2 C_5}{\partial x^2}+\frac{\partial^2 C_5}{\partial y^2}+\frac{\partial^2 C_5}{\partial z^2}\right)-R_5. \quad (39)$$

The boundary conditions for equation (39):

$$B_1\rho D\frac{\partial C_5}{\partial z}=Y_5, \quad (40.1)$$

$$B_1\rho D\frac{\partial C_5}{\partial z}=0, \quad (40.2)$$

$$B_2\rho D\frac{\partial C_5}{\partial z}=0, \quad (40.3)$$

$$B_{3.3}\rho D\frac{\partial C_5}{\partial x}=0, \quad (40.4)$$

$$B_{3.4}\rho D\frac{\partial C_5}{\partial x}=0, \tag{40.5}$$

$$B_{4.2}\rho D\frac{\partial C_5}{\partial y}=0, \tag{40.6}$$

$$B_{4.4}\rho D\frac{\partial C_5}{\partial y}=0. \tag{40.7}$$

The initial conditions for equation (39):

$$C_5\big|_{t=0}=C_{5.0}. \tag{41}$$

The mass balance equation:

$$\sum_{i=4}^{6}C_i=1. \tag{42}$$

The initial conditions for equation (42):

$$C_6\big|_{t=0}=C_{6.0}. \tag{43}$$

The expression for the mass reaction rate R_5 (Grishin and Shipulina, 2002):

$$R_5=k_5M_5T^{-2.25}\exp\left(-\frac{E_5}{RT_1}\right)\cdot\begin{cases}x_1^{0.25}x_2, x_1>0.05\\ x_1x_2, x_1\leq 0.05\end{cases}, \tag{44}$$

$$x_i=\frac{C_i}{\sum_{k=4}^{6}\frac{C_k}{M_k}M_i}, \tag{45}$$

As a result of computational experiments, the dependence of the ignition delay on the magnitude of the heat flux of focused solar radiation is obtained. It should be noted that the presence of moisture in the layer of forest fuel is ignored and humidity is assumed to be zero. Note that this assumption is consistent with the catastrophic fire danger scenarios in the forest. In connection with the observed and predicted climatic changes, the relevance of such studies will increase from year to year.

Table 3 shows the ignition delay depending on the heat flux of focused solar radiation and the known experimental data on the ignition of a layer of forest fuel by radiation.

Table 3. Ignition delay of forest fuel layer by radiant heat flux

Heat flux, q_s, kW/m²	Ignition delay (2D calculation), t_{ign}, s	Ignition delay (3D calculation), t_{ign}, s	Ignition delay (experiment (Kasperov and Goman, 2010), t_{ign}, s	Ignition delay (experiment), t_{ign}, s
15	99	105	No Ignition	
17	85	87	No Ignition	67±15
20	62	63	197-207	
25	33	34	93-100	
30	26	26	40-42	
35	21	21	27-30	
40	18	18	18-26	

Scenario modeling made it possible to establish a lower threshold for the heat flux of focused solar radiation, at which a layer of forest fuel ignites. Modeling was carried out in a full three-dimensional

Figure 11. Temperature distribution in the system "forest fuel layer-gas mixture" before and at the moment of ignition at q_s=15000 W/m²

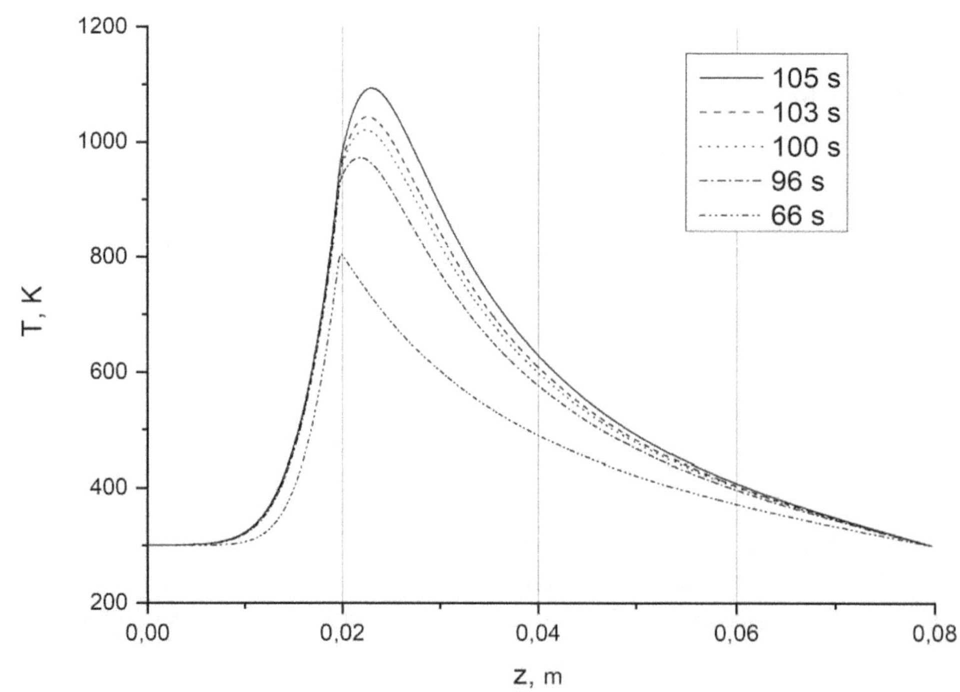

setting. A heat flux with a density of 15 kW/m^2 is the minimum value at which ignition of a layer of forest fuel is still possible. Thus, a relatively small 10-fold excess in the intensity of the flux of natural solar radiation can lead to the appearance of a source of a surface forest fire. The average deviation of the results of numerical modeling from experimental data was about 38%. The vertical temperature distribution in the center of the heat flux of the focused solar radiation is shown in Fig. 11 at various points in time.

It can be seen from the results that initially the forest fuel layer and the gas mixture heats inertly. This stage lasts until a temperature threshold of about 900 K is reached. Over time, in a certain vicinity of the surface of the forest fuel layer, the chemical reaction accelerates and the heat release increases in this zone. A peak characteristic of the ignition region is formed on the temperature curve (Fig. 11).

When the flux of focused solar radiation is equal to or greater than 30 kW/m^2, the differences in the ignition delay in three-dimensional and two-dimensional settings are about several 0.01 s. The greatest discrepancy between the results obtained from the two-dimensional and three-dimensional statements is noticeable at the lower threshold of the radiant heat flux, at which the ignition of the layer of forest fuel still occurs. Also, maximum differences between theoretical estimates and experimental data (Kasperov and Goman, 2010) are recorded in the region of low heat fluxes.

The main laws of heat and mass transfer, established in the case of two-dimensional formulation, are preserved for spatial modeling. Additional heat removal along the third coordinate and diffusion in the same direction lead to an increase in the ignition delay. Ignition criteria are met for the same conditions, but they come a little later.

It is of interest to study the conditions of ignition of a layer of forest fuel at small peripheral sizes of the zone of influence of focused solar radiation. Table 4 presents the ignition delays depending on the heat flux of radiation and the peripheral size of the exposure zone.

Table 4. Ignition delay depending on the peripheral size of the radiation exposure zone

Peripheral size d, mm	Ignition delay, t_{ign}, s				
	40 kW/m^2	35 kW/m^2	30 kW/m^2	25 kW/m^2	20 kW/m^2
10	26	35	48	69	108
8	35	49	70	104	159
6	59	84	126	No	No
4	154	No	No	No	No
2	No	No	No	No	No

As a result of the calculations, it was found that with the peripheral dimensions of the impact zone equal to and less than 10 mm, the value of t_{ign} increases with decreasing d. Such exposure zones can be considered point. Firstly, the ignition delay is longer than in areas of exposure. Secondly, ceteris paribus for the exposure zone with a peripheral size of 2 mm or less, it was found that ignition does not occur. The minimum deviation from the area impact zones is observed for a peripheral size of 10 mm and a heat flux of 40 kW/m^2. This is most likely due to the later attainment of ignition conditions. It takes more time for the fuel concentration to reach the required value. In addition, with such point-like zones of influence, the outflow of heat from this region as a result of heat conduction begins to play a significant role. Calculations showed that when heat fluxes are less than 20 kW/m^2, there is no ignition of a layer of forest fuel.

It should be noted that the results of computational experiments allow us to conclude the importance of the peripheral dimensions for zone of the thermal radiation influence. In (Kasperov and Goman, 2010), such data are not presented. It is only indicated that the test unit includes a radiation panel with a heating element. It is possible that the differences in the experimental data and theoretical results are also explained by the fact that in the experiment (Kasperov and Goman, 2010) such a parameter as the peripheral size of the zone of exposure to the radiant heat flux was not controlled. It is advisable to carry out an experiment in which the dependence of the ignition delay time on the peripheral dimensions of the exposure zone is taken into account. This is especially true for exposure areas with a small peripheral size.

CONCLUSION

Thus, this article presents a group of mathematical models for the ignition of forest fuels by focused solar radiation. Such models can be used to develop forest fire danger predicting software (CWFIS, 2020; EFFIS, 2020; WFAS, 2020; Podolskaya et al., 2011). A specific software implementation can be performed using modern computing systems (Matlab, 2020; MS Visual Studio, 2020; RAD Studio, 2020). It should be noted that previously shown a positive trend in reducing damage from forest fires when using forest fire danger predicting systems (Taylor and Alexander, 2006).

ACKNOWLEDGMENT

This work was implemented with the financial support of the Russian Foundation for Basic Research. [Project N 17-29-05093].

REFERENCES

Babrauskas, V. (2002). Ignition of wood: A review of the state of the art. *Journal of Fire Protection Engineering*, *12*(3), 163–189. doi:10.1177/10423910260620482

Babrauskas, V. (2003). *Ignition handbook: principles and applications to fire safety engineering, fire investigation, risk management, and forensic science*. Fire Science Publishers.

Baranovskiy, N. V., & Kuznetsov, G. V. (2017). *Forest fire occurrences and ecological impact prediction*. Publishing House of the Siberian Branch of the Russian Academy of Science. doi:10.15372/FOREST2017BNV

Canadian Wildland Fire Information System. (n.d.). Access: https://cwfis.cfs.nrcan.gc.ca/home

European Forest Fire Information System. (n.d.). Access: http://effis.jrc.ec.europa.eu

Grishin, A. M. (1997). *Mathematical modeling of forest fire and new methods of fighting them. Russia.* Publishing House of the Tomsk State University.

Grishin, A. M., & Shipulina, O. V. (2002). Mathematical model for spread of crown fires in homogeneous forests and along openings. *Combustion, Explosion, and Shock Waves, 38*(6), 622–632. doi:10.1023/A:1021127924913

Kasperov, G. I., & Goman, P. N. (2010) Investigation of the fire hazard properties of forest combustible materials of pine stands. *Transactions of BSTU. Ser. II, Forestry and Woodworking Industry, 18*, 337-340. (in Russian)

Kuznetsov, G. V. (1998). A mechanism of hyperthermal destruction of epoxy glass in gas flows at high pressure. *High Temperature, 36*, 70–74.

Kuznetsov, G. V., & Baranovskiy, N. V. (2013). Focused sun's rays and forest fire danger: New concept. *Proceedings of SPIE - The International Society for Optical Engineering, 8890*. 10.1117/12.2033929

Kuznetsov, G. V., & Sheremet, M. A. (2006). Conjugate heat transfer in a closed domain with a locally lumped heat-release source. *Journal of Engineering Physics and Thermophysics., 79*(1), 57–64. doi:10.100710891-006-0066-5

Kuznetsov, G. V., & Sheremet, M. A. (2009). Conjugate heat transfer in an enclosure under the condition of internal mass transfer and in the presence of the local heat source. *International Journal of Heat and Mass Transfer, 52*(1-2), 1–8. doi:10.1016/j.ijheatmasstransfer.2008.06.034

Kuznetsov, G. V., & Strizhak, P. A. (2008). Ignition of a single hot particle of liquid hydrocarbon fuels. [In Russian]. *Ivvestia of TPU., 312*, 5–9.

Kuznetsov, G. V., & Strizhak, P. A. (2008). Metal particles heated to high temperatures as sources of local ignitions of liquid substances. [in Russian]. *Fire Safety*, (4), 72–76.

Matlab. (2020). Available https://www.mathworks.com/products/matlab.html

Microsoft Visual Studio. (2020). Available https://visualstudio.microsoft.com (Accessed 03 April 2020)

Podolskaya, A.S., Ershov, D.V., & Shulyak, P.P. (2011). Application of the method for assessing the probability of occurrence of forest fires in ISDM-Rosleskhoz. *Modern Problems of Remote Sensing of the Earth From Space, 8*, 118 - 126. (in Russian)

Read, N., Duff, T. J., & Taylor, P. G. (2018). A lightning-induced wildfire ignition forecasting model for operational use. *Agricultural and Forest Meteorology, 253-254*, 233–246. doi:10.1016/j.agrformet.2018.01.037

Samarskiy, A. A. (1983). *Theory of difference schemes*. Nauka. (in Russian)

Samarskiy, A. A., & Vabishchevich, P. N. (2001). *Additive schemes for problems of mathematical physics*. Nauka.

Strakhov, V. L., Garashchenko, A. N., Kuznetsov, G. V., & Rudzinskii, V. P. (2001). Mathematical simulation of thermophysical and thermo chemical processes during combustion of intumescent fire-protective coatings. *Combustion, Explosion, and Shock Waves, 37*(2), 178–186. doi:10.1023/A:1017557726294

StudioR. A. D. (2020). Available https://www.embarcadero.com/ru/products/rad-studio

Taylor, S. W., & Alexander, M. E. (2006). Science, technology and human factors in fire danger rating: The Canadian experience. *International Journal of Wildland Fire*, *15*(1), 121–135. doi:10.1071/WF05021

Vilyunov, V. N. (1984). *Theory of ignition of condensed matter*. Science, Sib. Department.

WFAS Wildland Fire Assessment System. (n.d.). Access: http://www.wfas.net

Ye, J., Wu, M., Deng, Z., Xu, S., Zhou, R., & Clarke, K. C. (2017). Modeling the spatial patterns of human wildfire ignition in Yunnan province, China. *Applied Geography (Sevenoaks, England)*, *89*, 150–162. doi:10.1016/j.apgeog.2017.09.012

ADDITIONAL READING

Antonov, D. V., Volkov, R. S., Voitkov, I. S., Zhdanova, A. O., & Kuznetsov, G. V. (2018). Influence of Special Additives in a Water Aerosol on the Suppression of a Forest Fire with it. *Journal of Engineering Physics and Thermophysics.*, *91*(5), 1250–1259. doi:10.100710891-018-1855-3

Baranovskii, N. V., & Zakharevich, A. V. (2016). Conditions and Characteristics of the Ignition of a Typical Vegetable Combustible Material by a Local Energy Source. *Journal of Engineering Physics and Thermophysics.*, *89*(6), 1538–1544. doi:10.100710891-016-1524-3

Baranovskiy, N. V., Sherstnyov, V. S., Yankovich, E. P., Engel, M. V., & Belov, V. V. (2016) Focused sunlight factor of forest fire danger assessment using Web-GIS and RS technologies. Proceedings of SPIE - The International Society for Optical Engineering, 9688, article 968823, 6 P. 10.1117/12.2240378

Baranovskiy, N. V., & Zakharevich, A. V. (2015). *Experimental research of grassy rags ignition by heated up to high temperatures carbon particle*. Multifunctional Materials and Modeling., doi:10.1201/b18552

Baranovskiy, N. V., & Zakharevich, A. V. (2018). Mechanical, thermophysical and aerothermochemical processes at forest fuel ignition caused by the heated to high temperatures single metal particle. *International Review of Mechanical Engineering*, *12*(6), 522–532. doi:10.15866/ireme.v12i6.15065

Chen, T., Zhang, J., & Wu, J. (2016). Kinetic and energy production analysis of pyrolysis of lignocellulosic biomass using a three-parallel Gaussian reaction model. *Bioresource Technology*, *211*, 502–508. doi:10.1016/j.biortech.2016.03.091 PMID:27035484

Hameed, S., Sharma, A., Pareek, V., Wu, H., & Yu, Y. (2019). A review on biomass pyrolysis models: Kinetic, network and mechanistic models. *Biomass and Bioenergy*, *123*, 104–122. doi:10.1016/j.biombioe.2019.02.008

Kansa, E., Perlee, H., & Chaiken, R. (1977). Mathematical model of wood pyrolysis. *Combustion and Flame*, *29*, 311–324. doi:10.1016/0010-2180(77)90121-3

Liu, H., Wang, C., Zhao, W., Yang, S., & Hou, X. (2018). Pyrolysis characteristics and kinetic modeling of Artemisia apiacea by thermogravimetric analysis. *Journal of Thermal Analysis and Calorimetry*, *131*(2), 1783–1792. doi:10.100710973-017-6599-3

Thomas, J. C., Hadden, R. M., & Simeoni, A. (2017). Experimental investigation of the impact of oxygen flux on the burning dynamics of forest fuel bed. *Fire Safety Journal*, *91*, 855–863. doi:10.1016/j.firesaf.2017.03.086

KEY TERMS AND DEFINITIONS

Anthropogenic Load: Different human activities on forested territories lead to forest fire occurrence and characterized by presence of fire sources.

Focused Sunlight: Sunlight concentrated by natural or artificial lens.

Forest Fire: Uncontrolled aerothermochemical phenomenon characterized by step-by-step mechanism which includes following stages: inert heating, moisture evaporation, high temperature terpens evaporation, dry organic matter pyrolysis, flammable combustion, and smoldering.

Forest Fuel: It can be considered like dead and live forest fuel. Main types of forest fuel which can be involved in combustion during forest fire: ground forest fuel (needles, leaves and dry grass, small branches) and crown forest fuel (needles, small branches).

Ignition: Inflammation of forest fuel caused by definite source of high temperature or energy.

Ignition Delay: Time before flame flash after forest fuel heating.

Mathematical Simulation: The production of a computer model of forest fire conditions and prerequisites, especially for the purpose of study.

Monitoring: Monitoring refers to the periodic calculation of the parameters of forest fire danger with a portion of information available in real time.

Prediction: Under the prediction of forest fires is the calculation of the parameters of forest fire danger with a certain projection in advance in order to have enough time to anticipate an emergency. The calculation in this case is carried out in a mode ahead of the real time of the development of the catastrophe - the occurrence of a forest fire.

Section 4

Deterministic–Probabilistic Criteria to Predict Forest Fire Danger

Chapter 14
Forest Fire Probability Prediction Taking Into Account Anthropogenic Load, Lightning Activity, and Weather Conditions

ABSTRACT

A new probabilistic criterion for forest fire danger and a new methodology for determining the probability of forest fires taking into account thunderstorm activity and the level of anthropogenic load in a controlled area are considered. The results of a parametric study of the dependence of probability on meteorological conditions, thunderstorm activity, and anthropogenic load are also presented. The fire situation corresponds to the statistics of fire accidents in the territory of the Timiryazevskiy forestry of Tomsk region.

INTRODUCTION

Currently, the issue of forest fire danger assessment is relevant. The prediction of surface fires is of greatest importance, since more than 80% of all vegetation fires are surface fires. Almost all crown fires develop from surface fires (Bayham et al., 2020). An important role is played by the creation of a new methodology for predicting the occurrence of forest fires. The basis for the creation of such a methodology should be simple, but adequate to the physics of the process of forest fire ripening, mathematical models, as well as appropriate methodological, information and software.

An analysis of the forest fire danger prediction techniques existing in Russia and abroad shows that almost all the techniques have a weak physical basis and, as a rule, take only weather data into account. Thunderstorm activity and anthropogenic load are not properly taken into account (Grishin, 2002). It should be noted the Canadian, a number of South European methods (Viegas et al., 2000) and the Nest-

DOI: 10.4018/978-1-7998-7250-4.ch014

erov criterion (Nesterov, 1949). The advantages of these methods are the simplicity and a fairly good quality of the prediction, but within the territory on which a statistical analysis of forest fire incidents was carried out. The main disadvantage is the neglect of the real cause of forest fires, namely, the drying process of a layer of forest fuel on the underlying surface.

The existing network of weather stations (especially the regions of Siberia, the North and the Far East) does not allow us to talk about any normal coverage of the controlled territory. In this work, when developing a system for predicting the occurrence of forest fires, it is proposed to focus on the prospects for interaction with software that implements global and regional atmospheric models. It should be noted that the use of existing fire danger scales often leads to errors, since forest fire danger is actually estimated only by vegetation conditions (Telitsin, 1987). In addition, in the dry period, differences in the moisture content of forest fuel in the areas are smoothed out, and after 2-3 weeks there is practically no difference in the I and V classes (Telitsin, 1987). This information only confirms the conclusion that it is necessary to develop a new methodology for determining the probability of forest fires.

In the nuclear industry, for example, a probabilistic safety criterion is used (Grishin, 1999), and at present, the need to develop a similar criterion and the corresponding methodology has also arisen in forestry. An analysis of numerous literary sources devoted to the problem of forest fire danger predicting and forest taxation descriptions of specific forestry allows us to conclude that it is necessary to develop a forest fire danger prediction system that would have a spatial resolution at the level of the minimum forest taxation unit (site) and which would allow to obtain results in the system "forestry-quarter"', as the quarterly maps of forest areas, as a rule, do not have a geographic reference.

Simple mathematical models and an approximate analytical formula for determining the drying time of the forest fuel layer were previously developed (Grishin and Baranovskij, 2003). For practical use, the forest-taxation characteristics "completeness" and "bonitet" of the stand are integrated in the model (Baranovskiy and Grishin, 2002), which allows one to take into account the effect of screening of solar radiation by the canopy of the stand. In (Sofronov, 1970), it was noted that there is a dependence of the fraction of solar radiation penetrating under the canopy of the forest stand on these characteristics for various types of forests. It should be especially noted that for this it is enough to use the standard forest taxation descriptions that are available in each forestry.

The purpose of this work is to develop a new probabilistic criterion for forest fire danger and conduct a parametric study of the influence of anthropogenic load, thunderstorm activity and weather conditions on the probability of forest fires.

Background

The ecological system of forests of the Russian Federation occupies 1.2 billion hectares of territory and contains about 25% of the forest resources of the entire planet (Kuznetsov et al., 2005). Russian forests are not only an economic, but also an important ecological resource (Stryamets et al., 2020; Concept…, 2003). Global processes of regulating the state of the environment, biodiversity, climate, and river flows are significantly affected by the forests of the Russian Federation (Kuznetsov et al., 2005).

Tomsk region, especially its northern part, is a fairly typical forest-covered territory of the boreal forest zone. On her example, a fairly general description of the conditions for the occurrence of fires is possible. Forests of the region are located in the Ob river basin on exclusively flat territory with excess moisture and are of great environmental importance (Panevin, 2006). Relatively harsh climatic conditions determine a rather limited species composition of forests. The predominant breed in the Tomsk region is pine.

According to the Tomsk Aviation Forest Protection Base, for the period 1993-2002, 2363 forest fire accidents were recorded (Yanko, 2005). From 55 to 350 fires occur annually in the region. A feature of the forests of the Tomsk region is the presence of combustible material in all stands. Mostly surface fires develop in the region (98.5%). 1.1% of incidents and 12.5% of the burned-out area account for the crown fires, and underground fires occur even less frequently. The share of fires for anthropogenic reasons is quite stable over the years, and fires from a lightning discharge are cyclical in nature. Periods with massive thunderstorms give way to calmer ones. The firewood of the region's forests also varies significantly over the months of the fire danger season. The most "burning" months are June and July. The duration of the fire danger season according to the weather is from 137 to 161 days (Panevin, 2006).

One of the main causes of forest fires in many cases is lightning discharges of the cloud-to-ground class (Wang et al., 2020; Latham and Williams, 2001; Latham, 1991; Li et al., 2020; Uman, 1969). Extensive statistics on cloud-to-ground lightning discharges, which can be used to predict forest fire dangers in the boreal forest zone, are collected as part of the US National Lightning Detection Network (Cummins et al., 1998). This system can identify most cloud-to-ground lightning discharges in the United States and Canada with a spatial resolution of several kilometers and an accuracy of 1 ms in time. As a result of the system's operation, data on impact polarity, peak impact current, impact complexity (single or multi-impact) are archived (Lyons et al., 1998; Boccippio et al., 1995; Burke and Jones, 1996; Cummer and Inan, 1996; Huang et al., 1999).

Cloud-to-ground lightning discharges are recorded in other countries. For example (Soriano et al., 2005; Orville and Huffines, 2001), in Spain the proportion of single strikes was 53.6% and 89%, respectively, for negative and positive cloud-to-ground lightning discharges. The average current for the first strike of the negative and positive discharges was 23.5 kA and 35.3 kA, respectively. These statistics allow us to study the basic laws of ignition of deciduous and coniferous trees based on mathematical modeling. To predict forest fire danger from thunderstorm activity for a specific period, the results of analysis of the daily variation of cloud-to-ground thunderstorm discharges are of interest (Ateitio et al. 2001). Lightning discharges are also observed in Japan (Matsui et al., 2019; Tsunomura et al., 2016), France (Gallin et al., 2016), Brazil (Pinto et al., 2007; Paulucci et al., 2019; Pinto and Pinto, 2003; Pinto et al., 2003), Italy (Bernardi and Ferrari, 2004), and the Russian Federation (Wang et al., 2020; Adzhiev and Kuliev, 2018).

The Federal Service for Hydrometeorology and Environmental Monitoring plans to develop a network of direction finding for cloud-to-ground lightning discharges and it is advisable to develop physical and mathematical models for ignition of deciduous and coniferous trees as a result of a cloud-to-ground discharge.

A study of the spatial distribution of forest fires from thunderstorms in Ontario is significant (Podur et al., 2003). A rather long period was analyzed - from 1976 to 1998. Forest fires from thunderstorms account for 35% of all recorded fires in Canada, and the area covered by fire is 85% (Weber and Stocks, 1998). The total number of forest fires analyzed was 40,000 (17,000 of which occurred from thunderstorms) (Podur et al., 2003). Spatial statistics methods (SPP analysis) allowed authors to identify clusters with the highest probability of forest fires from thunderstorms. Important measures in SPP are spatial intensity (defined as the number of events per unit area) and NBS, the nearest neighbor statistic (determines how close events are to each other) (Podur et al., 2003).

Cluster analysis allows us to assess the spatial dynamics of forest fires from thunderstorms only in the conditions of invariance and stability of statistics on forest fire incidents. The objective existence of global climate changes and their contribution to local weather conditions will not allow to obtain adequate results on the ignition of forests. It is necessary to develop appropriate models for ignition of natural combustible materials that are currently not available. Cluster analysis was used to assess the probability of forest fires from thunderstorms in Central Spain (Nieto et al., 2006; Amatulli et al., 2007). General patterns characteristic of the province of Ontario were identified.

The need to use information about the availability of fire sources for assessing forest fire danger was noted by I.S. Melekhov (Melekhov, 1947). An increase in the level of urbanization, recreational load leads to changes in forest ecosystems, including as a result of forest fires. An analysis of the socio-psychological aspects of recreational visits to the forest and the occurrence of forest fires (Andreev and Larchenko, 1987). It has been established that with increasing distance from the settlement, the number of visits to the forest and forest fires decreases (Kurbatskiy, 1964; Melluma et al., 1982) according to the Rayleigh and Poisson distributions (Telitsin, 1984).

When studying population migration in geography, various mathematical models are used. Some of them were checked to describe the distribution of forest fires relative to settlements in rural areas (Andreev, 1986). It turned out that for this purpose, probability density functions of the lognormal distribution and a Pareto type function can be used. With the increase in the number of inhabitants, the probability of a fire near settlements increases. The distribution of fire sources over the territory and over time is similar to the distribution of forest fires and is reproduced annually with only a slight deviation from the long-term average (Telitsin, 1987).

The main anthropogenic sources of ignition are single particles of metals or nonmetals heated to high temperatures (Romanenkov and Levites, 1991). Such particles are formed and move over distances from several tens of centimeters to tens of meters as a result of wind blowing out extinguished bonfires, removal of hot particles from the front of the fire, welding or metal cutting. Forest fires are affected by social factors. For example, 93% of all fires in the national forests of Minnesota, Wisconsin and Michigan in 1986 occurred due to human faults (Cardille et al., 2001).

A significant GIS analysis of spatial and temporal patterns of forest fires of the 1950-1992 seasons that occurred for anthropogenic reasons in the forests of Vancouver Island (Canada). A total of 6329 forest fires were analyzed. Arc/Info and ArcView GIS software were used. The study area was divided into 36050 cells. The number of burnt cells is closely related to the proximity of country, paved roads and railways and decreases with increasing distance from them (especially from country roads). The number of burned out cells increases with the distance from camps and campsites, reaching a maximum at a distance of 5 km from their location, but decreases with subsequent removal. A large number (the difference with the neighboring territories by an order of magnitude or more) of cells burns out at a distance of more than 20 km from cobbled and railroads, Vancouver municipalities (Pew and Larsen, 2001).

GIS analysis reveals the spatial picture of forest fires for various anthropogenic reasons. However, it is not suitable for predicting forest fire danger. It is necessary to develop a probabilistic model for assessing forest fire danger for anthropogenic reasons, lightning activity and weather conditions.

From the analysis of the causes of fires, it follows that in countries with a low population density, the number of fires increases during the development of the territory. As you move away from settlements and transport routes, the number of fires decreases, and their average size naturally increases. Large-scale effects of forest fires from anthropogenic stress are possible (an example of the Mediterranean part of the Pyrenees peninsula in the period 1989-2003) (Benavent-Corai et al., 2007).

As a rule, to predict forest fire danger, data from stations for the registration of meteorological parameters (Kosarev and Andryushchenko, 2007; Filkov, 2005) or satellite-based TOVS sounding data

(Ponomarev et al., 2006) are used. For scenario modeling, reference information is used (Scientific-applied reference…, 1993).

It should be noted that in Siberia, in the North and the Far East of Russia, the network of weather stations is not sufficiently developed. However, in recent years, fairly developed mathematical models of the atmosphere (Tolstykh, 2001) and climate (Dymnikov et al., 2005) have already been used. The physical and mathematical approach currently in meteorology is becoming increasingly important. Mathematical methods make it possible to create complex models and predict the further development of atmospheric processes and are already widely used in the practice of operational weather predicting (Murphy and Winkler, 1984). The viability of short-term predictions is high and close to 90%. For models of medium-term weather predictioning, the semi-Lagrangian approach is widely used (Tolstykh, 2001). A three-dimensional model has been developed (Tolstykh, 1997). In addition, there are no obstacles to using the rough solution obtained by the global atmospheric model as the initial and boundary conditions for the non-hydrostatic model of the regional atmosphere (Esaulov and Starchenko, 2002). In combination with the methods for reconstructing the detailed structure of meteorological fields at the city and regional levels from the predicted large-scale values of meteorological fields (Chavro and Dmitriev, 2002), this can provide (for a specific territory) weather data with a higher spatial resolution than allowss a network of available weather stations (especially for regions with an undeveloped weather network).

In addition to global models, there are regional models of numerical weather predicting, for example, the MM5 (Lixiang et al., 2000) and WRF (Michalakes et al., 2001) models. The scope of regional models is extremely wide from scale of several meters to thousands of kilometers. It is necessary to develop a new prediction concept for forest fire danger, taking into account the possibility of using such models.

The process of drying the forest fuel layer under the influence of external conditions is important for assessing forest fire danger in a controlled area. In recent years, a number of physical and mathematical models have been developed for drying the forest fuel layer (Grishin and Baranovskij, 2003; Grishin and Filkov, 2005; Grishin et al., 2001a). The most complete of them (Grishin et al., 2001b) uses the basic concepts and methods of mechanics of continuous multiphase media and methods for solving the conjugate problems of heat and mass transfer taking into account the general mathematical model of forest fires (Grishin, 1997). The conjugate substitution and boundary layer model are used (the laminar regime of the steady flow is considered).

An analysis of forest fire danger predicting methods existing in Russia and abroad shows that almost all methods have a weak physical basis. Thus, there is no doubt the need to develop a new methodology for determining the probability of forest fires. For example, in the nuclear industry, a probabilistic safety criterion is used. Currently, the need to develop such a criterion and the corresponding methodology has arisen in forestry. As shown, the analysis of real forest taxation descriptions of specific forest tracts should be detailed at the level of the minimum forest taxation unit (site).

The existing network of weather stations (especially the regions of Siberia, the North and the Far East) does not provide any normal coverage of the controlled territory. Thus, it is necessary to develop a new concept for building a forest fire danger prediction system. In this work, when developing a system for predicting the occurrence of forest fires, it is proposed to focus on the prospects for interaction with developed mathematical models of the atmosphere and climate.

The forest fire danger scales currently used are usually tied to a specific region. Thus, it will be relevant to develop such a system of scales that would unambiguously and uniformly interpret the level of forest fire danger in any region in the form of scales of the probability of forest fires.

PROBABILISTIC CRITERION

The formula was obtained to predict forest fire probability for time interval *j* of a forest fire season using the basic principles of the probability theory (Baranovskiy, 2017; Baranovskiy and Zharikova, 2014):

$$P_j = [P(A)P(A_j / A)P(FF / A / A_j) + P(L)P(L_j / L)P(FF / L / L_j)]P_j(D) \quad (1)$$

It is necessary to use the definition of probabilities by means of the frequency of events and the statistical data of the specific forestry to determine all the parts in formula (1) ((Baranovskiy, 2017; Baranovskiy and Zharikova, 2014)):

$$P(A) \approx \frac{N_A}{N_{FS}}, \quad P(A_j / A) \approx \frac{N_{FD}}{N_{FW}}, \quad (2)$$

$$P(FF / A / A_j) \approx \frac{N_{FA}}{N_{FT}}, \quad (3)$$

$$P(L) \approx \frac{N_L}{N_{FS}}, \quad P(L_j / L) \approx \frac{N_{LH}}{N_{LD}}, \quad (4)$$

$$P(FF / L / L_j) \approx \frac{N_{FL}}{N_{FT}}, \quad (5)$$

Obviously, the more cases are considered for the specific forest, the more precise formulas (2) - (5) will be. All the parameters of a fire season have therefore to be registered every year.

P_j is the forest fire probability for interval *j* in the controlled forest area;
P(A) is the fire probability caused by the human activity;
$P(A_j/A)$ is the forest fire probability on week day *j*;
$P(FF/A,A_j)$ is the forest fire probability caused by the human activity in the forest area;
P(L) is the probability of dry thunderstorms in the forest area;
$P(L_j/L)$ is the probability of cloud-to-ground lightning discharge;
$P(FF/L,L_j)$ is the forest fire probability caused by the lightning provided that there is a dry thunderstorm in the forest area;
$P_j(D)$ is the probability that forest fuel layer will be dry;
index *j* corresponds to the day of a fire season.
N_A is the number of days during a fire season when anthropogenic load is enough for forest fuel ignition;
N_{FA} is the number of fires due to anthropogenic load;

N_{FT} is the total number of fires; N_L is the number of days when there was lightning (when dry thunderstorms occurred);
N_{FS} is the total number of days in a fire season;
N_{FL} is the number of fires due to lightning (when dry thunderstorms occurred);
N_{FD} is the number of fires on the specific day of week; N_{FW} is the total number of fires during week;
N_{LH} is the number of cloud-to-ground lightning discharges from 00:00 am;
N_{LD} is the total number of cloud-to-ground lightning discharges per day.

Thus, we can single out the probability that forest fuel reaches the state of forest fire maturity. In our case, the "maturation" of the forest fuel is determined by the drying process of the forest fuel layer under the influence of external conditions. This process can be quantitatively characterized by the drying time of the forest fuel layer. Under the drying time of the forest fuel layer is meant the period of time during which the moisture content of the layer reaches a critical value. For example, for pine needles it is 13%. The drying time of the forest fuel layer makes it possible to determine the probability of forest fires by weather conditions $P_j(D)$.

To determine the value of $P_j(D)$, it should be taken into account that, as a rule, tanning occurs in the daytime. Obviously, $P_j(D)$ takes its maximum value when the time interval up to the prediction hour (as a rule, the occurrence of most fires falls in the middle of the day) coincides with the drying time of the j-th time interval of the fire danger season in the i-th section of the controlled territory. The value of the fire season in days, as a rule, is known from statistical data on the forestry archives. As a result, for the quantity $P_j(D)$, the formula was obtained:

$$P_j(D) = \begin{cases} 0 \\ k_h \exp\left(-\left[\Delta \bar{\bar{t}}_j\right]^2\right) \end{cases}, \quad \Delta \bar{\bar{t}}_j = \frac{t_j - t_j^*}{t_j^*}, \tag{6}$$

where k_h is the correction factor for precipitation accounting, 0 in (6) corresponds to the case when there are no forest fuels on the forest area (the surface of roads, rivers, lakes, and water-saturated swamps).

In turn, after the forest fuel layer reaches the state of forest fire ripening, a forest fire can occur due to anthropogenic stress and thunderstorm activity. Moreover, these two events - the occurrence of a forest fire as a result of anthropogenic load and thunderstorm activity, are incompatible, i.e. cannot happen at the same time. Therefore, their probabilities must be summed up in accordance with the probability addition theorem. The following scenario of human behavior in the forest is selected - when a thunderstorm sets in, a person seeks to leave the forest territory or take refuge, i.e. there is no anthropogenic load when there is a thunderstorm (incompatible events).

Let us consider separately the term responsible for the probability of forest fires resulting from thunderstorm activity. The following formula was proposed:

$$P_j^L = P(L)P(L_j / L)P(FF / L, L_j)P_j(D). \tag{7}$$

First, forest fire ripening of the forest fuel layer occurs, then a forest fire can occur if there is thunderstorm activity, and it is also necessary to take into account the probability of thunderstorm activity. Since the dynamics of ground-based lightning discharges are of great importance, the probability of cloud-to-ground lightning discharges is introduced.

Let us consider a separate term responsible for the probability of forest fires resulting from anthropogenic load. The following formula was proposed:

$$P_j^A = P(A)P(A_j / A)P(FF / A, A_j)P_j(D) \tag{8}$$

Forest fire maturing of the forest fuel layer occurs, then a forest fire can occur provided that there is an anthropogenic load, and it is also necessary to take into account the probability of anthropogenic load. Since the change in the level of availability of fire sources during the week is of great importance, the probability of this event is introduced on a specific day of the week. As a result, the probability of forest fires in the controlled area is determined by the formula (1).

RESULTS AND DISCUSSION

Mathematical modeling of the probability of forest fires depending on various factors of forest fire danger for various options was carried out. A typical forest area on which a pine-type coniferous forest grows was considered. In the ground cover, the most part is the litter of pine needles with a small number of thin pine branches. Forest territory is in close proximity to the village, which leads to the predominance of anthropogenic pressure. Qualitatively, this correlates with the forest fire situation in the Timiryazevskiy local forestry of the Timiryazevskiy forestry of Tomsk region. Weather data on solar radiation were taken from the Aleksandrovskoye weather station and air temperature was taken from the Tomsk weather station (Scientific-applied reference…, 1993).

Option 1. The effect of solar radiation. Figures 1 and 2 show the diurnal variation in the density of solar radiation (data ((Scientific-applied reference…, 1993)) were used) for clear sky and medium cloud cover conditions, respectively. Figure 3 shows the dependence of the probability of a forest fire under clear sky conditions, and Figure 4 for conditions of moderate cloud cover. A clear sky means such conditions when the total cloud cover is not more than two points, the solar disk and the near-solar zone with a radius of 50 are free from clouds and their traces. Curves 1, 2, 3 reflect the probability for weather conditions in June, July and August, respectively.

A numerical study of the probability of a forest fire depending on the diurnal variation of the solar radiation flux density *qR(h)*, both for clear sky conditions and medium cloud conditions, was carried out using a mathematical model (Grishin and Baranovskij, 2003) to determine the drying time of the forest fuel layer. As can be seen from Figure 3, in clear skies, forest fire danger occurs in July, then in June and August, although the sums of total solar radiation in clear skies in descending order are as follows: 30.16 MJ/m^2 in June, 28.33 MJ/m^2 in July and 22.92 MJ/m^2 in August (Scientific-applied reference…, 1993). However, higher ambient temperatures in July contribute to faster drying of the forest fuel layer in July.

Under conditions of moderate cloud cover, the situation does not fundamentally change. The drying time of the forest fuel layer increases and the forest fire danger occurs later. The monthly distribution is

Figure 1. Daily variation in solar radiation density for clear sky conditions

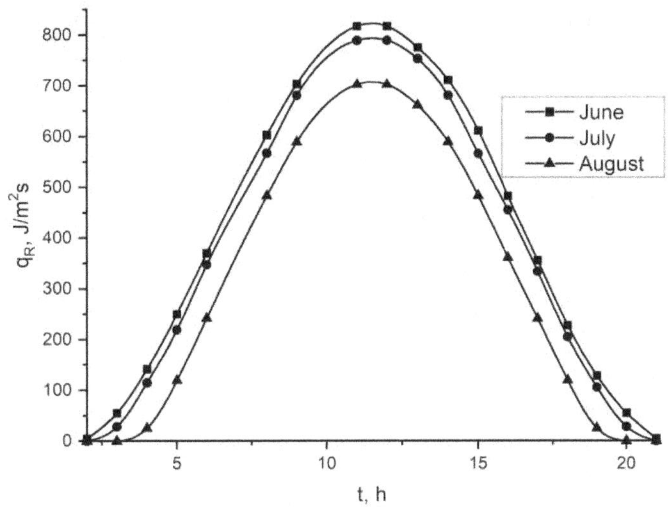

similar, except that the probability curves lie a little denser to each other, which is explained by a smaller difference in the amounts of solar radiation under average cloud conditions.

Figure 2. Daily variation in solar radiation density for conditions of moderate cloud cover

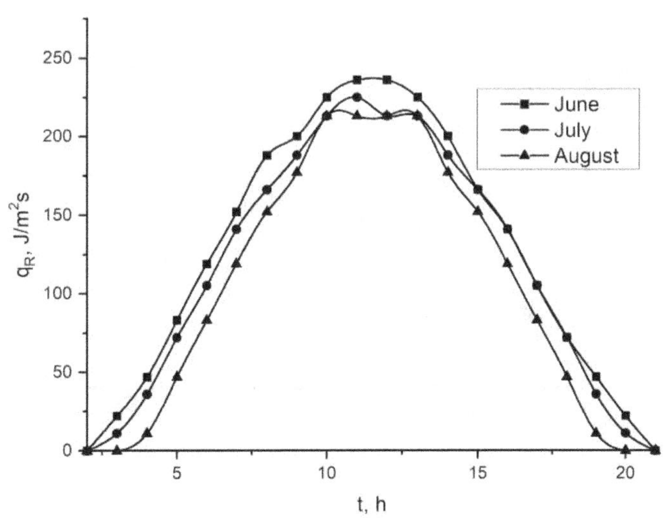

Option 2. The effect of ambient temperature. Figures 5–7 show the diurnal variation in air temperature for June, July, and August, respectively (the data (Scientific-applied reference…, 1993) was used). Figures 8–10 show the dependences of the probability of forest fire danger for June, July, and August, respectively. Curves 1, 2, 3 reflect the scenarios of low, medium, and high fire danger, respectively.

Figure 3. Probability of forest fire danger in clear sky

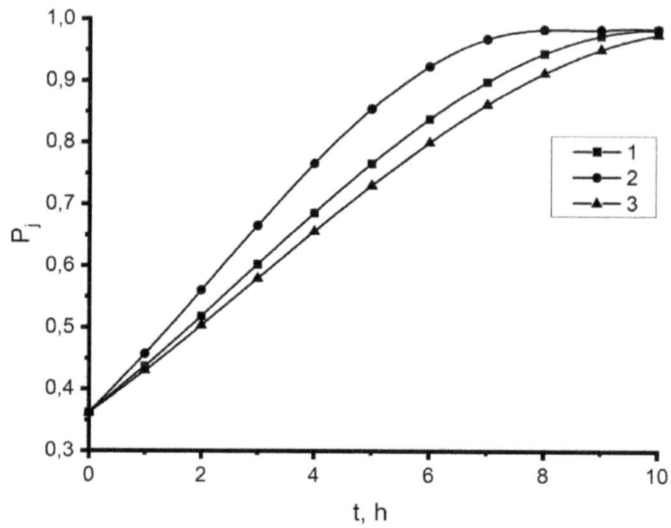

Figure 4. Probability of forest fire danger in moderate cloud conditions

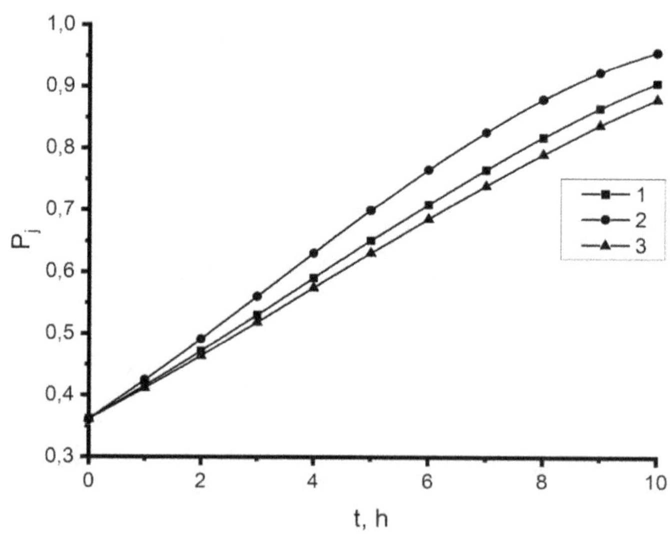

Figures 5-7 presents the daily trend of the ambient temperature. Curves 2 correspond to the average monthly temperature for June, July, and August and correspond to the scenario of average fire danger. Curves 1, 3 correspond to the low and high fire danger scenarios, respectively. The diurnal curves 1, 3 for definiteness were taken 3 °C lower and higher than the diurnal curves of the daily average monthly temperature. This choice is due to the fact that this ensures the excess of the average maximum temperature by several degrees (curve 3) and the minimum temperatures along curves 1 are also less than the average minimum by several degrees. Thus, these conditions may well take place in reality.

Figure 5. Daily variation of the ambient temperature (June) (Scientific-applied reference..., 1993) and the temperature of the gas phase in the forest fuel layer

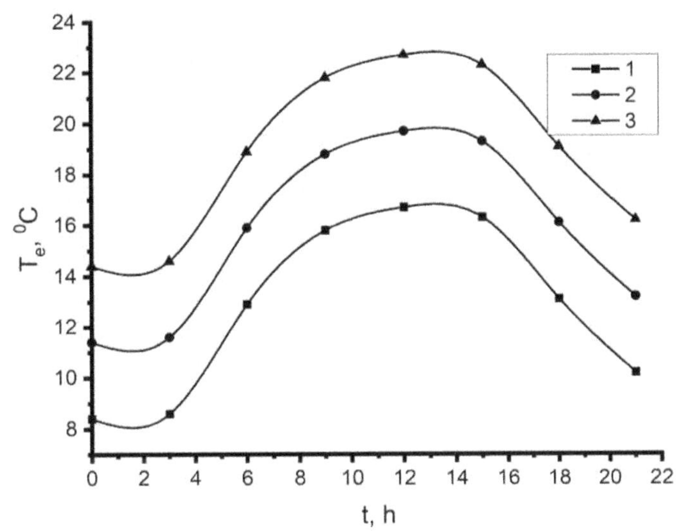

Figure 6. Daily variation of the ambient temperature (July) (Scientific-applied reference..., 1993) and the temperature of the gas phase in the forest fuel layer

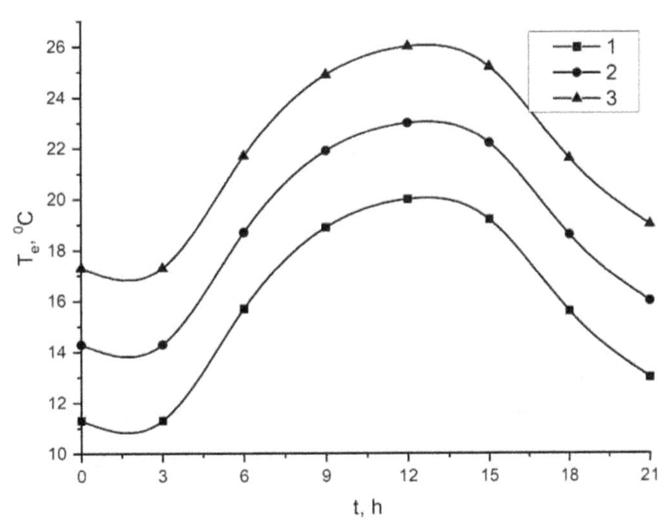

Analysis of the results of the author presented in Fig. 8-10 shows that higher ambient temperatures provide a faster onset of forest fire danger. In June and August, the effect of ambient temperature is almost the same. A logical result is a higher probability of forest fires in the high fire danger scenario and its lower value in the medium and low fire danger scenarios.

Figure 7. Daily variation of the ambient temperature (August) (Scientific-applied reference..., 1993) and the temperature of the gas phase in the forest fuel layer

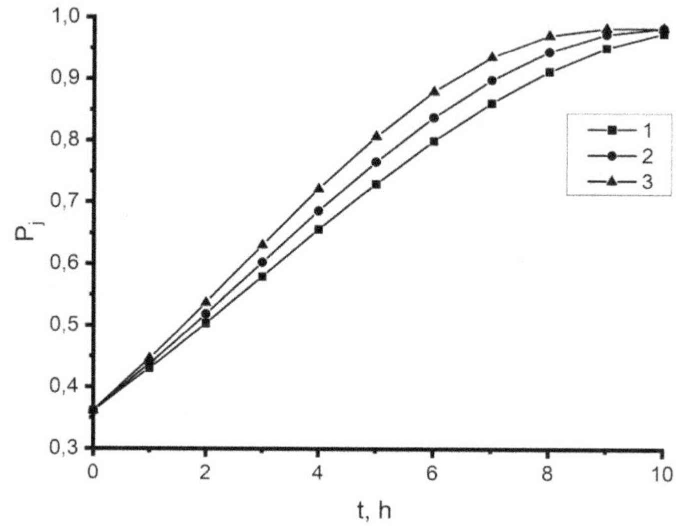

Figure 8. Probability of forest fire danger (June)

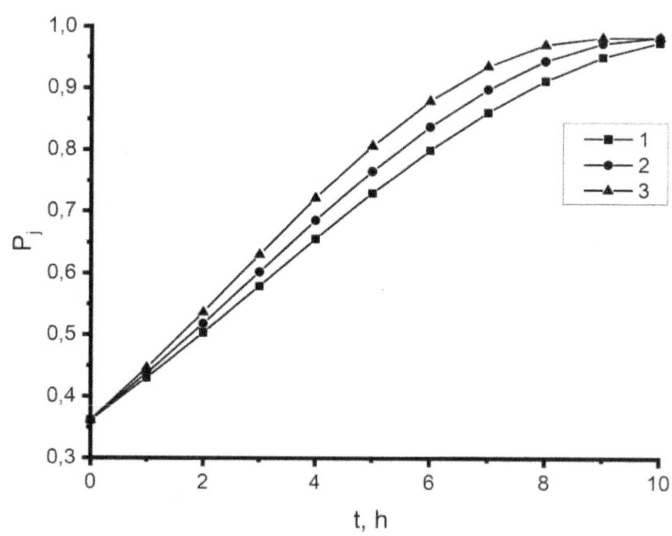

Option 3. Precipitation. Figure 11 shows the probability of forest fire danger in the absence of rainfall and in conditions of rainfall in an amount of 20 mm or more.

Figure 11 presents the results of determining the probability of a forest fire in the absence of precipitation (curve 1) and in case of 20 mm or more precipitation (curve 2). It should be noted that at the moment the accounting for precipitation is implemented in the method rather roughly, since with a 20

Figure 9. Probability of forest fire danger (July)

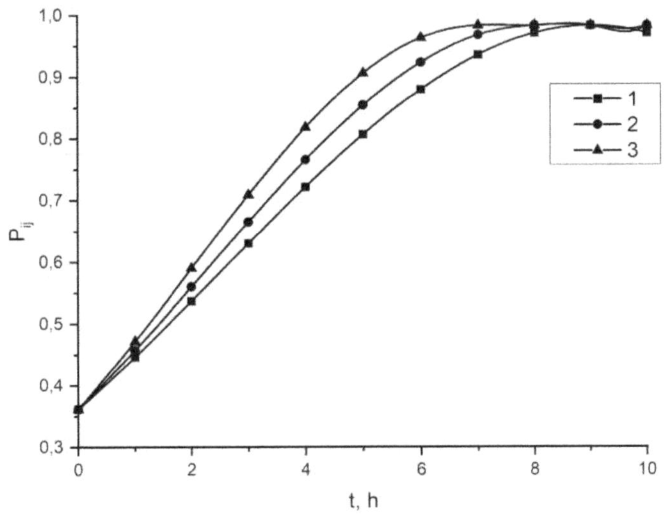

Figure 10. Probability of forest fire danger (August)

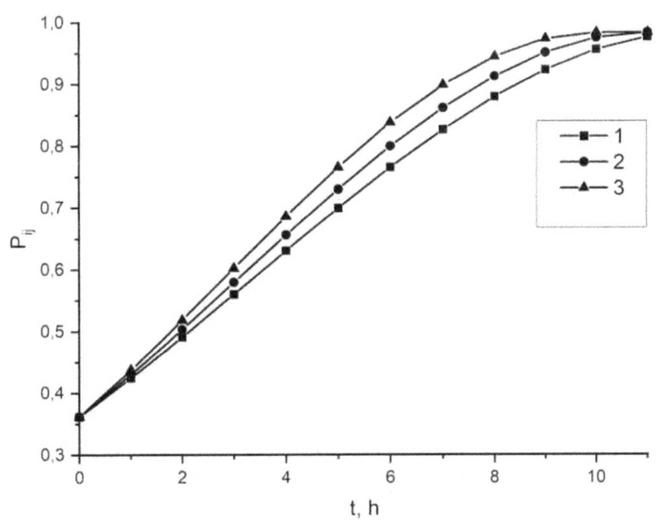

mm precipitation the probability assumes a zero value. However, as evidenced by the results (Larfeldt et al., 2000), 20 mm of precipitation is not always sufficient so that the forest area is not fire dangerous.

Option 4. The impact of anthropogenic load on the probability of a forest fire. Figure 12 shows the dependence of the probability of the presence of fire sources on the day of the week, and Figure 13 shows the forest fire danger dependencies for different days of the week. The change in the probability of the presence of fire sources reflects the change in the level of anthropogenic load in the Timiryazevskiy forestry of Tomsk region during the week.

Figure 11. Probability of forest fire danger and rainfall

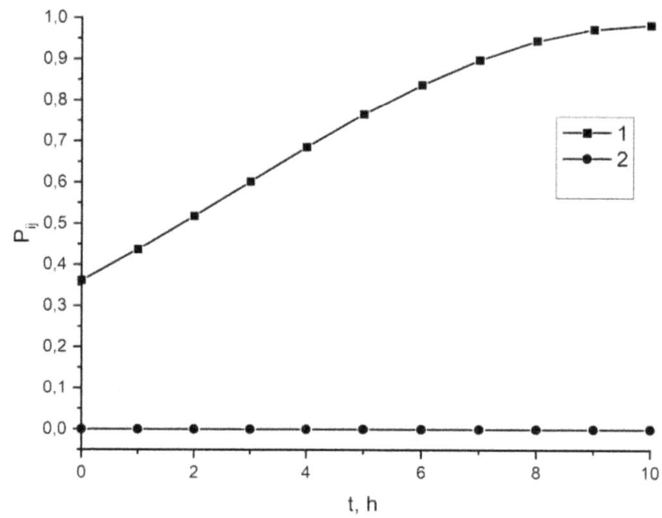

Figure 12. Probability of ignition sources

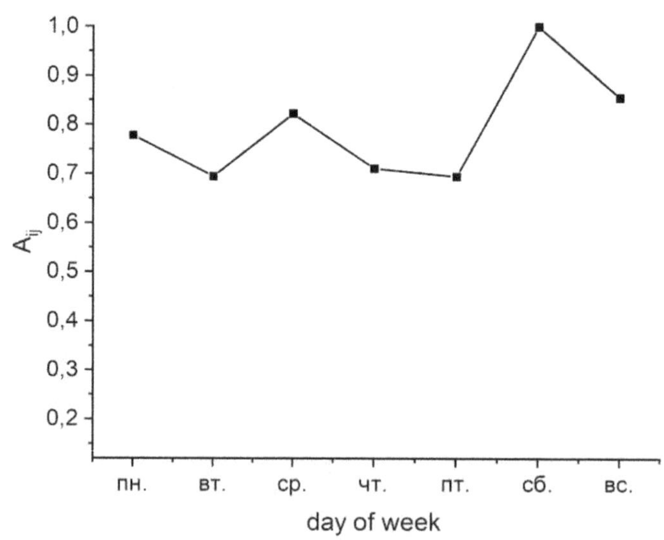

As a result of processing statistical data on forest fires in the Timiryazevskiy forestry of Tomsk region, the values of the probability of the presence of fire sources were obtained. A numerical study of the impact of anthropogenic load on the probability of forest fires has been carried out. Figure 13 shows that the probability of a forest fire significantly depends on a different level of anthropogenic load on different days of the week, which reflects the unevenness of the forest residents visiting the village during the week. It should be noted that due to an increase in the anthropogenic load on the forested territories adjacent to the settlements, this problem goes beyond the scope of physical and mathematical modeling

Figure 13. Probability of a forest fire during the week

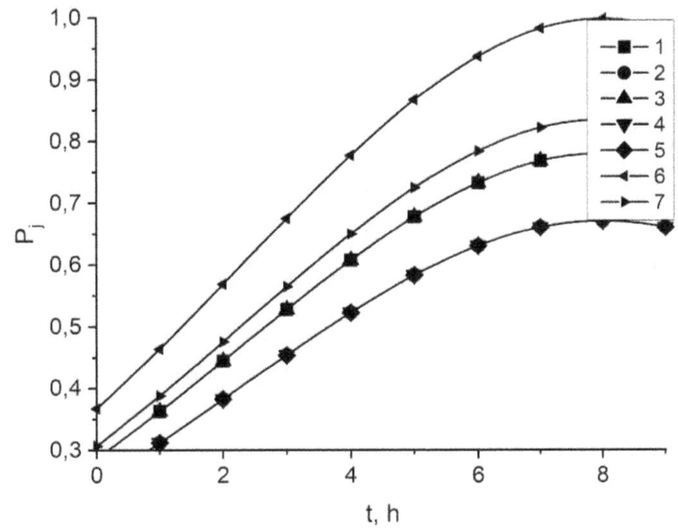

and mathematical statistics. In the future, the situation will only worsen and it is quite possible that it will be advisable to involve both sociologists and psychologists to solve this problem.

Option 5. Cloud-to-ground lightning discharges and the probability of forest fire danger. Figure 14 presents the dependence of the probability of cloud-to-ground lightning discharges on the time of day. Figures 15 and 16 show the dependences of the probability of forest fire danger at various values of the probability of occurrence of cloud-to-ground lightning discharges. This dependence was obtained using data on the distribution of the number of cloud-to-ground lightning discharges during the day.

Figure 14. Probability of cloud-to-ground lightning discharges during the day

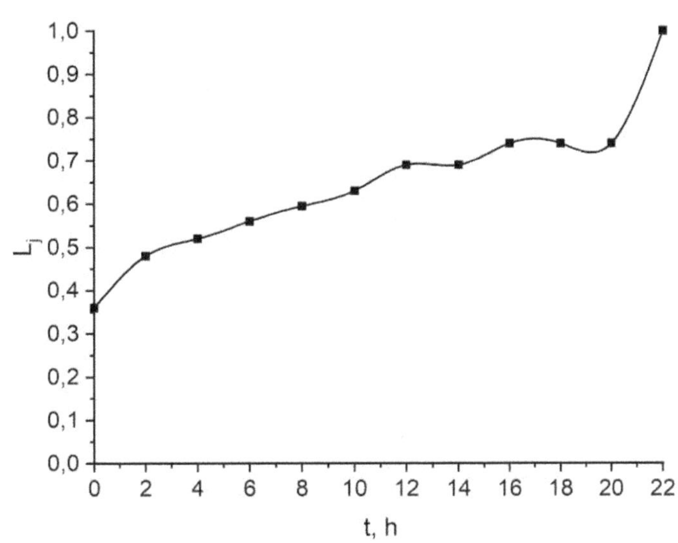

Figure 15. Probability of forest fire danger taking into account cloud-to-ground lightning discharges when anthropogenic load prevails

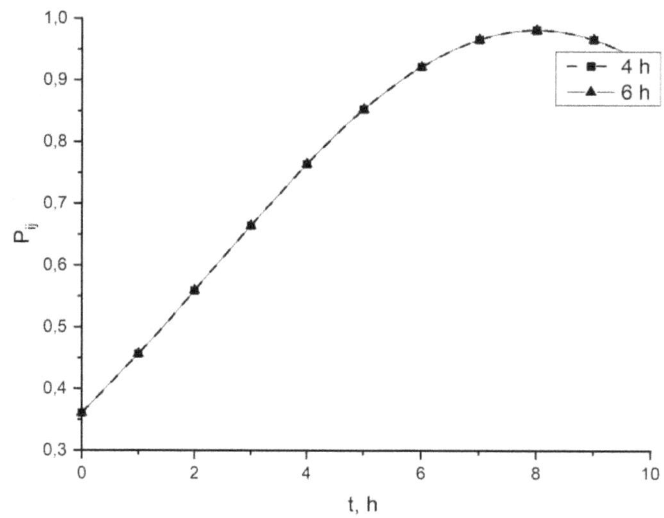

As can be seen from fig. 15, the prediction of the probability of a forest fire made by weather data for 4 and 6 hours practically does not differ, but it should be borne in mind that these calculations were carried out for the forestry, where forest fires resulting from anthropogenic pressure predominate. In fig. Figure 16 presents similar results, but for the forestry, where forest fires prevail as a result of thunderstorm activity. Here, the prediction results for the data obtained at 4 and 6 in the morning are different. Thus, based on the results of the calculations, it is possible to recommend meteorological parameters obtained at 5-6 a.m. as initial data, since this will allow taking into account most of the cloud-to-ground lightning discharges that occurred at night and in the early morning, and could potentially lead to foci

Figure 16. Probability of forest fire danger taking into account cloud-to-ground lightning discharges when thunderstorm prevails

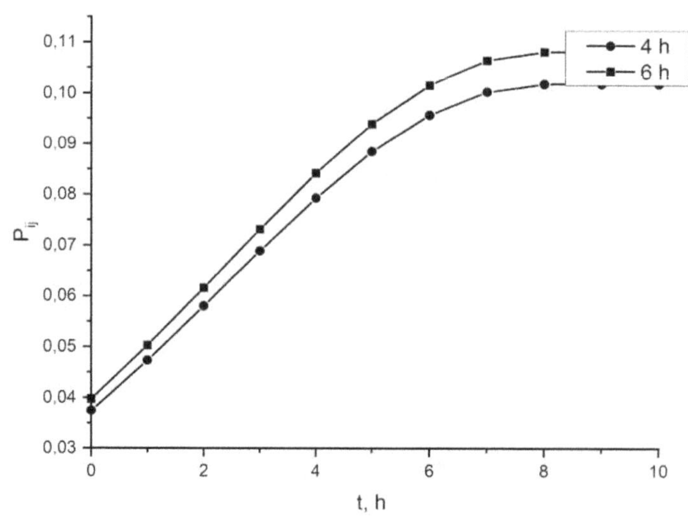

forest fires. Note that the number of days when dry thunderstorms occurred was taken to be equal to the number of days with thunderstorms, namely 22 (so many according to(Scientific-applied reference…, 1993) in June-August), which will give a somewhat overestimated estimate of the effect of thunderstorms.

Option 6. The effect of the initial moisture content of the forest fuel layer. Fig. 17 presents the dependence of the probability of forest fire danger for the initial moisture content of 40% and 25% (respectively, curves 1 and 2).

Figure 17. Probability of forest fire danger for initial moisture content of 40% and 25%

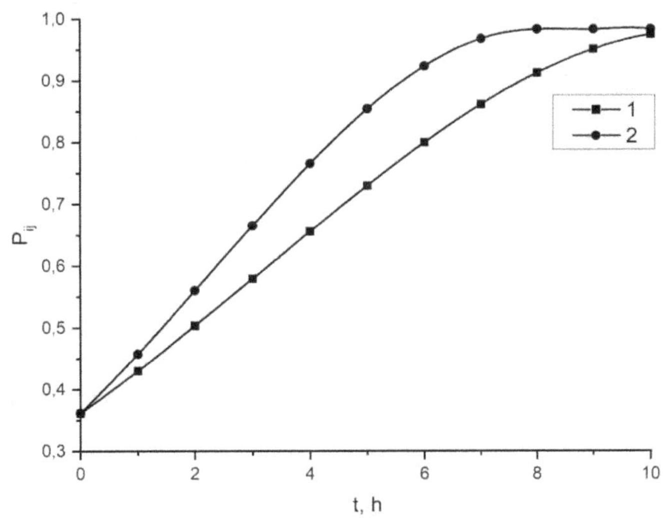

With an increase in the initial moisture content of the forest fuel layer, as follows from Figure 6.17, the drying time increases. This accordingly leads to the later initiation of forest fire danger.

Option 7. The probability of a forest fire becoming a city fire. When a forest fire occurs on the territory of an allotment that borders a settlement, it can go from a forest fire to a city fire. An example of this is the transition of a forest fire to an urban fire in 2000 in Los Alamos (USA). In (Fried et al., 2000), a formula was proposed for determining the probability of a forest fire moving into an urban fire:

$$P(WUI) = P(FF)P(WUI\,/\,FF), \tag{9}$$

where *P(WUI)* is the probability of a city fire, *P(FF)* is the probability of a forest fire, *P(WUI/FF)* is the probability of a city fire, provided that a forest fire occurs. *P(FF)* can be calculated by formula (1), and the conditional probability can be determined from statistics using the following formula $P(WUI\,/\,FF) \approx \frac{N_{WUI}}{N_{FT}}$, where N_{FT} is the total number of forest fires, N_{WUI} is the number of forest fires that have turned into urban fires. $(N_{WUI}, N_{FT}) = (5, 100)$ for the stand that rarely appears in statistics, $(N_{WUI}, N_{FT}) = (50, 100)$ for the stand that is often featured in forest fire statistics. In fig. 18, the curve with index 1 corresponds to the probability of a forest fire in the territory of the allotment bordering the village. Curve 2 corresponds to the situation when the stand is not often featured in statistics, curve 3

Figure 18. Probability of a forest fire becoming a city fire

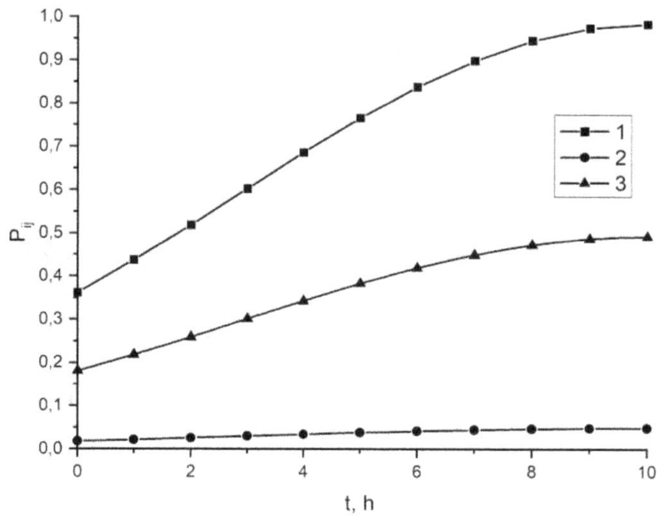

corresponds to the stand, the transition of a forest fire from which to a settlement is often recorded in statistics.

The results presented in Figure 18 are logical: the more often the highlight appeared in the statistics, the higher the conditional probability *P(WUI/FF)* and vice versa. It should be noted that formula (9) represents the simplest probabilistic model for the transition of a forest fire into an urban one. In the future, it will be advisable to introduce determinism into the definition of conditional probability *P(WUI/FF)* to simulate the process of transition of a fire flame front from a forest to a settlement, using the drying model to determine the probability of a forest fire. However, even in this simplest variant, the behavior of probability is qualitatively correct.

In addition to a quantitative description of the probability of forest fires for practical purposes, it seems appropriate to introduce a qualitative characteristic of forest fire danger in the form of scales. It is suggested to divide the probability variation interval [0.1] into five sub-intervals and matching them with the qualitative characteristics of a given level of forest fire danger and the corresponding regulations for forest protection services (table 1).

Forest fire danger scales are in compliance with the requirements for forest protection services.

The zero-dimensional mathematical formulation and the approximate analytical formula accurately determine the drying time of the forest fuel layer. The change in the probability of a forest fire adequately reflects the influence of input data, anthropogenic load and thunderstorm activity. The integration of the forest-taxation characteristic, the comparison of the scales of forest fire danger and the probability of forest fires open up the prospects for the practical application of this methodology for determining the probability of a forest fire.

As the calculation results show, with a decrease in the density of the solar radiation flux in a clear sky, the number of predicted forest fires decreases every day. These results are true, since with a decrease in the density of the solar radiation flux, the drying time of the forest fuel layer increases, the probability of forest fires decreases, and the predicted number of fires decreases, respectively. Under medium cloud conditions, the situation does not fundamentally change, only the curves of changes in the predicted number of forest fires lie more closely, which is explained by smaller differences in the density of the solar radiation flux.

Changes in ambient temperature also adequately affect changes in the number of projected forest fires. With decreasing ambient temperature, the drying time of the forest fuel layer increases, the probability of occurrence decreases, which leads to a decrease in the number of predicted forest fires in a controlled forest area every day.

In addition, parametric calculations were performed on the influence of the initial moisture content of the forest fuel layer, precipitation, anthropogenic load, cloud-to-ground lightning discharges on the predicted number of forest fires. As the analysis of the results showed, the technique adequately responds to changes in the initial moisture content and other parameters of the forest fire danger.

Table 1. Forest Fire Danger Scales

Fire danger class, P_j	Regulations for forest protection services (Schetinskiy, 2002)
P_jÎ[0, 0.2], (I) No fire danger	Ground-based patrols are carried out in places of flammable work, in order to monitor compliance with fire safety rules in the forest. Duty on fire observation towers (FOT) is not carried out.
P_jÎ(0.2, 0.4], (II) Low fire danger	Ground patrolling is carried out in areas assigned to classes I and II of natural fire danger, as well as in places of mass visits and rest of the population from 11 to 17 hours. Duty at the night vision gates and at reception points from 11 to 17 hours.
P_jÎ(0.4, 0.6], (III) Medium fire danger	Ground patrolling is carried out from 10 to 19 hours, in areas assigned to the first three classes of fire danger, and is especially intensified in the places of work and places most visited by the population. Duty at the FOT and at points of reception of reports from 10 to 19 hours, at points of reception of reports from 10 to 17 hours.
P_jÎ(0.6, 0.8], (IV) High fire danger	Ground patrolling is carried out from 8 to 20 hours in places of work, as well as in places visited by the population, regardless of the class of fire danger to which the sections are assigned. Duty hours on FOT are carried out during the whole daylight hours, at the points of reception of reports from 8 to 20 hours.
P_jÎ(0.8, 1], (V) Extreme fire danger	Ground patrolling is carried out throughout the daylight hours, and in the most dangerous places around the clock. To help the forest guard and temporary firemen, workers and employees of the forestry, a public asset and the police are involved in patrolling. Duty on FOT and at points of reception of reports are carried out, as in the IV class of fire danger.

CONCLUSION

This chapter proposes a new probabilistic criterion and a new deterministic-probabilistic model for calculating the probability of forest fires. In large forested areas, problems arise of large computational load and large amount of processed data. At the moment and in the near future, this is the main obstacle to obtaining a forest fire danger prediction in a mode ahead of the real time of the process development. This problem can be solved by applying supercomputers and parallel computing technologies.

This methodology can be extended with results on mathematical simulation of anthropogenic load on forested territories in the context of forest fire occurrence (Baranovskiy, 2020). Also clustering of cloud-to-ground lightning discharges can be used to determine characteristics of lightning activity over controlled forested territories (Belikova and Glebova, 2020). Mathematical simulation of forest fuel ignitions can be used to determine forest fire initiation conditions (Baranovskiy and Kuznetsov, 2017) with experimental assistance (Goman, 2020).

ACKNOWLEDGMENT

This work was implemented with the financial support of the Russian Foundation for Basic Research. [Project N 17-29-05093].

REFERENCES

Adzhiev, A. H., & Kuliev, D. D. (2018). Characteristics of Storm Activity and Parameters of Lightning Discharges in the South of the European Part of Russia. *Izvestiya. Atmospheric and Oceanic Physics*, *54*(4), 372–379. doi:10.1134/S0001433818040187

Amatulli, G., Perez-Cabello, F., & Du la Riva, J. (2007). Mapping lightning/human-caused wildfires occurrence under ignition point location uncertainty. *Ecological Modelling*, *200*(3-4), 321–333. doi:10.1016/j.ecolmodel.2006.08.001

Andreev, Yu. A. (1986). *Patterns of distribution of forest fires. Methods and means of combating forest fires*. VNIILM. (in Russian)

Andreev, Yu. A., & Larchenko, G. F. (1987). Socio-psychological aspects of recreational visits to the forest and the occurrence of fires. Forest fires and the fight against them. Moscow. [in Russian]. *VNIILM*, *1987*, 251–263.

Ateitio, J., Ezcurra, A., & Herrero, I. (2001). Cloud-to-ground lightning characteristics in the Spanish Bascue Country area during the period 1992—1996. *J. Atmos. and Solar-Terrest. Phys.*, *63*, 1005–1015.

Baranovskiy, N., & Zharikova, M. (2014). *A web-oriented geoinformation system application for forest fire danger prediction in typical forests of the Ukraine*. Lecture Notes in Geoinformation and Cartography. doi:10.1007/978-3-319-08180-9_2

Baranovskiy, N. V. (2017). Forest fire danger assessment using SPMD-model of computation for massive parallel system. *International Review on Modeling and Simulations*, *10*(3), 193–201. doi:10.15866/iremos.v10i3.10570

Baranovskiy, N. V. (2020). Mathematical Simulation of Anthropogenic Load on Forested Territories for Point Source. In N. Baranovskiy (Ed.), *Predicting, Monitoring, and Assessing Forest Fire Dangers and Risks* (pp. 64–88). IGI Global. doi:10.4018/978-1-7998-1867-0.ch003

Baranovskiy, N.V., & Grishin, A.M. (2002). Mathematical modeling of the drying of a layer of forest combustible materials under the canopy of a pine stand. *Computational Technologies, 7*(1), 320-324. (in Russian)

Baranovskiy, N. V., & Kuznetsov, G. V. (2017). Mathematical simulation of deciduous tree ignition by cloud-to-ground lightning discharge using large vessels approximation. *JP Journal of Heat and Mass Transfer*, *14*(4), 533–545. doi:10.17654/HM014040533

Bayham, J., Belval, E. J., Thompson, M. P., Dunn, C., Stonesifer, C. S., & Calkin, D. E. (2020) Weather, risk, and resource orders on large wildland fires in the western US. Forests, 11. doi:10.3390/f11020169

Belikova, M. Y., & Glebova, A. V. (2020). Cluster Analysis Algorithms for RS and WWLLN Data Processing. In N. Baranovskiy (Ed.), *Predicting, Monitoring, and Assessing Forest Fire Dangers and Risks* (pp. 151–181). IGI Global. doi:10.4018/978-1-7998-1867-0.ch007

Benavent-Corai, J., Rojo, C., Suarez-Torres, J., & Velasco-Garcia, L. (2007). Scaling properties in forest fire sequences: The human role in the order of nature. *Ecological Modelling*, *205*(3-4), 336–342. doi:10.1016/j.ecolmodel.2007.02.028

Bernardi, M., & Ferrari, D. (2004). Evaluation of the LLS efficiency effects on the ground flash density, using the Italian lightning detection system SIRF. *Journal of Electrostatics*, *60*(2-4), 131–140. doi:10.1016/j.elstat.2004.01.013

Boccippio, D. J., Williams, E. R., Heckman, S. J., Lyons, W. A., Baker, I. T., & Boldi, R. (1995). ELF transients, and positive ground strokes. *Science*, *269*(5227), 1088–1091. doi:10.1126cience.269.5227.1088 PMID:17755531

Burke, C. P., & Jones, D. L. (1996). On the polarity and continuing current in unusually large lightning flashes deduced from ELF events. J. Atmosp. and Solar-Terres. *Phys.*, *58*, 531–548.

Cardille, J. A., Ventura, S. J., & Turner, M. G. (2001). Environmental and social factors influencing wildfires in the Upper Midwest, United States. *Ecological Applications*, *11*(1), 111–127. doi:10.1890/1051-0761(2001)011[0111:EASFIW]2.0.CO;2

Chavro, A. I., & Dmitriev, E. V. (2002). *Methodology for reconstructing the detailed structure of meteorological fields on the urban and regional scales by their integrated characteristics. Program and abstracts of the international conference "ENVIROMIS-2002"*. Publishing House of Tomsk Central Scientific and Technical Institute. (in Russian)

Cummer, S. A., & Inan, U. S. (1996). Sprite-producing lightning using ELF radio atmospherics. *Geophysical Research Letters*, *24*(14), 1731–1734. doi:10.1029/97GL51791

Cummins, K. L., Murphy, M. J., Bardo, E. A., Hiscox, W. L., Pyle, R. B., & Pifer, A. E. (1998). A combined TOA/MDF technology upgrade of the U.S. national lightning detection network. *Journal of Geophysical Research*, *103*(D8), 9035–9044. doi:10.1029/98JD00153

Dymnikov, V. P., Lykosov, V. N., & Volodin, E. M. (2005). *Modeling climate and its changes. Modern Problems of Computational Mathematics and Mathematical Modeling* (Vol. 2). Nauka. (in Russian)

Esaulov, A.O., & Starchenko, A.V. (n.d.). *Implementation of the mesoscale model of atmospheric admixture transport on supercomputers. Program and abstracts of the international conference "ENVIROMIS-2002"*. Tomsk: Publishing House of Tomsk Central Scientific and Technical Institute. (in Russian)

Filkov, A. I. (2005). *Deterministic-probabilistic forest fire danger forecast system* (PhD Thesis). Tomsk: TSU. (in Russian)

Fried, J. S., Winter, G. J., & Gilless, J. K. (2000). Assessing the benefits of reducing fire risk in the Wildland-Urban Interface: A contingent valuation approach. *International Journal of Wildland Fire*, *9*(1), 9–19. doi:10.1071/WF99002

Gallin, L.-J., Farges, T., Marchiano, R., Coulouvrat, F., Defer, E., Rison, W., Schulz, W., & Nuret, M. (2016). Statistical analysis of storm electrical discharges reconstituted from a lightning mapping system, a lightning location system, and an acoustic array. *Journal of Geophysical Research, 121*, 3929–3953.

Goman, P. N. (2020). The Dynamics of Surface Forest Fire and Forest Fuel Ignition Under the Heat Radiation From the Fire Line. In N. Baranovskiy (Ed.), *Predicting, Monitoring, and Assessing Forest Fire Dangers and Risks* (pp. 1–47). IGI Global. doi:10.4018/978-1-7998-1867-0.ch001

Grishin, A. M. (1997). *Mathematical modeling of forest fire and new methods of fighting them. Russia*. Publishing House of the Tomsk University. (in Russian)

Grishin, A. M. (1999). *Mathematical modeling and prediction of disasters*. Publishing House of Tomsk University. (in Russian)

Grishin, A. M. (2002). *Modeling and forecasting disasters*. Publishing House of Tomsk University. (in Russian)

Grishin, A. M., & Baranovskij, N. V. (2003). Comparative analysis of simple models of drying of the layer of forest combustibles, including the data of experiments and natural observations. [in Russian]. *Inzhenerno-Fizicheskii Zhurnal., 76*(5), 166–169. doi:10.1023/B:JOEP.0000003234.63878.7b

Grishin, A. M., & Filkov, A. I. (2005). *Forecast of the occurrence and spread of forest fires*. Practice. (in Russian)

Grishin, A. M., Golovanov, A. N., Kataeva, L. Yu., & Loboda, E. L. (2001). Formulation and solution of the problem of drying a layer of forest combustible materials. [in Russian]. *Combustion and Explosion Physics., 37*, 65–76.

Grishin, A. M., Kataeva, L. Yu., & Loboda, E. L. (2001). Mathematical modeling of drying of a layer of forest combustible materials. Computational Technologies, 6(2), 140-144. (in Russian)

Huang, E., Williams, E., Boldi, R., Heckman, S., Lyons, W., Taylor, M., Nelson, T., & Wong, C. (1999). Criteria for sprites and elves based on Schumann resonance measurements. *Journal of Geophysical Research, 104*(D14), 1693–1694. doi:10.1029/1999JD900139

Kosarev, V. P., & Andryushchenko, T. T. (2007). *Forest meteorology with the basics of climatology: a Training manual.* St. Petersburg: Lan. (in Russian)

Kurbatskiy, N. P. (1964). *The problem of forest fires. Occurrence of forest fires*. Nauka. (in Russian)

Kuznetsov, V. I., Kozlov, N. I., & Khomyakov, P. M. (2005). *Mathematical modeling of forest evolution for forest management purposes*. LENAND. (in Russian)

Larfeldt, J., Leckner, B., & Melaaen, M. C. (2000). Modeling and measurements of the pyrolysis of large wood particles. *Fuel, 79*(13), 1637–1643. doi:10.1016/S0016-2361(00)00007-7

Latham, D., & Williams, E. (2001). *Lightning and forest fires. Forest Fires: Behavior and Ecological Effects*. Elsevier.

Latham, D. J. (1991). Lightning flashes from a prescribed fire-induced cloud. *Journal of Geophysical Research, 96*(D9), 17151–17157. doi:10.1029/91JD01808

Li, Y., Zhang, G., & Zhang, Y. (2020). Evolution of the Charge Structure and Lightning Discharge Characteristics of a Qinghai-Tibet Plateau Thunderstorm Dominated by Negative Cloud-to-Ground Flashes. Journal of Geophysical Research: Atmospheres, 125. doi:10.1029/2019JD031129

Lixiang, Z., Xiaoshan, Z., Yongzuo, L., & Jinggui, W. (2000). Case study on tropical cyclone track prediction with MM5. *J. Nanjing Institute of Meteorology.*, *23*, 73–80.

Lyons, W. A., Nelson, T. E., & Williams, E. R. (1998). Enhanced positive cloud-to-ground lightning in thunderstorms ingesting smoke from fires. *Science*, *282*(5386), 77–80. doi:10.1126cience.282.5386.77 PMID:9756479

Matsui, M., Michishita, K., & Yokoyama, S. (2019). Characteristics of Negative Flashes With Multiple Ground Strike Points Located by the Japanese Lightning Detection Network. IEEE Transactions on Electromagnetic Compatibility, 61, 751-758. doi:10.1109/TEMC.2019.2913661

Melekhov, I. S. (1947). *Nature of the forest and forest fires*. Arkhangelsk. (in Russian)

Melluma, A. Zh., Rungule, R. Kh., & Emsis, I. V. (1982). *Camping as a nature conservation problem*. Zinatne. (in Russian)

Michalakes, J., Chen, S., & Dudhia, J. (2001). Development of a next generation regional weather research and prediction model. In *Developments in Teracomputing: Proceedings of the Ninth ECMWF Workshop on the Use of High Performance Computing in Meteorology*. World Scientific.

Murphy, A. H., & Winkler, R. L. (1984). Probability predicting in meteorology. *Journal of the American Statistical Association*, *79*, 489–500.

Nesterov, V. G. (1949). *Forest combustibility and methods for its determination*. Goslesbumizdat. (in Russian)

Nieto, H., Aguado, I., & Chuvieco, E. (2006). Estimation of lightning-caused fires occurrence probability in Central Spain. *Forest Ecology and Management*, *234*(Suppl. 1.), S67. doi:10.1016/j.foreco.2006.08.095

Orville, R. E., & Huffines, G. R. (2001). Cloud-to-ground lightning in the United States: NLDN results in the first decade, 1989—98. *Monthly Weather Review*, *129*(5), 1179–1193. doi:10.1175/1520-0493(2001)129<1179:CTGLIT>2.0.CO;2

Panevin, V. S. (2006). *Forests and forestry of the Tomsk region*. Publishing House of Tomsk University. (in Russian)

Paulucci, T. B., França, G. B., Libonati, R., & Ramos, A. M. (2019). Long-Term Spatial–Temporal Characterization of Cloud-to-Ground Lightning in the Metropolitan Region of Rio de Janeiro. *Pure and Applied Geophysics*, *176*(11), 5161–5175. doi:10.100700024-019-02216-1

Pew, K. L., & Larsen, C. P. S. (2001). GIS analysis of spatial and temporal patterns of human-caused wildfires in the temperate rain forest of Vancouver Island, Canada. *Forest Ecology and Management*, *140*(1), 1–18. doi:10.1016/S0378-1127(00)00271-1

Pinto, I. R. C. A., & Pinto, O. Jr. (2003). Cloud-to-ground lightning distribution in Brazil. J. Atmosp. and Solar-Terrest. *Phys.*, *65*, 733–737.

Pinto, O. Jr, Pinto, I. R. C. A., & Diniz, J. H. (2003). A seven-year study about the negative cloud-to-ground lighting flash characteristics in Southeastern Brazil. *J. Atmosp. and Solar-Terrest. Phys.*, *65*, 739–748.

Pinto, O. Jr, Pinto, I. R. C. A., & Naccarato, K. P. (2007). Maximum cloud-to-ground lightning flash densities observed by lightning location systems in the tropical region: A review. *Atmospheric Research*, *84*(3), 189–200. doi:10.1016/j.atmosres.2006.11.007

Podur, J., Martell, D. L., & Csilagg, F. (2003). Spatial patterns lightning-caused forest fires in Ontario 1976—1998. *Ecological Modelling*, *164*(1), 1–20. doi:10.1016/S0304-3800(02)00386-1

Ponomarev, E. I., Ivanov, V. A., & Korshunov, N. A. (2006). Satellite data TOVS when solving the problem of forecasting thunderstorm fire danger in a forest. [in Russian]. *Geography and Natural Resources*, (1), 147–150.

Romanenkov, I. G., & Levites, F. A. (1991). *Fire protection of building structures*. Stroyizdat. (in Russian)

Schetinsky, E. A. (2002). Extinguishing Forest Fire: A Handbook for Forest Firefighters. Academic Press. (in Russian)

Scientific-applied reference on the climate of the USSR. (1993). *Series 3. Perennial data. Parts 1-6. Vol. 20. (Tomsk, Novosibirsk, Kemerovo regions and Altai Territory).* St. Petersburg: Gidrometeoizdat. (in Russian)

Sofronov, M. A. (1970). *Questions of forest pyrology*. ILID SB AS USSR. (in Russian)

Soriano, L. R., De Pablo, F., & Tomas, C. (2005). Ten-year study of cloud-to-ground lightning activity in the Iberian Peninsula. *J. Atmosp. and Solar-Terrestr. Phys.*, *67*, 1632–1639.

Stryamets, N., Elbakidze, M., Chamberlain, J., & Angelstam, P. (2020) Governance of non-wood forest products in Russia and Ukraine: Institutional rules, stakeholder arrangements, and decision-making processes. Land Use Policy, 94.

Telitsin, G. P. (1984). Studying the connection of forest attendance and the occurrence of fires. [in Russian]. *Forestry*, (1), 59–63.

Telitsin, G. P. (1987). *Method for determining the fire danger of a forest territory. Forest fires and the fight against them*. VNIILM. (in Russian)

The concept of development of forestry in Russia for 2003-2010. (2003). Moscow: MPR. (in Russian)

Tolstykh, M. A. (1997). *Global semi-Lagrangian atmospheric model based on compact finite-differences and its implementation on a parallel computer*. INRIA Research Report #3080, Theme 4 / INRIA. Domaine de Voluceau.

Tolstykh, M. A. (2001). Semi-Lagrangian high-resolution atmosphere model for numerical weather prediction. [in Russian]. *Meteorology and Hydrology*, (4), 5–15.

Tsunomura, S., Nishihashi, M., & Kusunoki, K. (2016). Electrical charge associated with cloud-to-ground lightning discharge in the Shonai area, Tohoku district, Japan. *Papers in Meteorology and Geophysics*, *66*(0), 25–37. doi:10.2467/mripapers.66.25

Uman, M. A. (1969). *Lightning*. McGraw-Hill.

Viegas, D. X., Bovio, G., & Ferreira, A. (2000). Comparative study of various methods of fire danger evaluation in Southern Europe. *International Journal of Wildland Fire*, *10*, 235–246.

Wang, X., Zhao, X., Cai, H., Liu, G., Liao, M., Qu, L., & Hu, S. (2020). Comparisons of optical characteristics of two upward lightning flashes triggered by a nearby positive cloud-to-ground lightning. Journal of Atmospheric and Solar-Terrestrial Physics, 198. doi:10.1016/j.jastp.2020.105193

Weber, M. G., & Stocks, B. J. (1998). Forest fires and sustainability in the boreal forests of Canada. *Ambio*, *27*, 545–550.

Yanko, I. V. (2005). *Pyrological assessment of the territory of the Tomsk region* (PhD Thesis). Tomsk: Tomsk State Pedagogical University. (in Russian)

ADDITIONAL READING

Adámek, M., Jankovská, Z., Hadincová, V., Kula, E., & Wild, J. (2018). Drivers of forest fire occurrence in the cultural landscape of Central Europe. *Landscape Ecology*, *33*(11), 2031–2045. doi:10.100710980-018-0712-2

Bowman, D. M. J. S., Moreira-Muñoz, A., Kolden, C. A., Chávez, R. O., Muñoz, A. A., Salinas, F., González-Reyes, Á., Rocco, R., de la Barrera, F., Williamson, G. J., Borchers, N., Cifuentes, L. A., Abatzoglou, J. T., & Johnston, F. H. (2019). Human–environmental drivers and impacts of the globally extreme 2017 Chilean fires. *Ambio*, *48*(4), 350–362. doi:10.100713280-018-1084-1 PMID:30128860

Clarke, H., Gibson, R., Cirulis, B., Bradstock, R. A., & Penman, T. D. (2019). Developing and testing models of the drivers of anthropogenic and lightning-caused wildfire ignitions in south-eastern Australia. *Journal of Environmental Management*, *235*, 34–41. doi:10.1016/j.jenvman.2019.01.055 PMID:30669091

Elia, M., Giannico, V., Lafortezza, R., & Sanesi, G. (2019). Modeling fire ignition patterns in Mediterranean urban interfaces. *Stochastic Environmental Research and Risk Assessment*, *33*(1), 169–181. doi:10.100700477-018-1558-5

Preisler, H. K., Brillinger, D. R., Burgan, R. E., & Benoit, J. W. (2004). Probability based models for estimation of wildfire risk. *International Journal of Wildland Fire*, *13*(2), 133–142. doi:10.1071/WF02061

Scheller, R., Kretchun, A., Hawbaker, T. J., & Henne, P. D. (2019). A landscape model of variable social-ecological fire regimes. *Ecological Modelling*, *401*, 85–93. doi:10.1016/j.ecolmodel.2019.03.022

Viedma, O., Urbieta, I. R., & Moreno, J. M. (2018). Wildfires and the role of their drivers are changing over time in a large rural area of west-central Spain. *Scientific Reports*, *8*(1), 17797. doi:10.103841598-018-36134-4 PMID:30542114

Ye, J., Wu, M., Deng, Z., Xu, S., Zhou, R., & Clarke, K. C. (2017). Modeling the spatial patterns of human wildfire ignition in Yunnan province, China. *Applied Geography (Sevenoaks, England)*, *89*, 150–162. doi:10.1016/j.apgeog.2017.09.012

You, W., Lin, L., Wu, L., Ji, Z., Yu, J., Zhu, J., Fan, Y., & He, D. (2017). Geographical information system-based forest fire risk assessment its spatiotemporal variability. *Ecological Indicators*, *77*, 176–184. doi:10.1016/j.ecolind.2017.01.042

KEY TERMS AND DEFINITIONS

Anthropogenic Load: Different human activities on forested territories lead to forest fire occurrence and characterized by presence of fire sources.

Cloud-to-Ground Lightning Discharge: An electrical discharge during a thunderstorm that occurs between a cloud and the earth's surface. It is a natural source of forest fires.

Drying: Drying is moisture evaporation from live of dead forest fuel under the environmental conditions.

Forest Fire: Uncontrolled aerothermochemical phenomenon characterized by step-by-step mechanism which includes following stages: inert heating, moisture evaporation, high temperature terpens evaporation, dry organic matter pyrolysis, flammable combustion and smoldering.

Forest Fuel: It can be considered like dead and live forest fuel. Main types of forest fuel which can be involved in combustion during forest fire: ground forest fuel (needles, leaves and dry grass, small branches) and crown forest fuel (needles, small branches).

Lightning Activity: An atmospheric phenomenon characterized by discharges of the cloud-to-cloud and cloud-to-ground class.

Mathematical Simulation: The production of a computer model of forest fire conditions and prerequisites, especially for the purpose of study.

Meteorological Parameters: Physical characteristics of local weather conditions in the forested area under consideration. Key parameters include ambient temperature, soil temperature, precipitation, wind speed, solar radiation, cloud cover, dew point temperature. These parameters are used for mathematical modeling of the drying of a layer of forest fuel.

Monitoring: Monitoring refers to the periodic calculation of the parameters of forest fire danger with a portion of information available in real time.

Prediction: Under the prediction of forest fires is the calculation of the parameters of forest fire danger with a certain projection in advance in order to have enough time to anticipate an emergency. The calculation in this case is carried out in a mode ahead of the real time of the development of the catastrophe - the occurrence of a forest fire.

Chapter 15
Forest Fire Probability Prediction Taking Into Account Different Reasons of Anthropogenic Load

ABSTRACT

This work is devoted to the creation of a probabilistic criterion for forest fire danger to take into account the various causes of anthropogenic load that lead to forest fires. Typical forested areas of the boreal zone are considered: Tomsk region (Russian Federation) and Vancouver Island (Canada). In addition, a description is given of a probabilistic criterion that takes into account the occurrence of a forest fire as a result of deliberate arson. The chapter presents the results of scenario modeling of forest fire danger. It is concluded that it is possible to modernize existing forest fire danger prediction systems in the USA, Canada, Southern Europe, Australia, and the Russian Federation.

INTRODUCTION

Both the consequences and the causes of forest fires are diverse. For practical purposes, among natural causes, it is sufficient to single out the occurrence of forest fires as a result of thunderstorm activity. Reasons such as volcanic activity and self-ignition of forest fuel can be neglected. The first reason in mind is the extremely small share in the total number of forest fires. The second one is for physical reasons, since self-ignition of the ground cover and further formation of the center of the surface forest fire is impossible (Kuznetsov and Baranovskiy, 2006). Anthropogenic causes of forest fires are quite diverse. A model should be developed for a differentiated assessment of forest fire danger for anthropogenic reasons, taking into account both the specifics of our state and attracting information from foreign sources (Gorev, 2004; Pew and Larsen, 2001).

A comparative analysis of Russian and foreign data shows that, at least in the Tomsk Region, the vast majority of man-made fires are caused by the fault of the local population and other reasons can be

DOI: 10.4018/978-1-7998-7250-4.ch015

neglected. However, on the territory of the Canadian island of Vancouver, the proportion of forest fires from other anthropogenic sources is high. Moreover, most fires have a mixed cause. Recent events in Spain allow us to talk about what needs to be included in the model and such a reason as arson. Therefore, in order to develop a universal geographically independent prediction model for forest fire danger as a result of anthropogenic load, all causes of fire outbreaks should be taken into account.

An analysis of the forest fire danger predicting methods existing in Russia and abroad shows that almost all methods have a weak physical basis (Baranovskiy and Kuznetsov, 2017). Thus, there is no doubt the need to develop a new methodology for determining the probability of forest fires. For example, in the nuclear industry, a probabilistic safety criterion is used (Grishin, 2002). Currently, the need to develop such a criterion and the corresponding methodology has arisen in forestry. As shown, the analysis of real forest taxation descriptions of specific forest stands should be detailed at the level of the minimum forest taxation unit (site). The forest fire danger scales currently used are usually tied to a specific region (Schetinskiy, 2002).

Thus, it will be relevant to develop such a system of scales that would unambiguously and uniformly interpret the level of forest fire danger in any region in the form of scales of the forest fires probability (Kuznetsov and Baranovskiy, 2009). A study of forest fires resulting from human activities on the territory of Vancouver Island (Canada) is significant (Pew and Larsen, 2001). The causes of forest fires from anthropogenic load on Vancouver Island are presented in Table 1. The causes of forest fires in the Tomsk Region are also quite diverse. Table 2 summarizes the causes of foci of forest fires in the region in 1993 – 2002 (Gorev, 2004).

Table 1. Causes of forest fires from anthropogenic load in Vancouver Island (Pew & Larsen, 2001)

Fire cause	Fire number	Burned area, ha	
		Total	Average (per fire)
Mixed	2567	17370	6,8
Recreation	1809	4462	2,5
Logging	1021	20927	20,5
Land clearing	512	966	1,9
Railway	189	276	1,5
Industrial objects	120	954	8,0
Motorways	104	2650	25,5
Undefined	7	8	1,2
Total	6329	47613	7,5

PROBABILISTIC CRITERION

In order to cover the countable number of anthropogenic causes of forest fires, the following events were defined: A_1 is the deliberate arson, A_2 is the negligent handling of fire, A_3 is agricultural harvest, A_4 is the influence of the railway, A_5 is the power transmission line, A_6 is the burning oil spill, A_7 is the

Table 2. Causes of forest fires in Tomsk region (Gorev, 2004)]

Fire cause	Fire number	
	Incidents	**%**
Population	1249	52,86
Lightning	891	37,71
Logging	12	0,51
Agricultural burning	3	0,13
High Temperature	1	0,04
Locomotive	9	0,38
Recreation fires	1	0,04
Power lines	1	0,04
Oil burning	2	0,08
Exhaust gas	1	0,04
Forestry objects	8	0,34
Rocket stage fall	2	0,08
Expedition	1	0,04
Unknown	182	7,70

influence highways, A_8 is the rocket stage fall, A_9 is the accidents at technological facilities located in forested areas, A_{10} is the reason has not been established.

Suppose that the occurrence of forest fires for various reasons there are independent events, that is, one reason does not depend on another. In this case, it is possible to record the probability of forest fires for anthropogenic reasons through the opposite event (Pakshirajan, 2013).

As a result, the probability of occurrence of forest fires on the basis of anthropogenic causes is determined by the following formula:

$$P(FF) = 1 - \prod_{i=1}^{10} (1 - P(FF_i)), \quad (1)$$

where FF_i is the occurrence of a forest fire for the i-th anthropogenic reason (independent event).

The probability of a forest fire for a specific anthropogenic cause is determined by the formula:

$$P(FF_i) = P(A)P(A_{j,i}/A)P(FF_i/A,A_{j,i}), \quad (2)$$

where $P(A)$ is the probability of anthropogenic load (visiting the forest area), $P(A_{j,i}/A)$ is the probability of the i-th source of fire of anthropogenic cause when visiting the forest territory on the j-th day of the week, $P(FF_i\ A,A_{j,i})$ is the probability of a forest fire on the j-th day for the i-th anthropogenic reason, which is determined by the formula $P(FF_i/\ A,A_{j,i})= P(D)P(I_i/\ D)$, where $P(D)$ is the probability that the forest fuel is sufficiently dry, $P(I_i/D)$ is the probability of ignition of the forest fuel by i -th source of anthropogenic load, provided that the forest fuel is sufficiently dry. The model can be quite simply extended to the general case of accounting for thunderstorm activity by introducing event (A_{11} is the thunderstorm activity).

$$P(FF_{11}) = P(L)P(L_k / L)P(FF_{11} / L, L_k), \tag{3}$$

where $P(L)$ is the probability of dry thunderstorms, $P(L_k/L)$ is the probability of a cloud-to-ground lightning discharge at the k-th hour of the day subject to the passage of a thunderstorm, $P(FF_{11}/L,L_k)$ is the probability of a forest fire at the k-th hour of the day subject to the passage of a thunderstorm, which is determined by the formula $P(FF_{11} / L,L_k)= P(D)P(I_L / D)$, where $P(D)$ is the probability that the forest fuel is sufficiently dry, $P(I_L/D)$ is the probability of ignition of the forest fuel by cloud-to-ground lightning discharge, provided that the forest fuel is sufficiently dry.

A typical forest area of the boreal zone is considered. Pine dominates in species composition. The ground cover for the most part consists of the litter of pine needles with a very small number of thin pine branches. The forested area is in proximity to the village, which leads to the predominance of anthropogenic pressure. To calculate the drying time of the forest fuel layer, reference data (Scientific-applied reference…, 1993) for the month of June were used. The scenario of anthropogenic load typical for the weekend is considered. For definiteness, the calculations were performed for Saturday (Matsenko et al., 1999).

The first scenario corresponds to the anthropogenic load on the territory of the Tomsk Region, when most fires occur due to the fault of the local population. According to the proposed system of reasons, this event was attributed to A_2 (negligent handling of fire). Table 3 presents the results of calculating the probability of forest fires at 1 p.m. using statistical data for the Tomsk region. In tables 3 and 4, the designation N_A is the number of days with anthropogenic load.

Table 3. Probability of forest fires (Tomsk region, Russian Federation)

N_A	P(FF)	P(FF$_1$)	P(FF$_2$)	P(FF$_3$)	P(FF$_4$)	P(FF$_5$)	P(FF$_6$)	P(FF$_7$)	P(FF$_8$)	P(FF$_9$)	P(FF$_{10}$)	P(FF$_{11}$)
90	0.552	0	0.491	0.001	0.003	0	0	0.0004	0	0.003	0.071	0.044
80	0.499	0	0.436	0.001	0.003	0	0	0.0004	0	0.002	0.063	0.044
70	0.446	0	0.382	0.001	0.002	0	0	0.0003	0	0.002	0.055	0.044
60	0.391	0	0.327	0.001	0.002	0	0	0.0003	0	0.002	0.047	0.044

The second scenario corresponds to the anthropogenic load on the territory of Vancouver Island, when various anthropogenic causes of forest fires are widely represented in statistics. Table 4 presents the results of calculating the probability of forest fires at 1 p.m. using statistics on Vancouver Island (Canada).

There is no information on forest fires from thunderstorms on Vancouver Island and for this reason, the level of anthropogenic load is higher in comparison with the Tomsk Region on the territory of this island. As a result, the total probability of forest fires is higher when all anthropogenic causes and thunderstorm activity are taken into account. The maximum contribution to the general level of forest fire danger in the Tomsk region is made by the careless handling of fire. The second reason for the level of contribution to the general fire danger is thunderstorm activity. The influence of other anthropogenic

Table 4. Probability of forest fire (Vancouver, Canada)

N_A	P(FF)	$P(FF_1)$	$P(FF_2)$	$P(FF_3)$	$P(FF_4)$	$P(FF_5)$	$P(FF_6)$	$P(FF_7)$	$P(FF_8)$	$P(FF_9)$	$P(FF_{10})$	$P(FF_{11})$
90	0.719	0	0.608	0.172	0.063	0	0	0.034	0	0.040	0.002	0
80	0.657	0	0.540	0.153	0.056	0	0	0.031	0	0.035	0.002	0
70	0.591	0	0.473	0.133	0.049	0	0	0.027	0	0.031	0.001	0
60	0.521	0	0.405	0.114	0.042	0	0	0.023	0	0.026	0.001	0

causes on the forest fire danger in the Tomsk region is minimal. An increase in the attendance of a forest area naturally influences an increase in the probability of forest fires in a controlled territory.

On the Canadian island of Vancouver, careless handling of fire also makes a major contribution to forest fire danger levels. However, unlike the Tomsk region, factors such as land clearing with the help of agricultural bollards (the second largest), the proximity of railways or highways, and the occurrence of emergency situations at technological facilities (industrial sources) have a noticeable effect. Presumably, this is explained by a higher standard of living (more cars) and a higher level of technical development (as a result, anthropogenic load increases on forested areas adjacent to industrial facilities).

The first versions of the deterministic-probabilistic forest fire danger prediction model took into account the anthropogenic load as a single common factor. As a result of this study, an improved version of the deterministic-probabilistic model for predicting forest fire danger was developed taking into account a differentiated assessment for anthropogenic reasons. A fairly complete set of anthropogenic causes of forest fires has been defined.

On the territory of the Tomsk region, the application of this technique may not be appropriate in these conditions. However, with an increase in the well-being of the population and the technical development of the Russian Federation (especially in the case of the development of remote territories), the demand for this model will grow. The results of model calculations for the Canadian island of Vancouver show that the current forest fire situation requires a differentiated assessment of forest fire danger for anthropogenic reasons. One should look for ways to disseminate positive experience in the field of forecasting forest fires in developed countries, such as the USA, Canada, countries of Southern Europe, Australia, and the Russian Federation (Deeming et al., 1977; Lee et al., 2002; EFFIS, 2020; Noble et al., 1980; Podolskaya et al., 2011). Taking into account information about the specific cause of anthropogenic forest fires made it possible to improve the existing methodology for predicting forest fire danger. Specification of the causes of forest fires may become the basis for new developments in the field of creating a system for assimilating data on the level of anthropogenic load in controlled forested areas (Baranovskiy, 2018; 2020).

MATHEMATICAL MODEL OF ARSON

In recent years, such an anthropogenic factor as deliberate arson has gained great importance. The events of recent years in the USA (California Department of Forestry and Fire Protection, 2020), Greece (WWF Russia, 2020) and Australia (Country Fire Authority, 2020) eloquently speak about this. A deterministic-probabilistic approach to forest fire danger predicting has been developed. Within the deterministic part, the process of drying and ignition of a forest fuel is simulated under the influence of weather conditions and the external influence factor. The probabilistic component allows to take into account, through the frequency of events, the statistical data of the forest fire retrospective.

In 2009, forest fires of unprecedented number of victims occurred in Australia (Country Fire Authority, 2020). As a result of the fires, about two hundred people died, some went missing. According to the Australian media, up to this point, there have not been so large-scale victims of fires. The most recent tragic events were the 1983 forest fires in Victoria, when 75 people were killed. The cause of the latest tragic events was deliberate arson.

In 2007, severe forest fires also occurred in Greece, one of the reasons for which was also intentional arson. The fires of 2007 became the most intense and destructive over the past century and a half and claimed the lives of almost a hundred people. The Greek government declared a state of emergency. WWF Greece believes that the main cause of fires was arson associated with the active development of private construction. Greece remains the only country in the European Union in which there is no state registration of land and forest land. Legislation thirty years ago allows landowners to build anywhere, provided that they own 0.4 hectares of land (WWF Russia, 2020). In addition, an amateur video featuring two arsonists was shown on Greek television. Most likely we are talking about the actions of pyromancers (Yurieva, 2020). These are people with a type of mental disorder, which is expressed in the desire to arrange the most spectacular fire possible (F63.1 according to the classification of mental disorders ICD-10).

In the USA, in 2007, forest fires raged in the state of California (from Santa Barbara on the Pacific coast to the Mexican border) (California Department of Forestry and Fire Protection, 2020). As a result of the fires, about 1,500 houses were destroyed, about 20 people died, dozens were injured. In 2007, state losses due to fires exceeded one billion dollars. In 2008, forest fires in southern California destroyed over 1,000 homes, including about 70 mansions in the elite village of Montequito. One of the reasons is deliberate arson.

In Spain, tense conditions with forest fires also regularly develop. Most fires are deliberate, as arsonists sell wood from burnt forests or gain building rights on scorched earth. In 2002 alone, the gendarmerie attracted 217 people on charges of arson.

Among the socio-psychological factors of intentional arson should be highlighted:

1. the welfare of people living in a wooded area causes envy and resentment of the lower social strata of society;
2. favorable conditions for the subsequent development of the scorched area by residential areas;
3. favorable conditions for the subsequent laying of roads, pipelines, power lines and the construction of industrial facilities;
4. imperfection of the legislation (the possibility of cutting on the burner and the purchase and sale of commercial wood after a fire at a more attractive price);
5. hooligan actions.

If you look at these reasons from the standpoint of probability theory (Pakshirajan, 2013), then the events-causes of arson (Janssens, 1991) can be distinguished: a) thirst for profit (i=1), b) manifestation of social inequality (i=2), c) hooligan actions (i=3).

A forested area can be characterized by the following criteria:

Figure 1. Scheme of causal relationships in the event of forest fire accidents as a result of deliberate arson

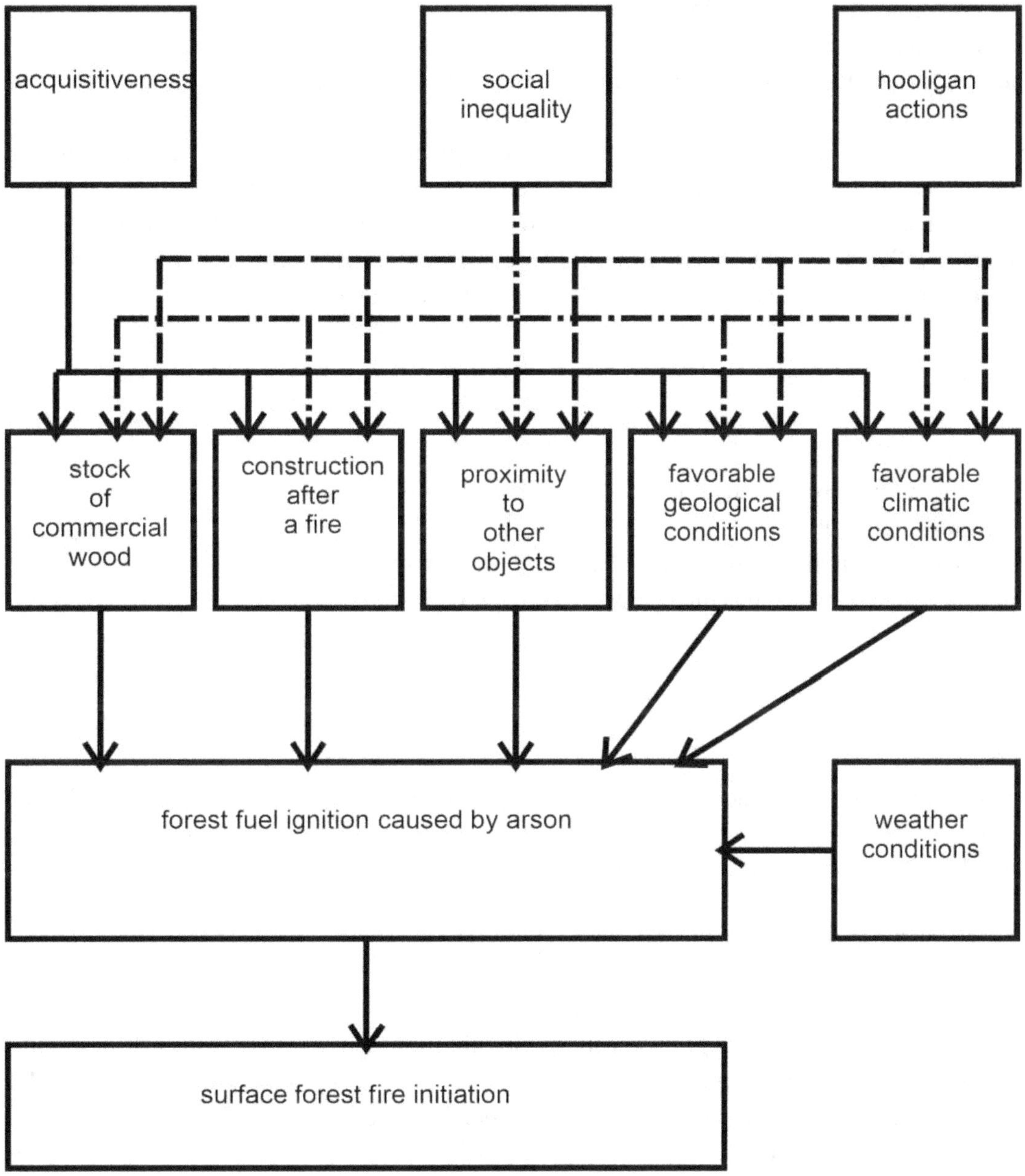

a) stock of commercial wood and its parameters; b) the value of the land; c) the proximity of any objects; d) favorable geological conditions; e) favorable climatic conditions.

From the standpoint of probability theory (Pakshirajan, 2013), events can be introduced that characterize the conditions on the forested area: a) on the forested area there is a supply of valuable commercial wood (j=1), b) a land plot of the forested area is available for construction after a fire (j=2) c) residential, infrastructural or industrial facilities are located near the forest site (j=3), d) favorable geological conditions on the forested area (j=4), e) favorable climatic conditions on the forested area (j=5). The scheme of causal relationships in the event of forest fire accidents as a result of deliberate arson is presented in Figure 1.

Assumptions: 1) elite development is carried out in wooded areas with a high level of recreational value and, accordingly, it is assumed that the forest is of high value in all respects; 2) elite development is carried out in places with favorable climatic and geological conditions; 3) the various causes of intentional arson are joint. When writing a formula to assess the probability of a forest fire, simplification is used that the causes of deliberate arson and the characteristics of the forested area are independent events.

Mathematically, the model for assessing the probability of forest fires as a result of deliberate arson is recorded through opposite events (by analogy with the integrated forest fire danger assessment):

$$P(FF) = 1 - \prod_{i=1}^{n} \left(1 - P(FF_i)\right), \tag{4}$$

where n is the number of reasons for intentional arson, $P(FF_i)$ is the probability of a forest fire for the i-th reason for intentional arson.

Since the characteristics of the forested area are joint and independent events, each probability of a forest fire for a specific reason is also determined through opposite events:

$$P(FF_i) = 1 - \prod_{j=1}^{m} \left(1 - P(FF_{ij})\right), \tag{5}$$

where m is the number of characteristics of the forested area, $P(FF_{ij})$ is the probability of forest fires in the presence of the j-th characteristic of the forested area for the i-th reason for deliberate arson. It is determined by the formula (the probability multiplication theorem is used):

$$P(FF_{ij}) = P(R_i)P(C_j)P(D)P(I_{ij} / R_i, C_j, D), \tag{6}$$

where $P(R_i)$ is the probability of the i-th cause of intentional arson, $P(C_j)$ is the probability of the j-th characteristic of the forested area, $P(D)$ is the probability that the forest fuel is sufficiently dry, $P(I_{ij}/R_i,C_j,D)$ is the conditional probability of forest fire for the i-th reason in the presence of the j-th characteristic of the forested area, provided that the forest fuel is sufficiently dry. The conditional probability $P(I_{ij}/R_i,C_j,D)$ can be estimated through the frequency of events according to the statistics of the forest fire retrospective:

$$P(I_{ij} / R_i, C_j, D) = \frac{N_{ij}^{FARC}}{N^{FT}}, \tag{7}$$

where N_{ij}^{FARC} is the number of fires from intentional arson for the i-th reason in the presence of the j-th characteristic of the forested area, N^{FT} is the total number of forest fires. Since the FF_{ij} events are joint, the maximum value (unit) of the probability $P(FF_i)$ takes when all the fires in the statistics occurred in territories for which all characteristics of the forest territory are present. Similarly, when the probabilities of forest fires for each reason are equal to one, the total probability of a forest fire resulting from deliberate arson will take the value one.

Model calculations should be carried out to determine the probability of forest fires resulting from deliberate arson to demonstrate the functionality of the model. A typical forested area on which a pine-type forest grows is considered. Ground cover consists mainly of litter of needles with a small inclusion of thin pine branches. The climatic conditions typical of summertime in a typical boreal territory of Western Siberia (Scientific-applied reference…, 1993) are considered. Since the statistics are scarce, and the investigation into the causes of many fires is not yet completed (California Department of Forestry and Fire Protection, 2020), scenario-based approach was used to assessing the forest fire danger as a result of deliberate arson. Consider the following scenarios: a) the causes of arson and the characteristics of the territory are fixed, meteorological parameters (precipitation) vary; b) there are all reasons for deliberate arson, the number of characteristics of a forested area varies; c) there are all the characteristics of the forested area, the number of reasons varies.

Scenario A is characterized by the following parameters of arson:

$$\begin{aligned} &P(R_1) = 1, \\ &P(R_2) = 1, \\ &P(R_3) = 1. \end{aligned} \tag{8}$$

This scenario is characterized by the presence of all the characteristics of the forested area:

$$\begin{aligned} &P(C_1) = 1, \\ &P(C_2) = 1, \\ &P(C_3) = 1, \\ &P(C_4) = 1, \\ &P(C_5) = 1. \end{aligned} \tag{9}$$

The influence of meteorological conditions was studied. For demonstration calculations, various values of such a parameter as the amount of precipitation were used. Calculations for five days, in which 0 mm, 3 mm, 12 mm, 18 mm and 21 mm fell, respectively, were carried out. The remaining meteorological parameters are identical.

The following model statistics on forest fires as a result of deliberate arson were used:

$$N_{ij}^{FARC} = \begin{bmatrix} 18 & 10 & 8 & 10 & 2 \\ 0 & 0 & 5 & 0 & 0 \\ 2 & 10 & 7 & 5 & 5 \end{bmatrix}. \tag{10}$$

Total number of forest fires N^{FT}=20. Table 5 presents the results of a numerical calculation of the probability of a forest fire resulting from a deliberate arson at 1 p.m. for scenario A.

Table 5. Probability of forest fires resulting from deliberate arson at 1 p.m. (Scenario A)

Day/precipitation	$P(D)$	$P(FFII_1)$	$P(FF_2)$	$P(FF_3)$	$P(FF)$
I, 0 mm	1.0	0.986	0.25	0.835	0.998
II, 3 mm	0.4	0.669	0.1	0.465	0.841
III, 12 mm	0.2	0.401	0.05	0.259	0.578
IV, 18 mm	0.1	0.219	0.025	0.137	0.343
V, 21 mm	0	0	0	0	0

Scenario B is characterized by the same meteorological conditions. For definiteness, catastrophic weather conditions are considered for which $P(D)$=1. There are all reasons for intentional arson:

$$\begin{aligned} &P(R_1) = 1, \\ &P(R_2) = 1, \\ &P(R_3) = 1. \end{aligned} \tag{11}$$

The number of characteristics of the forested area favoring arson varies.

Option 1. Forest territory is characterized only by the presence of a stock of valuable commercial timber:

$$\begin{aligned} &P(C_1) = 1, \\ &P(C_2) = 0, \\ &P(C_3) = 0, \\ &P(C_4) = 0, \\ &P(C_5) = 0. \end{aligned} \tag{12}$$

Model statistics on forest fires:

$$N_{ij}^{FARC} = \begin{bmatrix} 18 & 0 & 0 & 0 & 0 \\ 0 & 0 & 0 & 0 & 0 \\ 2 & 0 & 0 & 0 & 0 \end{bmatrix}. \tag{13}$$

Option 2. The forest area is characterized by the presence of a stock of valuable commercial timber and the availability of construction after a fire:

$$
\begin{aligned}
&P(C_1) = 1,\\
&P(C_2) = 1,\\
&P(C_3) = 0,\\
&P(C_4) = 0,\\
&P(C_5) = 0.
\end{aligned}
\tag{14}
$$

Model statistics on forest fires:

$$
N_{ij}^{FARC} = \begin{bmatrix} 18 & 10 & 0 & 0 & 0 \\ 0 & 0 & 0 & 0 & 0 \\ 2 & 10 & 0 & 0 & 0 \end{bmatrix}. \tag{15}
$$

Option 3. Forest territory is characterized by the presence of the first three factors:

$$
\begin{aligned}
&P(C_1) = 1,\\
&P(C_2) = 1,\\
&P(C_3) = 1,\\
&P(C_4) = 0,\\
&P(C_5) = 0.
\end{aligned}
\tag{16}
$$

Model statistics on forest fires:

$$
N_{ij}^{FARC} = \begin{bmatrix} 18 & 10 & 8 & 0 & 0 \\ 0 & 0 & 5 & 0 & 0 \\ 2 & 10 & 7 & 0 & 0 \end{bmatrix}. \tag{17}
$$

Option 4. Forest territory is characterized by the presence of all characteristics, except for favorable climatic conditions:

$$
\begin{aligned}
&P(C_1) = 1,\\
&P(C_2) = 1,\\
&P(C_3) = 1,\\
&P(C_4) = 1,\\
&P(C_5) = 0.
\end{aligned}
\tag{18}
$$

Model statistics on forest fires:

$$N_{ij}^{FARC} = \begin{bmatrix} 18 & 10 & 8 & 10 & 0 \\ 0 & 0 & 5 & 0 & 0 \\ 2 & 10 & 7 & 5 & 0 \end{bmatrix}. \tag{19}$$

Option 5. Forest territory is characterized by the presence of all factors:

$$\begin{aligned} &P(C_1) = 1, \\ &P(C_2) = 1, \\ &P(C_3) = 1, \\ &P(C_4) = 1, \\ &P(C_5) = 1. \end{aligned} \tag{20}$$

Model statistics on forest fires:

$$N_{ij}^{FARC} = \begin{bmatrix} 18 & 10 & 8 & 10 & 2 \\ 0 & 0 & 5 & 0 & 0 \\ 2 & 10 & 7 & 5 & 5 \end{bmatrix}. \tag{21}$$

Table 6 presents the results of calculating the probability of forest fires as a result of deliberate arson at 1 p.m. for various options according to scenario B.

Table 6. Probability of occurrence of forest fires as a result of deliberate arson at 1 p.m. (Scenario B)

Variant	$P(D)$	$P(FF_1)$	$P(FF_2)$	$P(FF_3)$	$P(FF)$
I	1.0	0.9	0	0.1	0.91
II	1.0	0.95	0	0.55	0.977
III	1.0	0.97	0.25	0.7	0.993
IV	1.0	0.985	0.25	0.78	0.997
V	1.0	0.986	0.25	0.835	0.998

For scenario C, all characteristics of the forested area are available. The number of reasons varies. For definiteness, catastrophic weather conditions are also considered for which $P(D)=1$.

$$
\begin{aligned}
&P(C_1) = 1,\\
&P(C_2) = 1,\\
&P(C_3) = 1,\\
&P(C_4) = 1,\\
&P(C_5) = 1.
\end{aligned}
\tag{22}
$$

Option 1. The main reason for arson is the thirst for profit:

$$
\begin{aligned}
&P(R_1) = 1,\\
&P(R_2) = 0,\\
&P(R_3) = 0.
\end{aligned}
\tag{23}
$$

Model statistics on forest fires:

$$
N_{ij}^{FARC} = \begin{bmatrix} 18 & 10 & 8 & 10 & 2 \\ 0 & 0 & 0 & 0 & 0 \\ 0 & 0 & 0 & 0 & 0 \end{bmatrix}. \tag{24}
$$

Option 2. The motivation for arson is the thirst for profit and social inequality:

$$
\begin{aligned}
&P(R_1) = 1,\\
&P(R_2) = 1,\\
&P(R_3) = 0.
\end{aligned}
\tag{25}
$$

Model statistics on forest fires:

$$
N_{ij}^{FARC} = \begin{bmatrix} 18 & 10 & 8 & 10 & 2 \\ 0 & 0 & 5 & 0 & 0 \\ 0 & 0 & 0 & 0 & 0 \end{bmatrix}. \tag{26}
$$

Option 3. All causes of arson are characteristic. The options for this option are the same as scenario 5 option B. Table 7 presents the results of calculating the likelihood of forest fires resulting from deliberate arson at 13 p.m. for various scenarios C.

Table 7. Probability of occurrence of forest fires as a result of deliberate arson at 13 pm (Scenario C)

Variant	$P(D)$	$P(FF_1)$	$P(FF_2)$	$P(FF_3)$	$P(FF)$
I	1.0	0.986	0	0	0.986
II	1.0	0.986	0.25	0	0.989
III	1.0	0.986	0.25	0.835	0.998

An analysis of the results of Scenario A (table 5) shows that the probability of forest fires due to deliberate arson is significantly affected by the probability of forest fires under weather conditions (the probability that forest fuel is sufficiently dry). The influence of meteorological factors determines the conditions under which the ignition of the forest fuel layer is possible when exposed to a fire source. Table 5 shows the effect of rainfall. A natural fact is a decrease in the probability of forest fires resulting from deliberate arson with an increase in the amount of precipitation. It was previously shown that an increase in ambient temperature and the flow of solar radiation increases the probability of forest fires under weather conditions (Baranovskiy, 2007) and, as a result, the probability of forest fires resulting from arson will increase. These findings are confirmed by the latest tragic events in Australia (Country Fire Authority, 2020). During this period, in places of large fires, dry and hot weather (over 40 °C) was observed.

Scenario B showed that under catastrophic meteorological conditions and the presence of all the causes of arson, an increase in the characteristics of the forest area favoring arson also increases the probability of forest fires resulting from deliberate arson (table 6). Under the extreme conditions of scenario B, the difference in the probability of forest fires resulting from deliberate arson only slightly differs from variant to variant. It can be assumed that the fire conditions in Spain and Greece are consistent with the extreme conditions of scenario B. In other real situations, of course, the differences can be more significant. It is advisable to develop a range of scenarios for an individual state or various regions of the country for a subtle consideration of the specifics of the territory.

Scenario C (table 7) shows another extreme case and shows that for an area that is attractive for arson in all respects, the probability of forest fires resulting from deliberate arson will increase with an increase in the number of causes of arson. It should be noted that in the conditions of insufficient statistical data to assess the probability of a specific reason, one can turn to the technology of socio-monitoring of the population to obtain a socio-psychological portrait of typical representatives of residents of specific territories. Then it will become possible to obtain a priori estimates of the probability of the causes of arson.

CONCLUSION

Arson has become an important and significant factor. The consequences of forest fires became especially tragic and significant in the 21st century. The social stratification of society, the imperfection of legislation, and the thirst for "easy" profit are the main reasons for deliberate arson. In this work, an important scientific and practical problem was solved by developing the prediction model for forest fire danger that was developed taking into account the factor of intentional arson. Model calculations were performed that demonstrate the functionality of the model proposed in this article. It should be noted that a generalized version of the model for differentiated prediction of forest fire danger has been developed and it is possible to implement this model in the methodology for predicting the forest fire danger of a particular state, it will require fine-tuning and tuning to take into account the specifics of the socio-economic development of this country.

Implementation in the form of program code in each case can be performed in various high-level languages, for example, in development systems such as MS Visual Studio, RAD Studio, Matlab (MSVS, 2020; RAD Studio, 2020; Matlab, 2020). In addition, comprehensive multidisciplinary studies can be carried out using the results presented in works (Deeming et al., 1977; Lee et al., 2002; EFFIS, 2020; Noble et al., 1980; Podolskaya et al., 2011) and articles (Apte et al., 2007; Bartalev et al., 2007; Bui et al., 2017; Brun et al., 2017; Carrio et al., 2019; Jacobson et al., 2006; Flemming et al., 2001; Martell, 2000; You et al., 2017).

ACKNOWLEDGMENT

This work was implemented with the financial support of the Russian Foundation for Basic Research. [Project N 17-29-05093].

REFERENCES

Apte, A., Hairer, M., Stuart, A. M., & Voss, J. (2007). Sampling the posterior: An approach to non-Gaussian data assimilation. *Physica D. Nonlinear Phenomena*, *230*(1-2), 50–64. doi:10.1016/j.physd.2006.06.009

Baranovskiy, N. V. (2007). *Mathematical modeling of the most probable scenarios and conditions of forest fires* (PhD Thesis). Tomsk: Tomsk State University. (in Russian)

Baranovskiy, N. V. (2018). Mathematical simulation of anthropogenic load from a linear source using partial differential equations in the context of the forest fires occurrence. *Advances in Differential Equations and Control Processes.*, *19*(3), 237–262. doi:10.17654/DE019030237

Baranovskiy, N. V. (2020). Mathematical Simulation of Anthropogenic Load on Forested Territories for Point Source. In N. Baranovskiy (Ed.), *Predicting, Monitoring, and Assessing Forest Fire Dangers and Risks* (pp. 64–88). IGI Global. doi:10.4018/978-1-7998-1867-0.ch003

Baranovskiy, N. V., & Kuznetsov, G. V. (2017). *Forest fire occurrences and ecological impact prediction: monograph*. Publishing House of the Siberian Branch of the Russian Academy of Science. doi:10.15372/FOREST2017BNV

Bartalev, S. A., Egorov, V. A., Loupian, E. A., & Uvarov, I. A. (2007). Multi-Year Circumpolar Assessment the Area Burnt in Boreal Ecosystems Using SPOT-Vegetation. *International Journal of Remote Sensing*, *28*(6), 1397–1404. doi:10.1080/01431160600840978

Brun, C., Artes, T., Cencerrado, A., Margalef, T., & Cortés, A. (2017). A High Performance Computing Framework for Continental-Scale Forest Fire Spread Prediction. *Procedia Computer Science*, *108C*, 1712–1721. doi:10.1016/j.procs.2017.05.258

Bui, D. T., Bui, Q.-T., Nguyen, Q.-P., Pradhan, B., Nampak, H., & Trinh, P. T. (2017). A hybrid artificial intelligence using GIS-based neural-fuzzy inference system and particle swarm optimization for forest fire susceptibility modeling at a tropical area. *Agricultural and Forest Meteorology*, *233*, 32–44. doi:10.1016/j.agrformet.2016.11.002

California Department of Forestry and Fire Protection Home Page. (2020). Available: http://www.fire.ca.gov

Carrió, D. S., Homar, V., & Wheatley, D. M. (2019). Potential of an EnKF Storm-Scale Data Assimilation System Over Sparse Observation Regions with Complex Orography. *Atmospheric Research*, *216*, 186–206. doi:10.1016/j.atmosres.2018.10.004

Country Fire Authority (Victoria state of Australia). (2020). Available: http://www.cfa.vic.gov.au (Accessed 07 April 2020)

Deeming, J. E., Burgan, R. E., & Cohen, J. D. (1977). *The National Fire Danger Rating System - 1978*. United States Department of Agriculture, Forest Service, General Technical Report INT-39, Intermountain Forest and Range Experiment Station, Odgen, UT.

European Forest Fire Information System. (2020). Available: http://effis.jrc.ec.europa.eu

Flemming, J., Reimer, E., & Stern, R. (2001). Data assimilation for CT-Modelling based on optimum interpolation. In *Preprints of 25th NATO/CCMS International Technical Meeting on Air Pollution Modelling and its Application*. Catholic University of Louvain.

Forest fires in Greece. (2020). *WWF News Russia*. Available: http://www.wwf.ru/resources/news/article/3200 (in Russian)

Gorev, G. V. (2004). *Evaluation of the climate predisposition of the territory to the occurrence of forest fires (on the example of the Tomsk region)* (PhD Thesis). Tomsk: Tomsk State University. (in Russian)

Grishin, A. M. (2002). *Modeling and forecasting of disasters*. Publishing House of Tomsk University. (in Russian)

Jacobson, A. R., Holzworth, R., Harlin, J., Dowden, R., & Lay, E. (2006). Performance Assessment of the World Wide Lightning Location Network (WWLLN) using the Los Alamos Sferic Array (LASA) as Ground Truth. *Journal of Atmospheric and Oceanic Technology*, *23*(8), 1082–1092. doi:10.1175/JTECH1902.1

Janssens, M. (1991). *Fundamental thermophysical characteristics of wood and their role in enclosure fire growth* (PhD Thesis). University of Gent, Belgium.

Kuznetsov, G. V., & Baranovskiy, N. V. (2006). Mathematical modeling of ignition of a layer of forest combustible materials by a particle heated to high temperatures [in Russian]. *Fire and explosion safety.*, *15*(4), 42–46.

Kuznetsov, G. V., & Baranovskiy, N. V. (2009). *Forest fire occurrences and ecological impact prediction*. Publishing House of the Siberian Branch of the Russian Academy of Science. (in Russian)

Lee, B. S., Alexander, M. E., Hawkes, B. C., Lynham, T. J., Stocks, B. J., & Englefield, P. (2002). Information systems in support of wildland fire management decision making in Canada. *Computers and Electronics in Agriculture*, *37*(1-3), 185–198. doi:10.1016/S0168-1699(02)00120-5

Martell, D. L. (2000). A Markov Chain Model of Day to Day Changes in the Canadian Forest Fire Weather Index. *International Journal of Wildland Fire*, *9*(4), 265–273. doi:10.1071/WF00020

Matlab. (2020). Available https://www.mathworks.com/products/matlab.html

Matsenko, V. V., Sokolov, A. Ya., Kalinin, S. I., Andriyanova, F. I., Andreeva, T. A., Ananyin, S. V., Krylov, M. N., & Kazantseva, L. V. (1999). Master Plan for the fire prevention of forests. Volume 1. Explanatory note. 5-99.14-17-PM. Barnaul: State Design and Research Institute "Rosgiproles", Altai Branch. (in Russian)

Microsoft Visual Studio. (2020). Available https://visualstudio.microsoft.com

Noble, I. R., Bary, G. A. A., & Gill, A. M. (1980). McArthur's fire danger meters expressed as equations. *Australian Journal of Ecology*, *5*(2), 201–203. doi:10.1111/j.1442-9993.1980.tb01243.x

Pakshirajan, R. P. (2013). Probability Theory. Springer.

Pew, K. L., & Larsen, C. P. S. (2001). GIS analysis of spatial and temporal patterns of human-caused wildfires in the temperate rain forest of Vancouver Island, Canada. *Forest Ecology and Management*, *140*(1), 1–18. doi:10.1016/S0378-1127(00)00271-1

Podolskaya, A.S., Ershov, D.V., & Shulyak, P.P. (2011). Application of the method for assessing the probability of occurrence of forest fires in ISDM-Rosleskhoz. *Modern Problems of Remote Sensing of the Earth From Space, 8*, 118 - 126. (in Russian)

Schetinskiy, E. A. (2002). *Extinguishing Forest Fire: A Handbook for Forest Firefighters*. VNIILM. (in Russian)

StudioR. A. D. (2020). Available https://www.embarcadero.com/ru/products/rad-studio

Scientific-applied reference on the climate of the USSR. Series 3. Perennial data. Parts 1-6.: Vol. 20. *Tomsk, Novosibirsk, Kemerovo regions and Altai Territory*. (1993). Gidrometeoizdat. (in Russian)

You, W., Lin, L., Wu, L., Ji, Z., Yu, J., Zhu, J., Fan, Y., & He, D. (2017). Geographical information system-based forest fire risk assessment its spatiotemporal variability. *Ecological Indicators*, *77*, 176–184. doi:10.1016/j.ecolind.2017.01.042

Yurieva, D. (2020). *Million for the pyromaniac. Greece is looking for those responsible for unprecedented forest fires.* Available: http://www.rg.ru/2007/08/28/nagrada.html (in Russian)

ADDITIONAL READING

Adab, H., Kanniah, K. D., & Solaimani, K. (2013). Modeling forest fire in the northeast of Iran using remote sensing and GIS techniques. *Natural Hazards*, *65*(3), 1723–1743. doi:10.100711069-012-0450-8

Al Janabi S., Al Shourbaji I., Salman M.A. (2017) Assessing the suitability of soft computing approaches for forest fires prediction // Applied Computing and Informatics, doi:10.1016/j.aci.2017.09.006

Amatulli, G., Perez-Cabello, F., & Du la Riva, J. (2007). Mapping lightning / human-caused wildfires occurrence under ignition point location uncertainty. *Ecological Modelling*, *200*(3-4), 321–333. doi:10.1016/j.ecolmodel.2006.08.001

Andreev, Yu. A. (1986). *Laws of distribution of forest fires. Methods and means of fighting forest fires.* VNIILM. (In Russian)

Baranovskiy, N. V. (2017). Forest fire danger assessment using SPMD-model of computation for massive parallel system. *International Review on Modelling and Simulations.*, *10*(3), 193–201. doi:10.15866/iremos.v10i3.10570

Benavent-Corai, J., Rojo, C., Suarez-Torres, J., & Velasco-Garcia, L. (2007). Scaling properties in forest fire sequences: The human role in the order of nature. *Ecological Modelling*, *205*(3-4), 336–342. doi:10.1016/j.ecolmodel.2007.02.028

Catry, F. X., Rego, F. C., Bacao, F. L., & Moreira, F. (2009). Modeling and mapping wildfire ignition risk in Portugal. *International Journal of Wildland Fire*, *18*(8), 921–931. doi:10.1071/WF07123

Diez-Delgado, R., Lloret, F., & Pons, X. (2004). Statistical analysis of fire-frequency models for Catalonia (NE Spain, 1975-1998) based on fire scar maps from Landsat MSS data. *International Journal of Wildland Fire*, *13*(1), 89–99. doi:10.1071/WF02051

Ganteaume, A., Camia, A., Jappiot, M., San-Miguel-Ayanz, J., Long-Fournel, M., & Lampin, C. (2013). A review of the main driving factors of forest fire ignition over Europe. *Environmental Management*, *9*(3), 651–662. doi:10.100700267-012-9961-z PMID:23086400

Jiménez-Ruano, A., Rodrigues Mimbrero, M., & de la Riva Fernández, J. (2017). Understanding wildfires in mainland Spain. A comprehensive analysis of the fire conditions in a climate-human context. *Applied Geography (Sevenoaks, England)*, *89*, 100–111. doi:10.1016/j.apgeog.2017.10.007

Poulos, H. M., Villanueva Díaz, J., Cerano Paredes, J., Camp, A. E., & Gatewood, R. G. (2013). Human Influences on fire regimes and forest structure in the Chihuahuan Desert Borderlands. *Forest Ecology and Management*, *298*, 1–11. doi:10.1016/j.foreco.2013.02.014

Ruffault, J., & Mouillot, F. (2017). Contribution of human and biophysical factors to the spatial distribution of forest fire ignitions and large wildfires in a French Mediterranean region. *International Journal of Wildland Fire*, *26*(6), 498–508. doi:10.1071/WF16181

Ye, J., Wu, M., Deng, Z., Xu, S., Zhou, R., & Clarke, K. C. (2017). Modeling the spatial patterns of human wildfire ignition in Yunnan province, China. *Applied Geography (Sevenoaks, England)*, *89*, 150–162. doi:10.1016/j.apgeog.2017.09.012

KEY TERMS AND DEFINITIONS

Anthropogenic Load: Different human activities on forested territories lead to forest fire occurrence and characterized by presence of fire sources.

Forest Fire: Uncontrolled aerothermochemical phenomenon characterized by step-by-step mechanism which includes following stages: inert heating, moisture evaporation, high temperature terpens evaporation, dry organic matter pyrolysis, flammable combustion and smoldering.

Mathematical Simulation: The production of a computer model of forest fire conditions and prerequisites, especially for the purpose of study.

Monitoring: Monitoring refers to the periodic calculation of the parameters of forest fire danger with a portion of information available in real time.

Prediction: Under the prediction of forest fires is the calculation of the parameters of forest fire danger with a certain projection in advance in order to have enough time to anticipate an emergency. The calculation in this case is carried out in a mode ahead of the real time of the development of the catastrophe - the occurrence of a forest fire.

Chapter 16
Mathematical Simulation of Anthropogenic Load on Forested Territories

ABSTRACT

Human activity causes forest fires near the municipalities and different transportation structures. It is suggested to define linear objects that caused human activity and the similarity of the human activity spread and heat conduction applicable for numerical simulation. The numerical parameter of the human activity is the virtual (possible) number of forest fires (VNF) near the linear object. One-dimensional and two-dimensional numerical models to evaluate the distribution of human activity from linear objects are presented. It is possible to mark out following linear objects: country roads, motorways, and railways. Dependence of the VNF for typical data from linear objects of human activity changed through the time and space are obtained. Conclusions are drawn about the patterns of human activity from linear objects. The prospects of further development are described for this direction to design software for forest fire danger prediction.

INTRODUCTION

Lightning activity and anthropogenic load are the primary reasons of forest fire initiation (Baranovskiy and Kuznetsov, 2017). The forest fires initiation caused by the lightning activity have been thoroughly examined (Baranovskiy et al., 2016; Baranovskiy and Kuznetsov, 2017; Preisler et al., 2004) by now compared to fires caused by anthropogenic load (Poulos et al., 2013; Amatulli et al., 2007). It should be noted that there are different approaches to assessing forest fire danger caused by anthropogenic load. As a rule, these are approaches that take into account statistics and the subsequent estimation of anthropogenic load as a stochastic process (Kurbatskiy, 1964; Bui et al., 2017). Moreover, the process of distribution (spread) of anthropogenic load is not modeled. An assessments of the expected number of fire incidents is made in a definite area (Kurbatskiy, 1964). It should be noted that sources of anthropogenic load of various types can be distinguished (Baranovskiy, 2006): point, linear, area. A linear source of

DOI: 10.4018/978-1-7998-7250-4.ch016

anthropogenic load is considered in this paper. It should be noted that statistics on forest fires provide data on the number of registered forest fires only for a small number of forest areas of definite forestry. However, this does not mean that the level of anthropogenic load is zero on other sites. It should be noted the importance of works on the evaluation of the inter-fire intervals (Diez-Delgaro et al., 2004). As a result, it is necessary to develop mathematical models that would describe the processes of distribution (spread) of anthropogenic load on a controlled forest-covered territory taking into account it's space-time dynamics.

Background

Forest fires cause ecological, economic and social damages (Grishin, 1997). An important link to prevent forest fire consequences is the prediction of their initiation (Grishin and Filkov, 2011; Al Janabi et al., 2017;Read et al., 2018). In Canada, the assessment of forest fire danger is carried out within the framework of the multi-component system of the Canadian Wildland Fire Information System (CWFIS) (CWFIS, 2020; Wotton, 2009; Gould et al., 2013; Lee et al., 2002; Martell, 2000; Pakshirajan, 2013). To predict forest fires probability, the Canadian Forest Fire Occurrence Prediction System is used for some Canadian areas.

The American Wildland Fire Assessment System (WFAS) has several modules for assessing potential fires (WFAS, 2020; Deeming et al., 1974). American system can predict lightning- and human-caused forest fires and final fire danger risk (Deeming et al., 1977). The application of the system is possible outside the United States in similar climatic and vegetative conditions. The fire danger indicators are expressed in dimensionless quantities. Dimensionless indices are constructed on the scales of various fractionality (hundred-point, twelve-point, five-point). For some models the scale is only partially used, less than 80%.These systems characterize the fire danger in general on the considered area, i.e. on tens and hundreds of thousands of hectares. This technique accurately predicts a fire danger, as it is built on the basis of statistics on forest fires over a period of more than 20 years.

The European Forest Fire Information System (EFFIS) originally was based on different South European indexes (EFFIS, 2020; Viegas et al., 2000). Recently, the Canadian Forest Fire Weather Index System was adopted to predict forest fire danger over the Europe. The information system for remote monitoring of forest fires (ISDM-Rosleskhoz) (ISDM-Rosleskhoz, 2020) was developed in 2003 and was executed in 2005 (Podolskaya et al., 2011). ISDM-Rosleskhoz is developed solve the following tasks (Podolskaya et al., 2011): 1) hot spot detection, registration and monitoring; 2) current evaluation of the characteristics of active forest fires; 3) assessment of forest fire risk. Recently, a probabilistic formulae to predict forest fires occurrence was designed taking into account lightning activity and anthropogenic load (Baranovskiy, 2015; Baranovskiy, 2017; Baranovskiy and Zharikova, 2014). Till now, only statistics were used in this criterion.

Urbanization and recreational load lead to changes in forests caused by human-induced fires (Melekhov, 1947). The processing of social and psychological reasons of recreational forest accretions and the occurrence of forest fires is conducted (Andreev and Larchenko, 1987). It has been established that the distance from the settlement increases, the number of visits to forests and forest fires decreases (Kurbatskiy, 1964; Melluma et al, 1982) according to the distributions of Rayleigh and Poisson (Telitsin, 1984). The distribution of ignitions in time and space is similar to the spatial and temporal distribution of forest fires (Telitsin, 1987; Matsenko et al., 1999).

The emergence of forest fires is influenced by social factors. For example, 93% of all fires in the national forests of Minnesota, Wisconsin and Michigan in 1986 occurred through human activity (Cardille et al., 2001). The number of factors under consideration included such characteristics as staff, density of roads and railways, distance to the territory not covered by forest, distance to the city, population density, etc. (Cardille et al., 2001). In forest areas with a higher population density, roads and railways, the level of fire danger is higher. Increasing the access of people to forest cover increases the forest fires occurrence probability.

The frequency and causes of fires are mainly due to the level of urbanization of the territory and, as a consequence, the level of anthropogenic load. The anthropogenic load on forest-covered territory is determined by the number of settlements and their population (Bui et al., 2017; Andreev, 2003). With an increase in population density, the severity of forests in relative values increases (Kurbatskiy, 1964). The influx of vacationers and hunters increases the summer maximum of fires, related to the distance from settlements and transport routes (Kurbatskiy, 1964; Ganteaumme and Guerra, 2018; Ye et al., 2017). The distribution of forest fires in connection with distance from transport routes is described by an exponential function (Telitsin, 1984).

To identify the anthropogenic component of the causes of the emergence of forest fires, data on the population of Tomsk region and the regression analysis method were used (Rawlings et al., 2015). The equation of the relationship between the number of forest fires and the population density has been obtained. The equation can be used for a preliminary assessment of the anthropogenic influence to the forested territory.

Estimates of the forest fires frequency are necessary for the work of forest fire services. The frequency of forest fires in Catalonia (Spain) was estimated from 1975 to 1998 according to the archives of maps of the burnt area (Diez-Delgado et al., 2004). Images from satellites made it possible to identify a fire with an area of more than 30 hectares and were used to characterize the spatial distribution of forest fires in Catalonia. Various models of the frequency of forest fires: Natural Fire Rotation Period (NFR), Poisson Process Model (PPM), Mean Fire Interval (MFI) were used to map the data for the specified period. The simplest of them NFR allows you to determine how many years it will take to burn the area similar to the area studied. Some forest areas were re-burned. A significant and positive correlation was found between the number of fires and the total burned area. A large number of forest fires concentrate simultaneously in one place during dry periods.

In work (Hong et al., 2018), a model of forest fire danger assessment is considered on the example of Dayu Country, China. The main goal of the work is to use the genetic algorithm (Haupt and Haupt, 2004) to obtain the optimal set of variables characterizing forest fire danger. The developed methodology was divided into the following stages: a) identification of the location of forest fires and the development of an inventory map; b) selection and classification of variables based on the natural break method (Chen et al., 2017), performing multicollinearity analysis (Dormann et al., 2013), estimating the weights of individual classes of variables using the Certainty Factor method, creating data for training and checking, normalizing variables; c) the use of a genetic algorithm; d) customization and execution of Random Forests (Ghimire et al., 2012) and Support Vector Machine (Duan et al., 2001) methods, the construction of a surface that demonstrates the danger of forest fires and the validation of the developed methodology. The variables take into account such characteristics as use of the territory, distance to the road network, distance to the river network. Satellite technologies (Aster, Landsat) were used as one of the data sources. The study found that the topography and type of vegetation play an important role in assessing forest fire danger.

In (Grala et al., 2017), an assessment of the influence of the human factor on the occurrence of forest fires in the Mississippi region (USA) is considered. This paper discusses the relative importance of temporal, spatial, and socioeconomic factors in the initiation of 52,532 forest fires as a result of human factors. The period of forest fires during 1991–2005 is considered. Forest fires caused by the use of equipment, children and burning debris were more frequent than fires near roads and railways. The model used 11 variables to determine the likelihood of seven types of forest fires due to human factors. The time variables used are Season and Weekend. Spatial variables were selected as a result of analysis using a geographic information system to estimate distances to highways and railways and settlements. Also considered a variable population. Socioeconomic factors such as income, poverty and unemployment were also taken into account. The transport network was correlated with forest fires due to human factors. In particular, it was found that increasing the distance from the road network by 1 km reduces the likelihood of a forest fire.

The work (Jimenez-Ruano et al., 2017) considers the problem of forest fires in the territory of Spain in the context of the interaction of climatic and anthropogenic factors. The following climatic data were used: monthly temperature and precipitation (Monthly Temperature Dataset of Spain and Monthly Precipitation Dataset of Spain) (Gonzalez-Hidalgo et al., 2011; Gonzalez-Hidalgo et al., 2015). These data are presented with a spatial resolution of 10 x 10 km. They are derived using the Spanish Meteorological Network for the period 1951-2010. The data was pre-processed using the nearest neighbor procedure. The following data was used as a human factor: territory use, population, and the Human Pressure Index. For data analysis, the principal component method (Flury, 1984) and the Varimax Rotation procedure (Kaiser, 1958) were used. Regression analysis methods were also used. The analysis showed that most fires occur in the summer, regardless of the region of study. It was found that the Human Pressure Index is a significant predictor of forest fires.

In (Hu et al., 2014), the main drivers of forest fires arising from thunderstorm activity and the human factor in the Great Xing'an Mountains region (China) were considered. The 40 year data for the period from 1967 to 2006, which were provided by the Heilongjiang Headquarters of Forest Fire Prevention, were used. This data includes information on the cause of the forest fire, the location and date of the forest fire, the size and date of extinguishing, the type of vegetation. All fires were divided into fires that occurred as a result of thunderstorm activity and the human factor. Meteorological data was also used. Precipitation, temperature, and the Drought Index have been selected as human factors. In accordance with (Liu et al., 2012a; Liu et al., 2012b), the population of this region increased from 30,000 to 116,000 over the period from 1964 to 1968. Between 1969 and 1986, 787,000 arrived, and 599,000 fell in this region. Between 1987 and 1997, the population remained relatively stable with an annual increase of 2%. During 1998 - 2011, the annual increase was -0.04%, which led to the number of 516,000 inhabitants by the beginning of 2011. Regression analysis was used to analyze the data to determine the trends of natural and anthropogenic factors. The maximum temperature during the fire season has a significant effect. After the Black Dragon Fire incident in 1987, local legislation was changed to reduce losses from forest fires. This led to a significant reduction in the frequency of forest fires due to human factors.

In (Ye et al., 2017), spatial patterns of forest fires resulting from human activity in the territory of Yunnan Province (China) are considered. The following predictors were selected as variables: distance from highways and railways, population density, tourism sites, and farms (Curt et al., 2016; Fusco et al., 2016; Zhang et al, 2016). Data on human settlements, transport infrastructure and other facilities were provided by the Yunnan Forestry Disaster Provincial Science and Technology Innovative Team. The road network was classified into two types: a) main roads; b) secondary roads. The first type was national roads, highways, provincial roads, railways. The second type was township roads, paths. These were linear objects. Also point sources were presented: urban settlements (city, country, township) and villages (village). In addition, water bodies were considered, including hydroelectric power stations and tourism sites. Also, farm areas were included in the analysis, as they were characterized by a high level of human activity and, as a result, the number of forest fires. To analyze the data, the main principles of the theory of probability were used, which were initially applied to the assessment of mineral reserves (Agterberg and Cheng, 2002). Forest fire mapping has been done in ArcGIS software. As a result, the risk was evaluated in 4 categories: Low, Medium, High, Very High predictive risk. It is established that the maximum human activity prevails within 5 km from roads and settlements. The greatest risk is due to roads, water bodies and farms. A probabilistic model for predicting the occurrence of a forest fire was developed.

The aim of the work is to study the distribution of human activity from linear objects on forest-covered territory for assessing forest fire danger on the basis of a deterministic-probabilistic approach.

MATHEMATICAL MODEL OF ANTHROPOGENIC LOAD

Recently, probabilistic formulae to predict forest fires occurrence was designed taking into account lightning activity and anthropogenic load. Probabilities can be estimated using frequency of events, for example (Baranovskiy, 2017; Baranovskiy and Zharikova, 2014):

$$P(A_j / A) \approx \frac{\mathrm{N_{FD}}}{\mathrm{N_{FW}}}, \qquad (1)$$

Where $P(A_j/A)$ is the probability of a high temperature source presence on the jth day; $\mathrm{N_{FD}}$ is the number of human-caused fires on a specific day of the week; $\mathrm{N_{FW}}$ is the total number of human-caused fires during a week.

The next step is the space-time forecast of the fields of anthropogenic parameters. Of great importance is the probability of fire sources, which is determined by the expression of formula (1). Still, it is impossible to calculate the number of forest fires caused by the human activity throughout the week and on a definite day of the week with high spatial resolution. VNF is a numerical characteristic of human-caused forest fires. It is necessary to introduce a grid that will cover the forest territory adjacent to the transport route passing through the forest. The human activity at the initial time is characterized by a zero level. Human activity is considered as a deterministic process. It is postulated that human activity is similar to heat conduction. This let to formulate the following model of anthropogenic load.

The solution area is shown in Fig. 1.

It is possible to denote N* as one of following parameters ($N_{A,}$ $N_{FA,}$ $N_{FT,}$ $N_{FS,}$ N_{FD}, N_{FW}) that correspond to human activity. Equation of spatial-temporal dynamics of anthropogenic load:

$$\frac{\partial N^*}{\partial t} = \frac{\partial}{\partial x}\left(A\frac{\partial N^*}{\partial x}\right) + \frac{\partial}{\partial y}\left(A\frac{\partial N^*}{\partial y}\right), \qquad (2)$$

Figure 1. Solution area

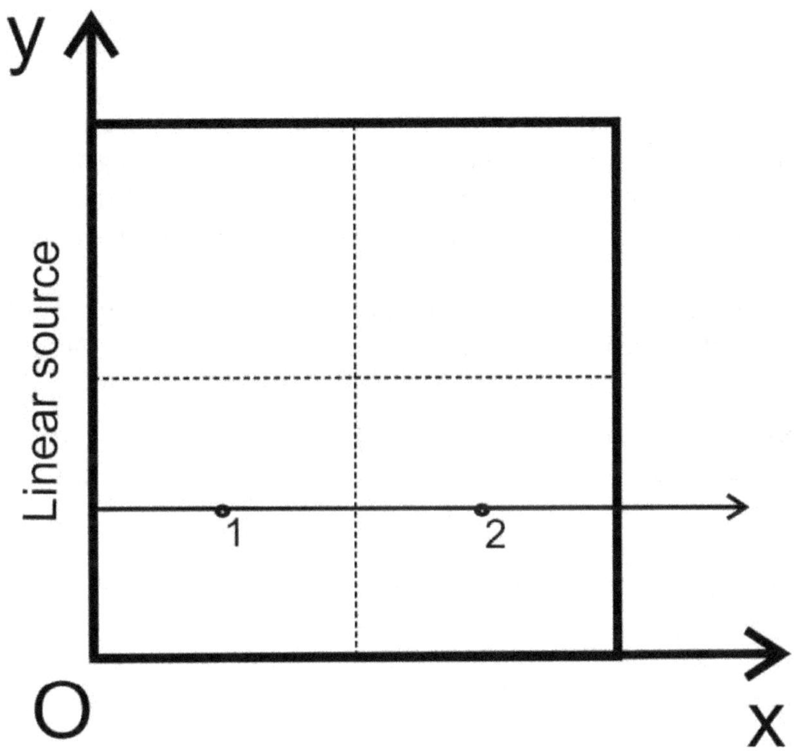

The boundary conditions, for example, of the first type:

$$N^*(0, y, t) = N^*_{x0}, \quad (3)$$

$$N^*(x_{end}, y, t) = 0, \quad (4)$$

$$N^*(x, 0, t) = 0, \quad (5)$$

$$N^*(x, y_{end}, t) = N^*_{y_{end}}, \quad (6)$$

Initial conditions:

$$N^*(x, y)\Big|_{t=0} = N^*_0(x, y), \quad (7)$$

Equation for one-dimensional statement:

$$\frac{\partial N^*}{\partial t} = \frac{\partial}{\partial x}\left(A \frac{\partial N^*}{\partial x} \right), \tag{8}$$

Where N* is the VNF; t is temporal coordinate; x, y are spatial coordinates; A is the coefficient that characterizes the propagation velocity of ignition sources. Method for solving heat conduction problems can be used for numerical solution of this problem in a complete formulation, which is described by non-stationary partial differential equations of the parabolic type (Samarskii and Vabishchevich, 1995a; 1995b).

Processing of the spatial and temporal patterns of forest fires of 1950-1992, which caused by human activity in the territory of Vancouver Island, Canada, is indicative (Fig. 2) (Pew and Larsen, 2001).

Table 1. Forest fires caused by anthropogenic load in Vancouver island (Pew and Larsen, 2001)

Fire Cause	Number of Fires	Burnt	Area, ha
		Total	Average (for fire)
Mixed	2567	17370	6,8
Recreation	1809	4462	2,5
Logging	1021	20927	20,5
Land clearing	512	966	1,9
Railway	189	276	1,5
Industrial sources	120	954	8,0
Routes	104	2650	25,5
Undefined	7	8	1,2
Total	6329	47613	7,5

The analysis was subjected to 6,329 forest fire ignitions (Table 1, Figure 2). Software Arc/Info and ArcView GIS were used. The area of the study was divided into 36050 cells.

The number of burnt cells is closely related to the proximity of country roads, motorways and railways and decreases with increasing distance from them (especially from country roads) (see Fig. 2.a). A large number (the difference with neighboring territories is an order of magnitude or more) of the cells burns at a distance of more than 20 km from motorways and railways (see Fig. 2.b,c).

RESULTS AND DISCUSSION

Figure 2 shows that the maximum number of forest fires is typical for country roads. The number of fires is less for motorways and railways. This is due to the fact that human behavior on such transport routes is different. It is suggested the following scenarios. Initially, the scenario of forest fires near the railways should be considered. As a rule, the railway passes through the forest and only occasionally there are large settlements along the route of the train. This is typical for fast trains and long-distance trains. As

Figure 2. The number of burned cells in the forest-covered territory of Vancouver, depending on the distance to the infrastructure (Pew and Larsen, 2001): a) country roads; b) motorways; c) railways

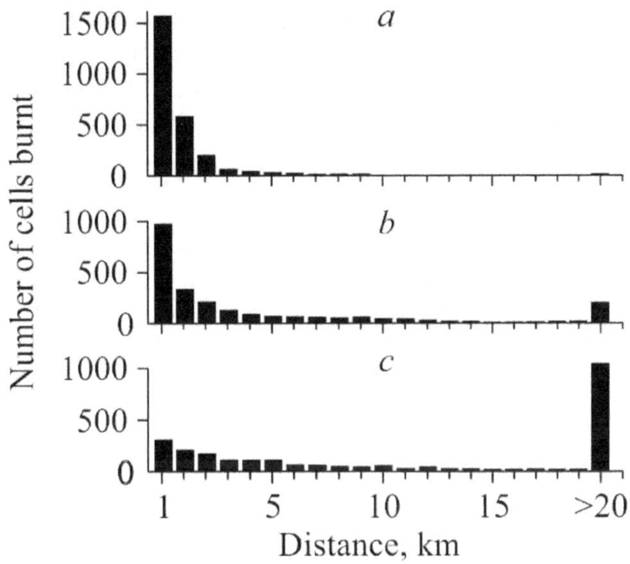

for suburban areas, they are full of stopping areas, on which electric trains stop. Thus, the number of people visiting the forest is small. Either near the railway, forest fires occur as a result of a locomotive's spark, sparks from under the wheels of a train, or particles heated to high temperatures (cigarette butts). Although the last source in connection with the introduction of a ban on smoking in trains is gradually ceasing its influence. Next the behavior of a person on motorways should be described. The situation is somewhat analogous to the railways in the context of the periodic appearance of settlements on the route of the car. However, unlike the train, the car can stop between settlements (be they large or small settlements). This causes a greater number of forest fires. In conclusion, one should describe the scenario of human behavior on country roads. It is these roads that are used by loggers, for agricultural and recreational purposes. As a result, the maximum value of the virtual number of forest fires should be used in the boundary conditions of the first type for country roads. Moderate value should be used for motorways. And the minimum value of the virtual number of forest fires should be set for the railways. The depth of persons penetration into the forest can be taken into account by varying the coefficient of anthropogenic load A. It should be assumed that the calculation of the spread of anthropogenic load should begin from 5-6 am on local time. It should be expected that the maximum anthropogenic load will occur for a period of time from 14 to 16 hours local time. This is indirectly confirmed by statistics on forest fires. Timiryazevskiy forestry (Tomsk region) statistics were used as a reference data, summarized for the period from 2008 to 2010 (Data, 2010).

Thus, the following initial data for numerical investigation are used. The final calculation time is 16 hours starting at 6 am, that is, a calculation within 10 hours. The virtual number of forest fires is 20 for country roads, 15 for motorways and 10 for railways. The coefficient of anthropogenic load was based on the solution of the quasi-inverse problem with the aim of determining it by comparing the virtual number of forest fires at a given distance from the transport route. It should be noted that the question of the initial data requires a separate study on the analysis and compilation of statistics for a particular territory using geospatial analysis methods in GIS systems (Adab et al., 2013; You et al., 2017).

To demonstrate calculation procedure, N_{FD} parameter was chosen as a virtual number of fires N^*. Figure 3 shows the distribution of the virtual number of forest fires in relation to the distance from country roads at different times. Figures 4 and 5 show, respectively, the dependence of the virtual number of forest fires on time at different distances from the country road and the virtual number field of forest fires in the plane near the country road. Last pictures were obtained as a result of mathematical model (2) – (8) implementation.

Figure 3. The distribution of the virtual number of forest fires, depending on the distance from the country road at different times

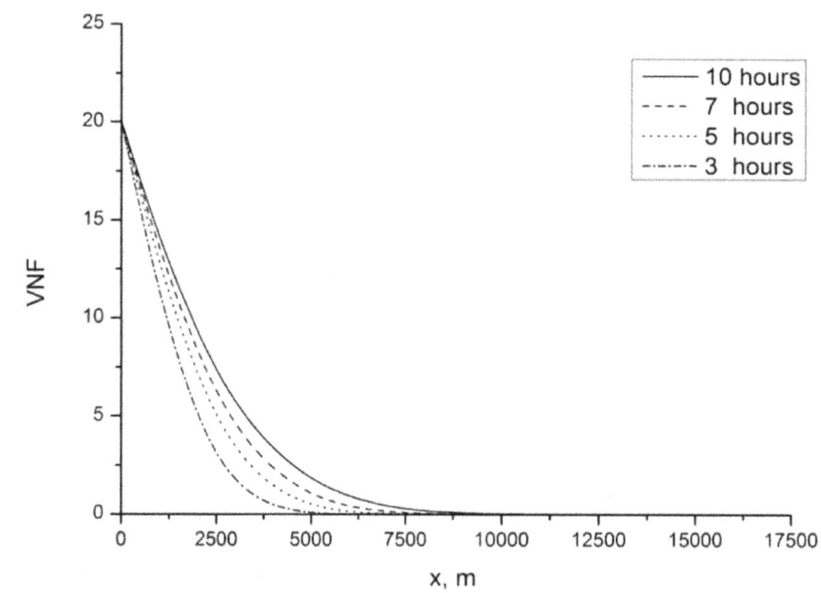

Figure 6 shows the distribution of the virtual number of forest fires as a function of the distance from motorways at different times. Figures 7 and 8 show, respectively, the dependence of the virtual number of forest fires on time at different distances from the motorways and the virtual number of forest fires in the plane near the motorway.

Figure 9 shows the distribution of the virtual number of forest fires as a function of the distance from the railways at different times. Figures 10 and 11 show, respectively, the dependence of the virtual number of forest fires on time at different distances from the railway and the virtual number of forest fires in the plane near the railway.

Analysis of the results shows that for all types of transport routes (country road, motorway and railway) over time, the virtual number curve of forest fires penetrate into the adjacent forest area. Taking into account the model statistics (typical for the Timiryazevskiy forestry of the Tomsk region), the non-zero level of the virtual number of forest fires is calculated at some distance from the transport routes. For country roads, this distance is approximately equal to 5 kilometers. For motorways and railways, this

Figure 4. Dependence of the virtual number of forest fires on time at different distances from the country road: 1 – 3 km; 2 – 6 km; 3 – 9 km; 4 – 12 km

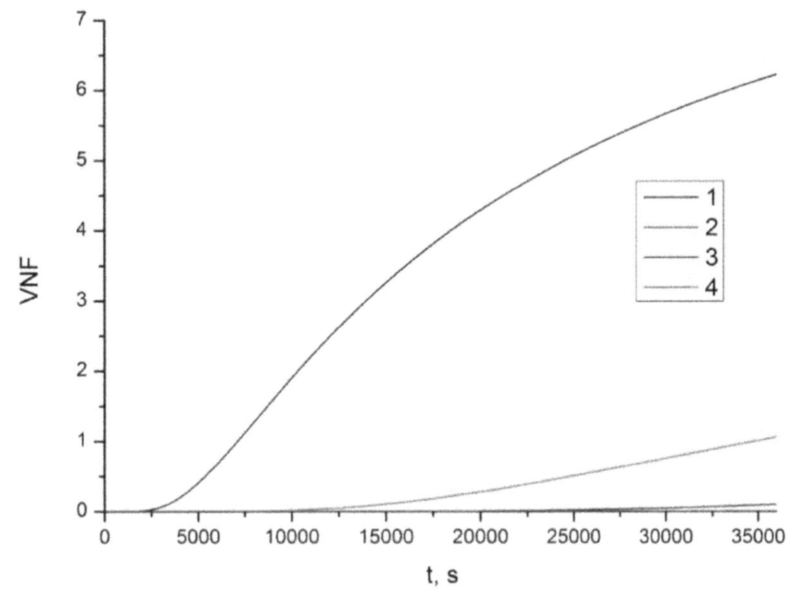

distance corresponds to 15 kilometers. Such distribution of anthropogenic load corresponds to previously proposed scenarios of human behavior near various transport routes, which are similar to data (Data, 2010).

Figure 5. Field of virtual number of forest fires for 16-00 hours for anthropogenic load caused by the influence of a country road

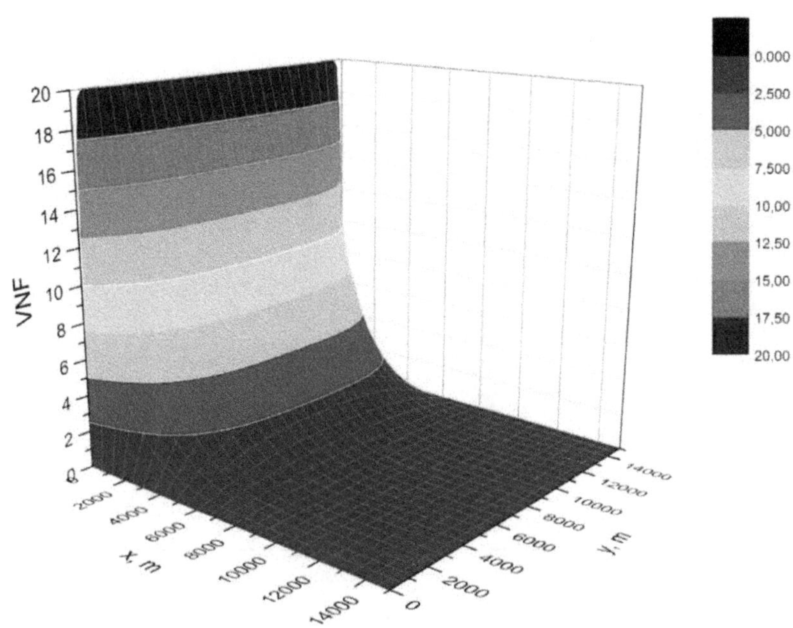

Figure 6. The distribution of the virtual number of forest fires as a function of the distance from the motorway at different times

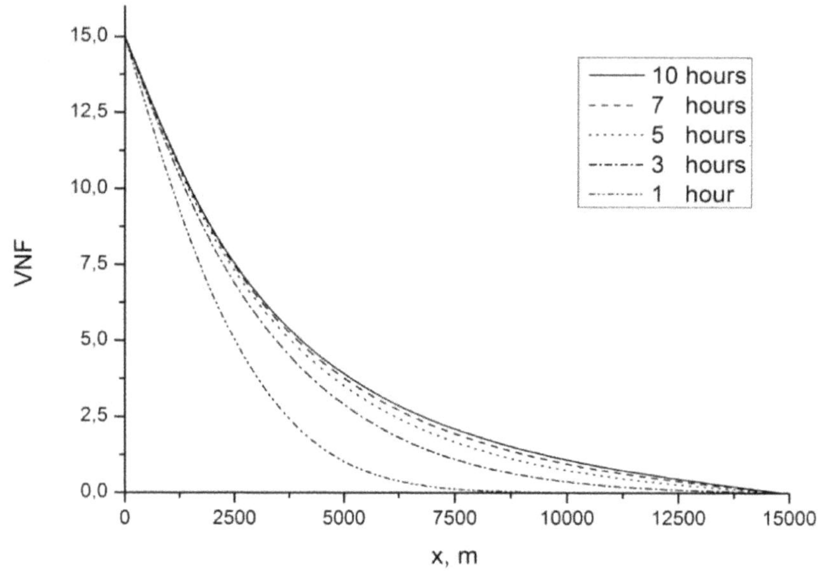

For different transport routes, a different type of behavior of the virtual number of forest fires over a horizontal coordinate is characteristic, depending on the time elapsed since the onset of anthropogenic impact on forest cover. Curves describing the virtual number of forest fires for motorways and railways lie closer to each other than similar curves for country road.

Figure 7. Dependence of the virtual number of forest fires on time at different distances from the motorway: 1 – 3 km; 2 – 6 km; 3 – 9 km; 4 – 12 km

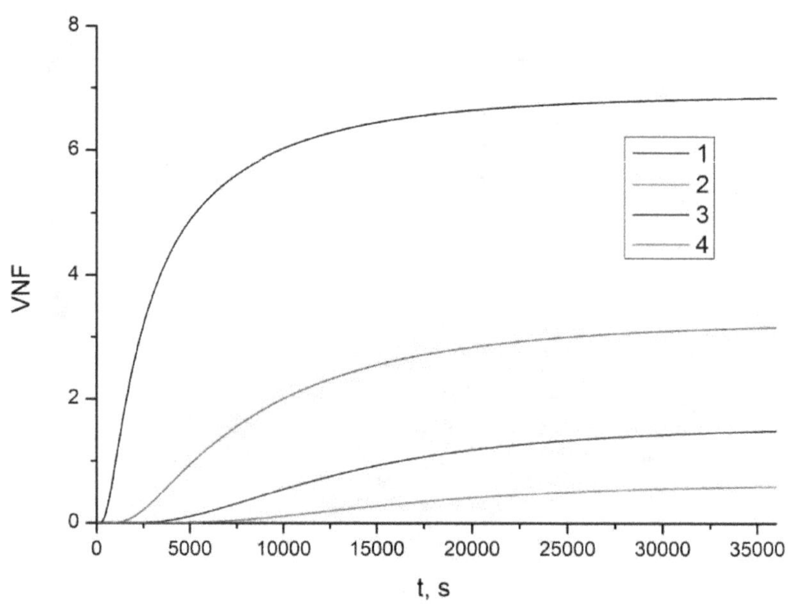

Figure 8. The field of the virtual number of forest fires for 16-00 hours for anthropogenic load, caused by the influence of a motorway

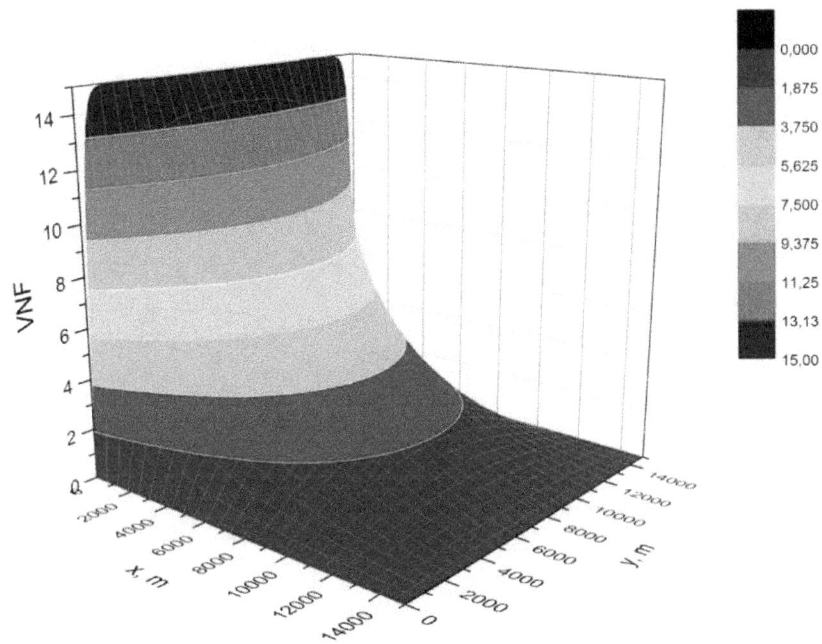

It can be concluded that the most intensive growth of the virtual number of forest fires is noted at distances small from the linear source of anthropogenic load. Large values of gradients are characteristic

Figure 9. Distribution of the virtual number of forest fires in relation to the distance from the railway at different times

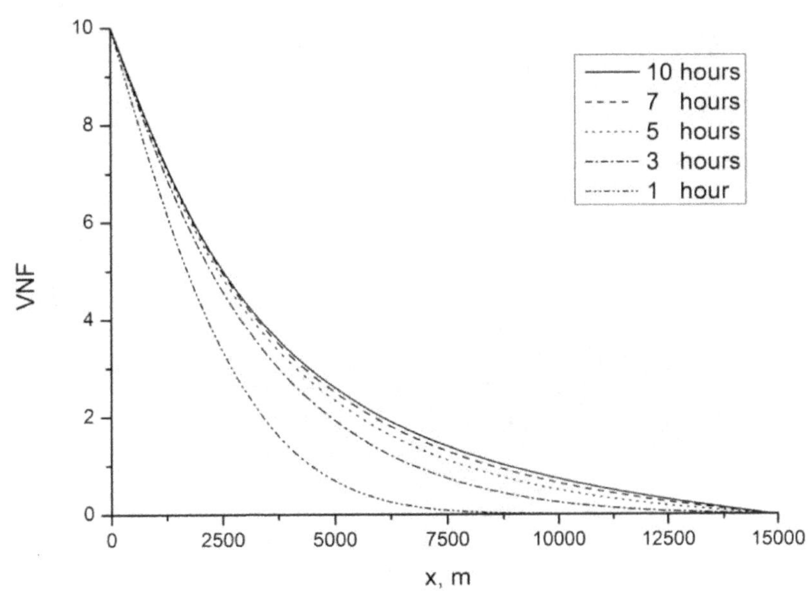

Figure 10. Dependence of the virtual number of forest fires on time at different distances from the railway: 1 – 3 km; 2 – 6 km; 3 – 9 km; 4 –12 km

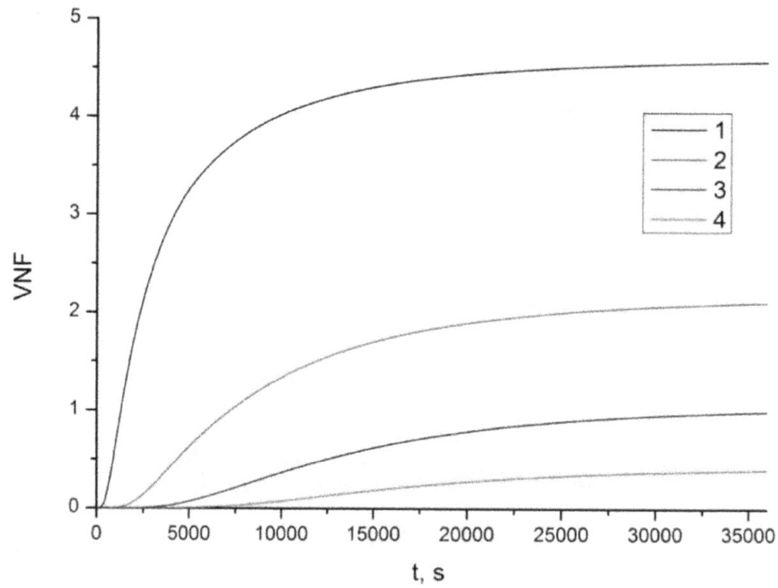

for motorways and railways. Curves of the virtual number of forest fires for country roads at a small distance from the source of anthropogenic load are close in absolute values to zero.

Figure 11. Field of virtual number of forest fires for 16-00 hours for anthropogenic load caused by the influence of a railway

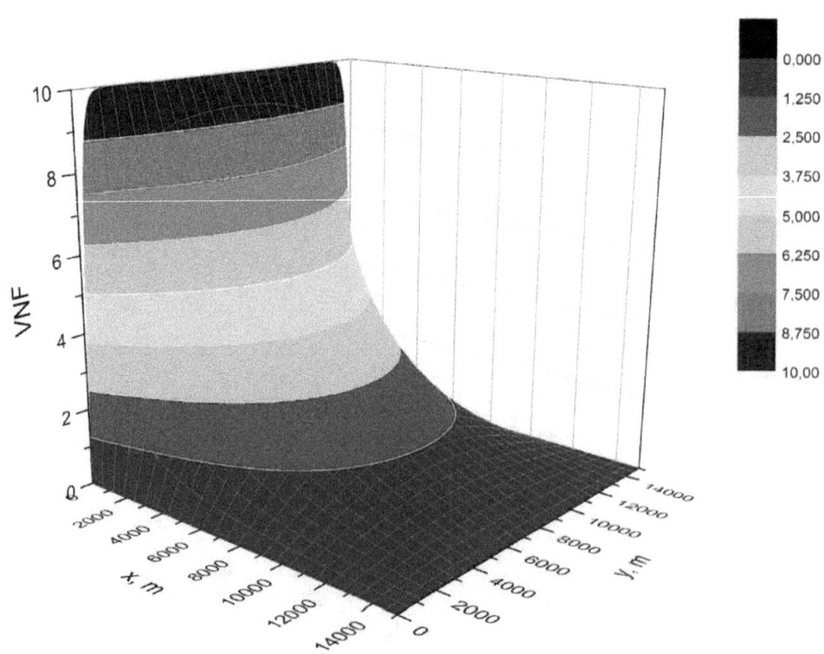

Two-dimensional distributions of the virtual number of forest fires are similar for all types of linear source of anthropogenic load. Quantitatively, the distribution along the coordinate, measured along the normal to the transport path, corresponds to one-dimensional distributions. A model source that is sufficiently extended along the second coordinate is considered, but the same level of anthropogenic load along a linear source is given. In a real situation, this may not be fulfilled, since near the populated area there may be a combined effect of linear and point sources of anthropogenic load. Then it is necessary to set the diminishing value of the virtual number of forest fires in the boundary conditions of the first type in order to use the approximation of a linear source of anthropogenic load. The maximum value of the virtual number of forest fires should be set near a populated area in this case.

It should be noted that at present there are tools to control the number of passing cars on a certain section of the transport route. These devices can be used to collect statistical and monitoring information on traffic density on the road. As a result, these data can be associated with an estimate of the virtual number of forest fires on the border of a forest adjacent to a linear source of anthropogenic load.

In the framework of this study, a model concept for the creation and development of a national forest fire danger prediction system is proposed (Baranovskiy, 2015). The concept presupposes the creation and development of a national forest fire danger prediction system with the aim of improving the environmental, economic security of the state through monitoring and managing the level of fire danger taking into account the main significant factors. The main objectives of the concept follow from the requirements formulated above for the modern system of forecasting forest fire danger. It should be noted that the availability of input data for computational models is very important. Within the framework of the present concept, it is proposed to unite into a single system (distributed hardware and software complex) as an information and computing core (parallel program complex for a multiprocessor computer system) (Baranovskiy, 2017; Baranovskiy, 2015b; Brun et al., 2017, and systems to detect cloud-to-ground lightning discharges (Jacobson et al., 2006). Strengthening the influence of the anthropogenic factor obliges to devote attention to the development of a system for the data assimilation on the level of anthropogenic load in order to assess its spatial-temporal dynamics when creating forest fire danger forecast complexes. The technological base of the model concept is shown in Fig. 12.

New systems can be created for monitoring, forecasting and controlling the level of forest fire danger caused by anthropogenic load on the basis of the developed mathematical model. It will be possible to develop hardware and software systems for automated tracking of the impact of anthropogenic load from a linear source on contiguous forested areas using this model. Such systems of ecological and mathematical control over the condition of forest can be supplemented by geoinformation subsystems for visualization of initial and forecast or monitoring information by means of electronic maps (Baranovskiy and Zharikova, 2014; Adab et al., 2013; You et al., 2017; Sakellariou et al., 2017).

It is possible to compare the expected functional characteristics of the forest fire danger assessment system due to anthropogenic load on the basis of the developed mathematical model with known analogs. First, it should be noted that the components of the Canadian system for forecasting ignitions from lightning and anthropogenic loads have been developed only for some areas of Canada. And the development of a generalized system, even for the whole of Canada, is difficult because the system was not based on mathematical models that used to simulate chemical and thermophysical processes. Among the merits should be noted the successful application of Canadian subsystems at the local level for some regions of the country. It should also be noted that the use of the Canadian forest fire danger forecast system had a positive economic effect (Taylor and Alexander, 2006).

The American system is built on the basis of a large number of amendments and indices, which make it possible to obtain data on the level of forest fire danger, including those caused by human activity. The weak physical basis of this system should also be attributed to shortcomings. However, the US system is successfully applied within the state. It is difficult to say about the prospects for the applicability of the American system directly to the territory of another state. It will be necessary to adjust all the corrections and adjust the indices in order to take into account the local characteristics of the territory and forest fire situation of another state. The European system is based on the Canadian subsystem and, accordingly, has similar advantages and disadvantages.

The Russian ISDM-Rosleskhoz system already uses elements of a deterministic-probabilistic assessment of forest fire danger taking into account thunderstorm activity and anthropogenic load (Podolskaya et al., 2011). The system builds maps of the probability of forest fire occurrence on the territory of the Russian Federation. Minimum detail of the territory is the area of responsibility (monitoring) of one meteorological station. It should be noted that in 2010 there were catastrophic forest fires in Central Russia (Devisilov, 2010).. Then, large forest areas and some settlements were destroyed. Human victims were marked in the fire-dangerous season of 2010. It is known that a retrospective analysis of forest fire danger in the ISDM-Rosleskhoz system showed that 48% of the fires of 2010 were detected with a probability in the range of 0.8 - 1. That is, the timely introduction of probabilistic criteria (Baranovskiy, 2017; Baranovskiy and Zharikova, 2014) in the practice of protecting forests from fires could reduce the economic and social damage to 50%. The main drawback of the ISDM-Rosleskhoz system lies in the lack of sufficient initial information on the causes of fires in forest areas. The network of lightening stations is quite rarefied in Siberia and the Far East in comparison with the European part. However, the use of the ISDM-Rosleskhoz system in conjunction with the monitoring of "Avialesoohrana" makes it possible to significantly reduce the consequences of forest fires on the territory of the Russian Federation.

As a result, it may be concluded that the modernization of existing systems is advisable (CWFIS, 2020; WFAS, 2020; EFFIS, 2020; ISDM-Rosleskhoz, 2020) by including anthropogenic load forecasting subsystems based on the results of this study. First of all, this concerns the Russian ISDM-Rosleskhoz system. But modernization of the North American and European system is also possible. In principle, probabilistic criterion can be considered as a certain index of forest fire danger, which is based on a quantitative estimate of the virtual number of forest fires for the modernization of the Canadian and European systems. Or to normalize one in the range 0 - 100 points for integration with the American system.

CONCLUSION

The ultimate goal of research in the area of anthropogenic load on forests should be a subsystem for the data assimilation on the level of anthropogenic load. Such a subsystem should perform an objective analysis of the data. By analogy with meteorological data assimilation systems (Apte et al., 2007; Srivas et al., 2017), it can perform the analysis and forecast stages.

Thus, it is proposed a predictive mathematical model for the system of data assimilation on the level of anthropogenic load. Subsequent studies will present a parametric analysis of this model, as well as methods for determining the coefficient A of model. The use of this model in the forest fire danger prediction system will increase the level of detail when taking into account such an anthropogenic load factor. The use of this model will provide a hybrid deterministic-probabilistic forecast of the forest fire occurrence from anthropogenic load, when the spread of anthropogenic load is modeled as a deterministic process, but the probability of a forest fire is calculated.

Evaluation of the advantages and disadvantages of the existing forest fire danger forecasting systems (including those caused by anthropogenic load) shows the prospects of their further exploitation and modernization using the developed mathematical model for determining the fields of the virtual number of forest fires in a controlled forest-covered area. As a result, modern hardware and software systems can be created using the developed mathematical model, high-performance computing and geo-information technologies, as well as tools for controlling automobile flows (Robionika, 2020). Joint application of mathematical models to predict virtual number of forest fires from linear and point sources can be mathematical basis of future developments in forest fire danger prediction (Baranovskiy, 2020).

ACKNOWLEDGMENT

This work was implemented with the financial support of the Russian Foundation for Basic Research. [Project N 17-29-05093].

REFERENCES

Adab, H., Kanniah, K. D., & Solaimani, K. (2013). Modeling forest fire in the northeast of Iran using remote sensing and GIS techniques. *Natural Hazards*, *65*(3), 1723–1743. doi:10.100711069-012-0450-8

Agterberg, F. P., & Cheng, Q. (2002). Conditional independence test for weight-of-evidence modeling. *Natural Resources Research*, *11*(4), 249–255. doi:10.1023/A:1021193827501

Al Janabi, S., Al Shourbaji, I., & Salman, M. A. (2017). *Assessing the suitability of soft computing approaches for forest fires prediction*. Applied Computing and Informatics. doi:10.1016/j.aci.2017.09.006

Amatulli, G., Perez-Cabello, F., & Du la Riva, J. (2007). Mapping lightning/human-caused wildfires occurrence under ignition point location uncertainty. *Ecological Modelling*, *200*(3-4), 321–333. doi:10.1016/j.ecolmodel.2006.08.001

Andreev, Y. A. (2003). *Influence of anthropogenic and natural factors on the occurrence of fires in forests and settlements* (Dissertation). VNIIPO EMERCOM of Russia, Moscow, Russian Federation. (in Russian)

Andreev, Y. A., & Larchenko, G. F. (1987). *Socio-psychological aspects of recreational forest visits and the emergence of fires. Forest fires and the fight against them.* VNIILM. (in Russian)

Apte, A., Hairer, M., Stuart, A. M., & Voss, J. (2007). Sampling the posterior: An approach to non-Gaussian data assimilation. *Physica D. Nonlinear Phenomena*, *230*(1-2), 50–64. doi:10.1016/j.physd.2006.06.009

Baranovskiy, N., Krechetova, S., Belikova, M., & Perelygin, A. (2016). WWLLN data cluster analysis methods for lightning-caused forest fires monitoring. *Iranian Journal of Electrical and Computer Engineering*, *6*(6), 3112–3120. doi:10.11591/ijece.v6i6.12780

Baranovskiy, N., & Zharikova, M. (2014). A web-oriented geoinformation system application for forest fire danger in the typical forests of the Ukraine. Lecture Notes in Geoinformation and Cartography, 13-22. () doi:10.1007/978-3-319-08180-9_2

Baranovskiy, N. V. (2006). *Mathematical foundations of the system of data assimilation on the level of anthropogenic load on a controlled forest-covered territory*. Science at the Turn of the Millennium "Progressive technologies of development": 3rd International scientific-practical conference, Tambov, Russian Federation. (in Russian)

Baranovskiy, N. V. (2015a). Project of the eurasian segment of the new system of the forest fire risk prediction based on information and computer technologies. *Journal of Automation and Information Sciences*, *47*(3), 40–56. doi:10.1615/JAutomatInfScien.v47.i3.40

Baranovskiy, N. V. (2015b). Algorithms for parallelizing a mathematical model of forest fires on supercomputers and theoretical estimates for the efficiency of parallel programs. *Cybernetics and Systems Analysis*, *51*(3), 471–480. doi:10.100710559-015-9738-5

Baranovskiy, N. V. (2017). Forest fire danger assessment using SPMD-model of computation for massive parallel system. *International Review on Modeling and Simulations*, *10*(3), 193–201. doi:10.15866/iremos.v10i3.10570

Baranovskiy, N. V. (2020). Mathematical Simulation of Anthropogenic Load on Forested Territories for Point Source. In N. Baranovskiy (Ed.), *Predicting, Monitoring, and Assessing Forest Fire Dangers and Risks* (pp. 64–88). IGI Global. doi:10.4018/978-1-7998-1867-0.ch003

Baranovskiy, N. V., & Kuznetsov, G. V. (2017). *Forest fire occurrences and ecological impact prediction*. Publishing House of the Siberian Branch of the Russian Academy of Science. doi:10.15372/FOREST2017BNV

Baranovskiy, N. V., & Kuznetsov, G. V. (2017). Coniferous tree ignition by cloud-to-ground lightning discharge using the approximation of "ideal" crack in bark. *JP Journal of Heat and Mass Transfer*, *14*(1), 173–186. doi:10.17654/HM014010173

Brun, C., Artes, T., Cencerrado, A., Margalef, T., & Cortés, A. (2017). A High Performance Computing Framework for Continental-Scale Forest Fire Spread Prediction. *Procedia Computer Science*, *108C*, 1712–1721. doi:10.1016/j.procs.2017.05.258

Bui, D. T., Bui, Q.-T., Nguyen, Q.-P., Pradhan, B., Nampak, H., & Trinh, P. T. (2017). A hybrid artificial intelligence using GIS-based neural-fuzzy inference system and particle swarm optimization for forest fire susceptibility modeling at a tropical area. *Agricultural and Forest Meteorology*, *233*, 32–44. doi:10.1016/j.agrformet.2016.11.002

Canadian Wildland Fire Information System. (n.d.). Access: https://cwfis.cfs.nrcan.gc.ca/home

Cardille, J. A., Ventura, S. J., & Turner, M. G. (2001). Environmental and Social Factors of Influencing Wildfires in the Upper Midwest, United States. *Ecological Applications*, *11*(1), 111–127. doi:10.1890/1051-0761(2001)011[0111:EASFIW]2.0.CO;2

Chen, W., Xie, X., Wang, J., Pradhan, B., Hong, H., Tien Bui, D., Duan, Z., & Ma, J. (2017). A comparative study of logistic model tree, random forest and classification and regression tree models for spatial prediction of landslide susceptibility. *Catena*, *151*, 147–160. doi:10.1016/j.catena.2016.11.032

Curt, T., Frejaville, T., & Lahaye, S. (2016). Modelling the spatial patterns of ignition causes and fire regime features in southern France: Implications for the prevention policy. *International Journal of Wildland Fire*, *25*(7), 785–796. doi:10.1071/WF15205

Data archive on forest fires in the Timiryazevskiy forestry of the Tomsk region. (2010). Department of Forestry of Tomsk Region. (in Russian)

Deeming, J. E., Burgan, R. E., & Cohen, J. D. (1977). *The National Fire Danger Rating System - 1978*. United States Department of Agriculture, Forest Service, General Technical Report INT-39, Intermountain Forest and Range Experiment Station, Odgen, UT.

Deeming, J. E., Lancaster, J. W., Fosberg, M. A., Furman, W. R., & Schroeder, M. J. (1974). *The National Fire Danger Rating System*. United States Department of Agriculture, Forest Service, Research Paper RM-84, Rocky Mountain Forest and Range Experiment Station, Fort Collins, CO.

Devisilov, V. A. (2010). Russian forest is asking the protection! *Safety in Technosphere*, (6), 3–7.

Diez-Delgado, R., Lloret, F., & Pons, X. (2004). Statistical analysis of fire-frequency models for Catalonia (NE Spain, 1975-1998) based on fire scar maps from Landsat MSS data. *International Journal of Wildland Fire*, *13*(1), 89–99. doi:10.1071/WF02051

Dormann, C. F., Elith, J., Bacher, S., Bushmann, C., Carl, G., Carre, G., Marquez, J. R. G., Gruber, B., Lafourcade, B., Leitao, P. J., Munkemuller, T., McClean, C., Osborne, P. E., Reineking, B., Schroder, B., Skidmore, A. K., Zurell, D., & Lautenbach, S. (2013). Collinearity: A review of methods to deal with it and a simulation study evaluating their performance. *Ecography*, *36*(1), 27–46. doi:10.1111/j.1600-0587.2012.07348.x

Duan, K., Keerthi, S., & Poo, A. (2001). *Evaluation of simple performance measures for tuning SVM hyperparameters (Technical report)*. National University of Singapore. Department of Mechanical Engineering.

European Forest Fire Information System. (n.d.). Access: http://effis.jrc.ec.europa.eu

Flury, B. N. (1984). Common principal components in K groups. *Journal of the American Statistical Association*, *79*, 892–898.

Fusco, E. J., Abatzoglou, J. T., Balch, J. K., Finn, J. T., & Bradley, B. A. (2016). Quantifying the human influence on fire ignition across the western USA. *Ecological Applications*, *26*(8), 2390–4201. doi:10.1002/eap.1395 PMID:27907256

Ganteaume, A., & Guerra, F. (2018). Explaining the spatio-seasonal variation of fires by their causes: The case of southeastern France. *Applied Geography (Sevenoaks, England)*, *90*, 69–81. doi:10.1016/j.apgeog.2017.11.012

Ghimire, B., Rogan, J., Galiano, V. R., Panday, P., & Neeti, N. (2012). An evaluation of bagging, boosting and random forests for land-cover classification in CapeCod, Massachusetts, USA. *GIScience & Remote Sensing*, *49*(5), 623–643. doi:10.2747/1548-1603.49.5.623

Gonzalez-Hidalgo, J. C., Brunetti, M., & de Luis, M. (2011). A new tool for monthly precipitation analysis in Spain: MOPREDAS database (monthly precipitation trends December 1945 – November 2005). *International Journal of Climatology*, *31*(5), 715–731. doi:10.1002/joc.2115

Gonzalez-Hidalgo, J. C., Pena-Angulo, D., Brunetti, M., & Cortesi, N. (2015). MOTEDAS: A new monthly temperature database for mainland Spain and the trends in temperature (1951 – 2010). *International Journal of Climatology*, *35*(15), 4444–4463. doi:10.1002/joc.4298

Gould, J. S., Patriquin, M. N., Wang, S., McFarlane, B. L., & Wotton, B. M. (2013). Economic evaluation of research to improve the Canadian forest fire danger rating system. *Forestry*, *86*(3), 317–329. doi:10.1093/forestry/cps082

Grala, K., Grala, R. K., Hussain, A., Cooke, W. H. III, & Varner, J. M. III. (2017). Impact of human factors on wildfire occurrence in Mississippi, USA. *Forest Policy and Economics*, *81*, 38–47. doi:10.1016/j.forpol.2017.04.011

Grishin, A. M. (1997). *Mathematical modeling of forest fire and new methods of fighting them*. Publishing House of the Tomsk State University.

Grishin, A. M., & Filkov, A. I. (2011). A deterministic-probabilistic system for predicting forest fire hazard. *Fire Safety Journal*, *46*(1-2), 56–62. doi:10.1016/j.firesaf.2010.09.002

Haupt, L. R., & Haupt, S. E. (2004). *Practical genetic algorithms* (2nd ed.). John Wiley & Sons Inc.

Hong, H., Tsangaratos, P., Ilia, I., Liu, J., Zhu, A.-X., & Xu, C. (2018). Applying genetic algorithms to set the optimal combination of forest fire related variables and model forest fire susceptibility based on data mining models. The case of Dayu Country, China. *The Science of the Total Environment*, *630*, 1044–1056. doi:10.1016/j.scitotenv.2018.02.278 PMID:29554726

Hu, T., & Zhou, G. (2014). Drivers of lightning- and human-caused fire regimes in the Great Xing'an Mountains. *Forest Ecology and Management*, *329*, 49–58. doi:10.1016/j.foreco.2014.05.047

Information system for remote monitoring of forest fires ISDM-Rosleskhoz. (n.d.). Access: https://nffc.aviales.ru/main_pages/index.shtml

Jacobson, A. R., Holzworth, R., Harlin, J., Dowden, R., & Lay, E. (2006). Performance Assessment of the World Wide Lightning Location Network (WWLLN) using the Los Alamos Sferic Array (LASA) as Ground Truth. *Journal of Atmospheric and Oceanic Technology*, *23*(8), 1082–1092. doi:10.1175/JTECH1902.1

Jimenez-Ruano, A., Rodrigues Mimbrero, M., & de la Riva Fernandez, J. (2017). Understanding wildfires in mainland Spain. A comprehensive analysis of fire regime features in a climate-human context. *Applied Geography (Sevenoaks, England)*, *89*, 100–111. doi:10.1016/j.apgeog.2017.10.007

Kaiser, H. F. (1958). The varimax criterion for analytic rotation in factor analysis. *Psychometrika*, *28*(3), 187–200. doi:10.1007/BF02289233

Kurbatskiy, N. P. (1964). *Fires of the taiga, patterns of their occurrence and development. Dissertation.* Institute of Forest of the USSR Academy of Sciences. (in Russian)

Kurbatskiy, N. P. (1964). *The problem of forest fires. The emergence of forest fires.* Nauka. (in Russian)

Lee, B. S., Alexander, M. E., Hawkes, B. C., Lynham, T. J., Stocks, B. J., & Englefield, P. (2002). Information systems in support of wildland fire management decision-making in Canada. *Computers and Electronics in Agriculture*, *37*(1-3), 185–198. doi:10.1016/S0168-1699(02)00120-5

Liu, Q., Gao, P., & Wang, Z. (2012). Great Xing'an Mountains: The relationship between population growth and economic development. *Statist Consult*, *6*, 17–19.

Liu, Z. H., Yang, J., Chang, Y., Weisberg, P. J., & He, H. S. (2012). Spatial patterns and drivers of fire occurrence and its future trend under climate change in a boreal forest of Northeast China. *Global Change Biology*, *18*(6), 2041–2056. doi:10.1111/j.1365-2486.2012.02649.x

Martell, D. L. (2000). A Markov Chain Model of Day to Day Changes in the Canadian Forest Fire Weather Index. *International Journal of Wildland Fire*, *9*(4), 265–273. doi:10.1071/WF00020

Matsenko, V. V., Sokolov, A. Y., & Kalinin, S. I. (1999). General plan of the anti-fire arrangement of forests. Vol. 1. Explanatory note. 5-99.14-17-PM. State Design and Survey Institute Rosgiproles, Altai Branch. (in Russian)

Melekhov, I. S. (1947). *Nature of forest and forest fires*. Nauka. (in Russian)

Melluma, A. Z., Rungule, R. H., & Emsis, I. V. (1982). *Recreation on nature as a nature protection problem*. Zinatne. (in Russian)

Moving car detector project. (n.d.). *Robionika*. Access: http://www.robionika.ru/projects/4 (in Russian)

Pakshirajan, R. P. (2013) Probability Theory. Springer. doi:10.1007/978-93-86279-54-5

Pew, K. L., & Larsen, C. P. S. (2001). GIS analysis of the spatial and temporal patterns of human-caused wildfires in the temperate rain forest of Vancouver Island, Canada. *Forest Ecology and Management*, *140*(1), 1–18. doi:10.1016/S0378-1127(00)00271-1

Podolskaya, A.S., Ershov, D.V., & Shulyak, P.P. (2011). Application of the method for assessing the probability of occurrence of forest fires in ISDM-Rosleskhoz. *Modern Problems of Remote Sensing of the Earth From Space, 8*, 118 - 126. (in Russian)

Poulos, H. M., Villanueva Díaz, J., Cerano Paredes, J., Camp, A. E., & Gatewood, R. G. (2013). Human Influences on fire regimes and forest structure in the Chihuahuan Desert Borderlands. *Forest Ecology and Management*, *298*, 1–11. doi:10.1016/j.foreco.2013.02.014

Preisler, H. K., Brillinger, D. R., Burgan, R. E., & Benoit, J. W. (2004). Probability based models for estimation of wildfire risk. *International Journal of Wildland Fire*, *13*(2), 133–142. doi:10.1071/WF02061

Rawlings, J. O., Pantula, S. G., & Dickey, D. A. (2015). *Applied Regression Analysis. A Research Tool. Springer Texts in Statistics book series (STS)*. Springer.

Read, N., Duff, T. J., & Taylor, P. G. (2018). A lightning-induced wildfire ignition forecasting model for operational use. *Agricultural and Forest Meteorology*, *253-254*, 233–246. doi:10.1016/j.agrformet.2018.01.037

Sakellariou, S., Tampekis, S., Samara, F., Sfougaris, A., & Christopoulou, O. (2017). Review of state-of-the-art decision support systems (DSSs) for prevention and suppression of forest fires. *Journal of Forestry Research*, *28*(6), 1107–1117. doi:10.100711676-017-0452-1

Samarskii, A. A., & Vabishchevich, P. N. (1995a). Computational Heat Transfer: Vol. 1. *Mathematical Modeling*. Wiley.

Samarskii, A. A., & Vabishchevich, P. N. (1995b). Computational Heat Transfer: Vol. 2. *The Finite Difference Method*. Wiley.

Srivas, T., de Callafon, R. A., Crawl, D., & Altintas, I. (2017). Data Assimilation of Wildfires with Fuel Adjustment Factors in farsite using Ensemble Kalman Filtering. *Procedia Computer Science*, *108C*, 1572–1581. doi:10.1016/j.procs.2017.05.197

Taylor, S. W., & Alexander, M. E. (2006). Science, technology and human factors in the fire danger: The Canadian experience. *International Journal of Wildland Fire*, *15*(1), 121–135. doi:10.1071/WF05021

Telitsin, G. P. (1984). Study of the connection between forest attendance and the occurrence of fire. [in Russian]. *Лесоведение*, 59–63.

Telitsin, G. P. (1987). *Method for Determining the Fire Hazard of Forest Territory. Forest Fires and Fighting them*. VNIILM. (in Russian)

Viegas, D. X., Bovio, G., Ferreira, A., Nosenzo, A., & Sol, B. (2000). Comparative study of various methods of fire hazard evaluation in Southern Europe. *International Journal of Wildland Fire*, *10*, 235–246.

WFAS Wildland Fire Assessment System. (n.d.). Access: http://www.wfas.net

Wotton, B. M. (2009). Interpreting and using outputs from the Canadian Forest Fire Danger Rating System in research applications. *Environmental and Ecological Statistics*, *16*(2), 107–131. doi:10.100710651-007-0084-2

Ye, J., Wu, M., Deng, Z., Xu, S., Zhou, R., & Clarke, K. C. (2017). Modeling the spatial patterns of human wildfire ignition in Yunnan province, China. *Applied Geography (Sevenoaks, England)*, *89*, 150–162. doi:10.1016/j.apgeog.2017.09.012

You, W., Lin, L., Wu, L., Ji, Z., Yu, J., Zhu, J., Fan, Y., & He, D. (2017). Geographical information system-based forest fire risk assessment its spatiotemporal variability. *Ecological Indicators*, *77*, 176–184. doi:10.1016/j.ecolind.2017.01.042

Zhang, Y., Lim, S., & Sharples, J. J. (2016). Modelling spatial patterns of wildfire occurrence in South-Eastern Australia. *Geomatics, Natural Hazards & Risk*, *7*(6), 1800–1815. doi:10.1080/19475705.2016.1155501

ADDITIONAL READING

Adámek, M., Jankovská, Z., Hadincová, V., Kula, E., & Wild, J. (2018). Drivers of forest fire occurrence in the cultural landscape of Central Europe. *Landscape Ecology*, *33*(11), 2031–2045. doi:10.100710980-018-0712-2

Baranovskiy, N. (2018). Mathematical model to simulate anthropogenic loading around point source in one-dimensional statement. International Multidisciplinary Scientific GeoConference Surveying Geology and Mining Ecology Management, SGEM, 18(3.2), 857-862

Bowman, D. M. J. S., Moreira-Muñoz, A., Kolden, C. A., Chávez, R. O., Muñoz, A. A., Salinas, F., González-Reyes, Á., Rocco, R., de la Barrera, F., Williamson, G. J., Borchers, N., Cifuentes, L. A., Abatzoglou, J. T., & Johnston, F. H. (2019). Human–environmental drivers and impacts of the globally extreme 2017 Chilean fires. *Ambio*, *48*(4), 350–362. doi:10.100713280-018-1084-1 PMID:30128860

Clarke, H., Gibson, R., Cirulis, B., Bradstock, R. A., & Penman, T. D. (2019). Developing and testing models of the drivers of anthropogenic and lightning-caused wildfire ignitions in south-eastern Australia. *Journal of Environmental Management*, *235*, 34–41. doi:10.1016/j.jenvman.2019.01.055 PMID:30669091

Elia, M., Giannico, V., Lafortezza, R., & Sanesi, G. (2019). Modeling fire ignition patterns in Mediterranean urban interfaces. *Stochastic Environmental Research and Risk Assessment*, *33*(1), 169–181. doi:10.100700477-018-1558-5

Grishin, A. M., & Filkov, A. I. (2011). A deterministic-probabilistic system for predicting forest fire hazard. *Fire Safety Journal*, *46*(1-2), 56–62. doi:10.1016/j.firesaf.2010.09.002

Scheller, R., Kretchun, A., Hawbaker, T. J., & Henne, P. D. (2019). A landscape model of variable social-ecological fire regimes. *Ecological Modelling*, *401*, 85–93. doi:10.1016/j.ecolmodel.2019.03.022

Viedma, O., Urbieta, I. R., & Moreno, J. M. (2018). Wildfires and the role of their drivers are changing over time in a large rural area of west-central Spain. *Scientific Reports*, *8*(1), 17797. doi:10.103841598-018-36134-4 PMID:30542114

Wotton, B. M. (2009). Interpreting and using outputs from the Canadian Forest Fire Danger Rating System in research applications. *Environmental and Ecological Statistics*, *16*(2), 107–131. doi:10.100710651-007-0084-2

Yashwanth, B. L., Shotorban, B., Mahalingam, S., & Weise, D. R. (2015). An investigation of the influence of the heating modes on ignition and pyrolysis of woody wildland fuel. *Combustion Science and Technology*, *187*(5), 780–796. doi:10.1080/00102202.2014.973948

Ye, J., Wu, M., Deng, Z., Xu, S., Zhou, R., & Clarke, K. C. (2017). Modeling the spatial patterns of human wildfire ignition in Yunnan province, China. *Applied Geography (Sevenoaks, England)*, *89*, 150–162. doi:10.1016/j.apgeog.2017.09.012

You, W., Lin, L., Wu, L., Ji, Z., Yu, J., Zhu, J., Fan, Y., & He, D. (2017). Geographical information system-based forest fire risk assessment its spatiotemporal variability. *Ecological Indicators*, *77*, 176–184. doi:10.1016/j.ecolind.2017.01.042

KEY TERMS AND DEFINITIONS

Anthropogenic Load: Different human activities on forested territories lead to forest fire occurrence and characterized by presence of fire sources.

Forest Fire: Uncontrolled aerothermochemical phenomenon characterized by step-by-step mechanism which includes following stages: inert heating, moisture evaporation, high temperature terpens evaporation, dry organic matter pyrolysis, flammable combustion, and smoldering.

Linear Source: A linear source of anthropogenic load means an extended linear object, from which the distribution of anthropogenic load is directed perpendicular to the depth of the forested area.

Mathematical Simulation: The production of a computer model of forest fire conditions and prerequisites, especially for the purpose of study.

Monitoring: Monitoring refers to the periodic calculation of the parameters of forest fire danger with a portion of information available in real time.

Prediction: Under the prediction of forest fires is the calculation of the parameters of forest fire danger with a certain projection in advance in order to have enough time to anticipate an emergency. The calculation in this case is carried out in a mode ahead of the real time of the development of the catastrophe - the occurrence of a forest fire.

Virtual Number of Forest Fires: Possible (calculated) number of forest fire around source of anthropogenic load which equal to real number of forest fire in some selected sites.

About the Author

Nikolay Viktorovich Baranovskiy received Specialist Degree in Mechanics from Tomsk State University in 2000. He received PhD Degree in Physics and Mathematics from Tomsk State University in 2007. His current position is Associate Professor at Tomsk Polytechnic University. His current interests include forest fire danger and ecological impact prediction, parallel computing and spatial data analysis using GIS and RS technologies. Prof. Baranovskiy is a Member of International Association of Wildland Fire, Association for Fire Ecology, International Association for Fire Safety Science, and International Association of Engineers. He is an author of more than 150 articles and papers in Russian and International Journals.

Index

S

V

www.ingramcontent.com/pod-product-compliance
Lightning Source LLC
LaVergne TN
LVHW060425300926
849162LV00006B/1525